International Conference on Building Energy Conservation, Thermal Safety and Environmental Pollution Control (ICBTE 2020)

E3S Web of Conferences Volume 212 (2020)

Brest, Belarus
29 - 30 October 2020

Editors:

N. Sheshko
U. Navaseltsau

ISBN: 978-1-7138-2095-6

Printed from e-media with permission by:

Curran Associates, Inc.
57 Morehouse Lane
Red Hook, NY 12571

Some format issues inherent in the e-media version may also appear in this print version.

No changes have been made to the content of these proceedings. There may be changes to pagination and minor adjustments for aesthetics.

Printed with permission by Curran Associates, Inc. (2021)

For additional information, please contact EDP Sciences – Web of Conferences
at the address below.

EDP Sciences – Web of Conferences
17, Avenue du Hoggar
Parc d'Activité de Courtabœuf
BP 112
F-91944 Les Ulis Cedex A
France

Phone: +33 (0) 1 69 18 75 75
Fax: +33 (0) 1 69 28 84 91

contact-edps@webofconferences.org

Additional copies of this publication are available from:

Curran Associates, Inc.
57 Morehouse Lane
Red Hook, NY 12571 USA
Phone: 845-758-0400
Fax: 845-758-2633
Email: curran@proceedings.com
Web: www.proceedings.com

TABLE OF CONTENTS

The 2020 International Conference on Building Energy Conservation, Thermal Safety and Environmental Pollution Control (ICBTE 2020)

Brest, Belarus, October 29 - 30, 2020

Chairmen:

Bahanovich A. G. – Doctor of Technical Sciences, Associate Professor, Rector of BrSTU (Belarus).

Deputy Chairman:

Shalabyta N. N. – Candidate of Technical Sciences, Associate Professor, Vice-rector for Scientific Work, BrSTU (Belarus).

Cai Xinli – Doctor, Professor, Vice president of Anhui Jianzhu University (China).

Members of the Scientific Committee:

Volchek A. A. – Doctor of Geological Sciences, Professor, Dean of the Faculty of Engineering Systems and Ecology, BrSTU (Belarus).

Semenyuk S. M. – Candidate of Technical Sciences, Associate Professor, Dean of the Faculty of Civil Engineering, BrSTU (Belarus).

Zhu Shuguang – Doctor, Professor, Dean of School of Environment and Energy Engineering, Anhui Jianzhu University (China).

Tur V. V. – Doctor of Technical Sciences, Professor, Head of the Department of Concrete Technology and Building Materials, BrSTU (Belarus).

Poita P. S. – Doctor of Technical Sciences, Professor, Department of Geotechnics and Transport Communications, BrSTU (Belarus).

Lyu Kefeng – Doctor, Lecturer, Teacher of Anhui Jianzhu University (China).

Mazhaysky Yu. A. – Doctor of Agricultural Sciences, Professor, Chief Research Worker of the Meshchersky Branch of the All-Russian Scientific Research Institute of Water Engineering and Melioration named after A.N. Kostyakov, Director General of the Meshchersky Scientific and Technical Center, Honorary Worker of the Russian Agro-Industrial Complex (Russia).

Jeznach J. – Doctor of Technical Sciences, Professor, Professor of Warsaw University of Natural Sciences – SGGW, Member of the Presidium and Academic Secretary of the Committee for Agronomical Sciences of the Polish Academy of Sciences (Poland).

Gavardashvili G. V. – Doctor of Technical Sciences, Professor, Director of the Institute of Water Management named after Ts. Mirtskhulava of the Georgian Technical University (Georgia).

Gribauskene V. – Doctor of Technical Sciences, Professor, Kaunas University of Applied Sciences in Forestry and Environmental Engineering – KMAIK (Lithuania).

Burlibaev M. Zh. – Doctor of Technical Sciences, Professor, Deputy General Director for Scientific Work of the Kazakhstan Agency for Applied Ecology, Academician of the Academy of Water Management of the Russian Federation (Kazakhstan).

Bogdasarov M. A. – Doctor of Geological and Mineralogical Sciences, Professor, Head of the Department of Geography and Environmental Management, Educational Establishment "Brest State University named after A.S. Pushkin" (Belarus).

Derkach V. N. – Doctor of Technical Sciences, Associate Professor, Director of the branch of RUE "BelBSR Institute", Scientific and Technical Center (Belarus).

edp sciences

E3S Web of Conferences

Statement of Peer review

In submitting conference proceedings to E3S Web of Conferences, I certify to the Publisher that I adhere to the **Policy on Publishing Integrity** of the journal in order to safeguard good scientific practice in publishing.

1. All articles have been subjected to peer review administered by the proceedings editors.
2. Reviews have been conducted by expert referees, who have been requested to provide unbiased and constructive comments aimed, whenever possible, at improving the work.
3. Proceedings editors have taken all reasonable steps to ensure the quality of the materials they publish and their decision to accept or reject a paper for publication has been based only on the merits of the work and the relevance to the journal.

Title, date and place of the conference

The 2020 International Conference on Building Energy Conservation, Thermal Safety and Environmental Pollution Control (ICBTE 2020).
October 29-30, 2020
Brest, Belarus
Brest State Technical University

Proceedings editor(s):

Sheshko N, Brest State Technical University
Navasettsau U., Brest State Technical University

Date and editor's signature

17, avenue du Hoggar - PA de Courtabœuf – BP 112 - 91944 Les Ulis Cedex A (France)
Tél. : 33 (0)1 69 18 75 75 - Fax : 33(0)1 69 07 45 17 – www.edpsciences.org

THE APPLICATION OF OZONE TO REDUCE THE COLORING INTENSITY OF AQUEOUS SOLUTIONS OF DYES USED IN THE TEXTILE INDUSTRY

Sergey Belov[1*], *Grigoriy* Naumchik[1]
[1]UO BrSTU, Brest, Republic of Belarus

Abstract. The article discusses the mechanisms of interaction of ozone with aqueous solutions of organic dyes. Two different dyes with different types of chromophore systemwere investigated: azo dye and thiazine dye. A detailed methodology for performing experimental studies makes it possible to accurately dose ozone into the reaction mixture. The data of experimental studies are presented indicating the effectiveness of the use of ozone for the destruction of dyes in aqueous solutions. The specific doses of ozone were established to reduce the color intensity of the studied dye solutions to standard values.
Keywords: Ozone, dyes, destruction, color intensity, wastewater, textile industry.

Introduction

Wastewater from textile factories is characterized by a high content of pollutants. They contain up to 1000 *mg / L* of suspended solids and are characterized by high chemical oxygen demand (COD) which is determined by the content of organic pollutants in colloidal and soluble forms [1, 2, 3]. However, one of the most characteristic indicators of wastewater pollution from textile enterprises is their color intensity by the degree of dilution [1, 4, 5].

The color of waste water in the production of textiles is mainly due to the widespread use of organic dyes. Depending on the specialization of the enterprise, fibers, fabrics and finished products are dyed during the production process. A significant part of the dyes (from 10% to 50%) during the dyeing process remains in the waste technological solutions and in the wash water formed after washing the dyed items [1, 6, 7]. As a result, the intensity of color in terms of the degree of dilution in the wastewater of textile enterprises using dyes in the production cycle can reach 1: 1000. At the same time, the intensity of coloring in terms of the degree of dilution in domestic and industrial waste water

* Corresponding author: siarhei.bialou@gmail.com

discharged both into the municipal sewer network and into natural water bodies should not exceed 1:20 [1, 4].

Synthetic dyes currently used for dyeing textile materials are persistent organic compounds that are difficult to biooxidize [1, 4, 8]. Therefore, when entering municipal wastewater these compounds pass practically unchanged through biological treatment facilities and enter natural water bodies having a negative impact on their ecosystem [9]. The following methods can be used to remove dyes that cause the color of wastewater: reagent coagulation, electrocoagulation, sorption, barometric methods, destructive methods (oxidation, reduction, UV treatment) [1, 8]. Each method has its own advantages and disadvantages and should be applied based on the conditions of a specific production. However, one of the most versatile methods is ozonation [10, 11, 12, 13]. Unlike reagent coagulation and electrocoagulation, ozonation does not form precipitates, unlike sorption,ozonation does not generate spent regeneration solutions compared to barometric methods so ozonation is more reliable and easy to operate [14, 15, 16].

The main value of ozone for removing pollutants, in contrast to other reagents, lies in its high reactivity which makes it possible to destroy practically all water-soluble organic dyes using ozonation. In this case, ozone exhibits its high oxidizing properties at any pH value [17]. Ozone interacts most vigorously with unsaturated bonds [18]:

ethylene $>C=C<$, azomethine $>C=N-$, azo $-N=N-$, thiocarbonyl $>C=S$.

However, with a carbonyl bond $>C=O$ the reaction is not very active.

When developing a technology for purifying waste water from dyes, the reaction with an azo bond is of greatest interest, since most dyes used in light industry are azo dyes in terms of chromophore structure [19]. Literature sources indicate that the interaction of ozone with azo bonds occurs with the breaking of the double bond and the formation of intermediate addition products which are instantly hydrolyzed in an aqueous medium with the final breaking of the azo bond and the formation of oxy compounds [18]. This reaction mechanism, in principle, can be similar to the interaction of ozone with an ethylene bond [20]:

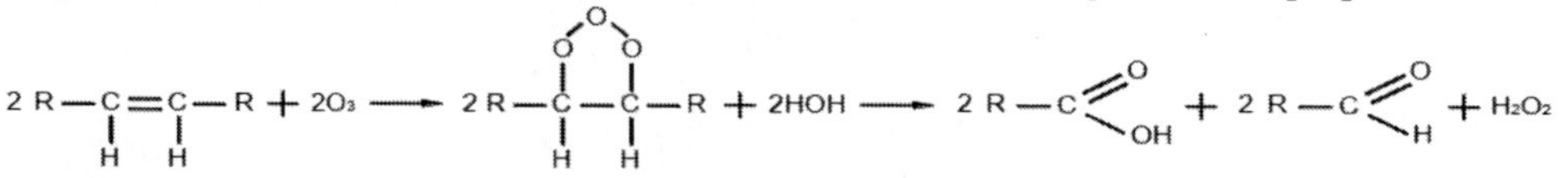

However, it was found when searching in the literature that the mechanism of the reaction of ozone with azo bonds of dyes in aqueous solutions has not been studied in detail.

The interaction of ozone with aromatic compounds has been studied in more detail which is also of great theoretical importance for determining the mechanism of interaction of ozone with dyes in aqueous solutions. All synthetic dyes contain aromatic residues in their molecules. Therefore, various derivatives of aromatic hydrocarbons are formedduring the decomposition of dyes. These compounds, as a rule, are not colored or weakly colored but they are resistant to biological oxidation and are highly pathogenic for aquatic ecosystems. The interaction of ozone with aromatic compounds also occurs through the mechanism of the addition of ozone to the double bond and subsequent hydrolysis of the resulting intermediate compound [6].

The relevance of the development of effective methods for the removal of dyes from the wastewater of the textile industry has recently increased in connection with the increasing requirements for the protection of the surrounding aquatic environment. [7, 21]

We investigated in this work the efficiency of using ozone to reduce the color intensity of model solutions prepared on the basis of two dyes with different chemical structures of the chromophore.

We studied model solutions of the following dyes:

1) Acid metal complex azo dye "Lanasyn Black M-DL" produced by the Swiss company "Clariant". The dye was obtained at OJSC Brest Stocking Plant as one of the most widely used in production and difficult to remove in wastewater treatment.
2) Thiazine dye "Methylene blue". This dye, by its nature, is not an azo dye, it is highly resistant to oxidants since it itself has weak oxidizing properties. This dye has a relatively low molecular weight and a very high extinction coefficient [22, 23].

The chemical formulas of the above dyes are shown in Figure 1.

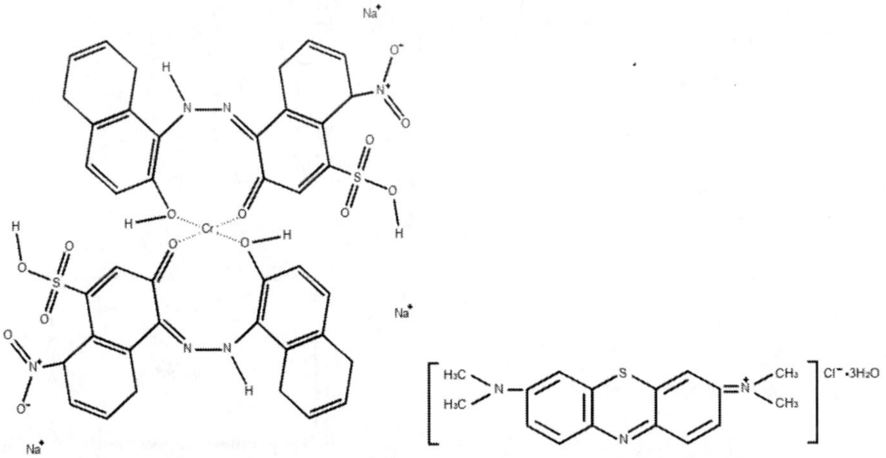

Lanasyne Black M-DL Methylene blue

Figure 1 - Chemical formulas of the studied dyes

Method

We used solutions of dyes "Lanasyn Black M-DL" and "Methylene blue" in distilled wateras an object of research. The concentration of the initial solutions was 1 g / L. Working solutions of dyes "Lanasyn Black M-DL" with a concentration of 30 mg / L (color intensity at a dilution ratio of 1: 200) and "Methylene blue" with a concentration of 4 mg / L (color intensity at a dilution rate of 1: 400)were prepared from these solutions by the dilution method.

The technique presented in the article [24] was usedto determine the intensity of the color by the degree of dilution.

The method in which gaseous ozone is introduced into the aqueous solution of the treated substance was not used since this method of introducing ozone does not allow to accurately determine the amount of ozone that has reacted. It is impossible when bubbling a solution to determine how much ozone has dissolved in an aqueous solution in the bubble column and reacted with substances dissolved in this solution and how much ozone remained in the gas phase and got into the ozone destructor.The dose of ozone when it is introduced into the treated aqueous solution by bubbling is a value that depends on the average diameter of the bubbles coming out of the gaseous ozone disperser, the temperature of the treated aqueous solution which affects the ozone solubility, the contact time of the solution with ozone bubbles which depends on the height of the column of the treated solution, concentration of ozone in the supplied gas mixture, duration of the ozonation process and other factors. Ozone was introduced by the method of its precise dosage in the form of an aqueous solutionto exclude the influence of the above factors on the efficiency of ozonation. An aqueous solution of ozone was obtained in a glass column filled with distilled water by bubbling water with an ozone-oxygen mixture. Distilled water was bubbled through a porous dispersant. In this way, an aqueous solution of ozone dissolved in distilled water was obtained. The scheme of the installation for the dissolution of ozone in water is shown in Figure 2.

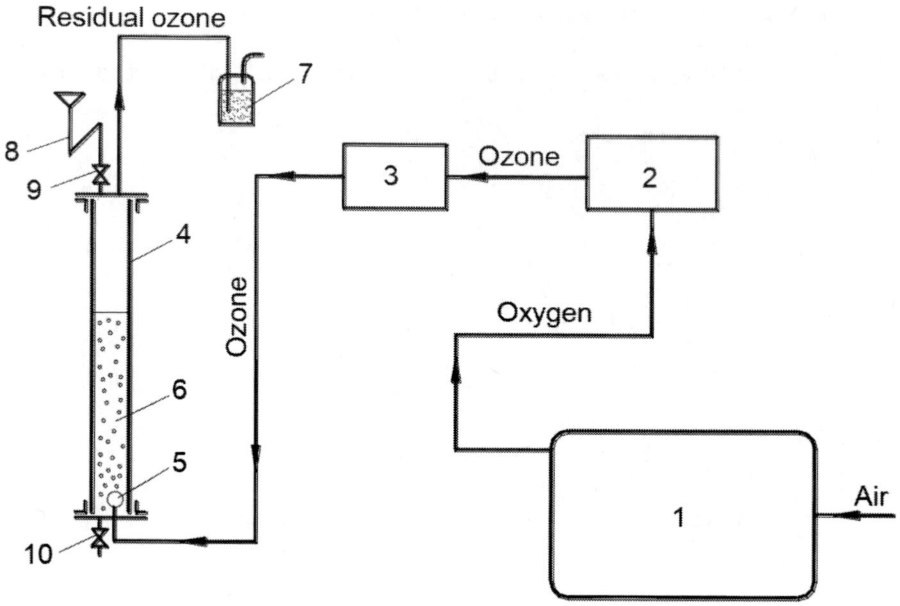

Figure 2. - Scheme of the device for obtaining an ozone solution in distilled water.
1 – oxygen concentrator Atmung OXY 6000; 2 –ozonizer PLATON 10/2; 3 – ozone meter MEDOZON 254/5; 4 – contact column; 5 – dispersant of the ozone-oxygen mixture; 6 – distilled water with dissolved ozone; 7 – destructor of residual ozone; 8 – hydraulic valve; 9 – tap for pouring distilled water; 10 - tap for sampling ozonized distilled water.

Oxygen (oxygen concentration 95%) produced by oxygen concentrator 1 enters ozonizer 2 where ozone is produced. Ozone-oxygen mixture is fed into the contact column 4 through a porous disperser 5 while distilled water 6 is saturated with ozone. Further, the ozone-oxygen mixture depleted in ozone enters the destructor 7 which is a Drexel flask filled with a packing that degrades ozone. Ozonized distilled water 6 from the contact column 4 is taken using the valve 10. New portions of distilled water are introduced into the contact column 4 through the hydraulic valve 8 using the valve 9.

1.5 liter flask with a ground glass stopper was used as a reactor for the interaction of ozone with an aqueous solution of the treated substance.The experiment was carried out as follows. A working solution with a volume of 500 ml was poured into a 1.5 *L* reactor flask. To carry out the experiment, ozonized water with a volume of about 1 liter was taken from the column into a special beaker then the concentration of ozone dissolved in distilled water was determined using the MEDOZON 245/8 (Zh-30) ozonometer. For this purpose, ozonized water was taken from the middle of the beaker using a pipette and poured into a quartz cuvette of an ozone meter. The average value of four determinations was taken as the ozone concentration in the ozonized water introduced into the reaction. Next, the volume of ozonized water was calculated which must be poured into the processed sample in order to provide a given dose of ozone.After that, the calculated volume of ozonized water was measured with a measuring cylinder and added to the treated dye solution in the reactor flask. If the volume of the solution mixture in the coble-reactor thus obtained was less than 1 *L* then distilled water was added to the reactor flask in order to bring the volume of the mixture in it to exactly 1 *L*. Thus, it was achieved that when ozone was introduced in the form of an aqueous solution, a twofold dilution of the treated dye solution with distilled water always occurred.Taking into account the above, the initial color intensity in terms of the degree of dilution of the test solutions is two times lower than that of working solutions because at a zero dose of ozone 500 ml of distilled water must be added to 500 ml of the working solution in order to obtain a mixture volume of 1 liter. Therefore, the initial concentration of the test solution "LanasynBlack M-DL" by the degree of dilution was 1: 100 and for the solution "Methylene blue" - 1: 200 (at zero dose of ozone). Then, the reactor flask was sealed with a ground glass stopper and vigorously shaken for 20 minutes. The hermetically sealed reactor flask provided almost 100% use of the ozone introduced into it for the destruction of the dye under study which was in the reactor flask. Further, using the technique described in article [24], the color intensity was determined from the degree of dilution of the solution obtained as a result of ozonation.

The effect of reducing the color intensity $Э_{сниж.\_ок}$ is determined from the ratio:

$$Э_{сниж.\_ок} = \frac{K_{исх} - K_{обр}}{K_{исх}} \cdot 100 \quad \% \qquad (1)$$

$K_{исх}$ - multiplicity of dilution of the original dye solution, corresponding to the color intensity according to the degree of dilution of the original dye;

$K_{обр}$ - the multiplicity of dilution of the dye solution treated with ozone, corresponding to the color intensity according to the dilution rate.

The multiplicity of dilution is the number of volumes of distilled water that must be added to the test sample until the color completely disappears in a layer 20 cm high. The "Coloring-monitoring" program described in article [24] when analyzing the absorption spectrum of the test solution gives exactly the dilution factor on the basis of which the intensity of the color is determined by the degree of dilution. The multiplicity of dilution is the reciprocal of the color intensity in terms of the dilution degree, i.e. if the multiplicity of dilution is 100 then the color intensity according to the dilution degree is 1: 100.

In Figure 3, the effectiveness of ozone is expressed through the effect of color reduction $Э_{сниж.\_ок}$ determined from the ratio:

$$Э_{сниж.\_ок} = \frac{O_{исх} - O_{обр}}{O_{исх}} \cdot 100 \quad \% \qquad (2)$$

$O_{исх}$ - the number of volumes of distilled water that must be added to the sample of the initial dye solution until the color completely disappears in a layer 20 cm high.

$O_{обр}$ - the number of volumes of distilled water that must be added to the sample of the dye solution treated with ozone until the color disappears completely in a layer 20 cm high.

Spectrophotometric studies were carried out on an SF-2000 spectrophotometerto study the process of destruction of the dye under the influence of ozone. We used quartz cuvettes K10 with an optical path length of 10 *mm*. The spectra of solutions of dyes in distilled water were studied before and after treatment with various doses of ozone in the visible range (390 ... 760 *nm*). The scanning step was 0.1 *nm*when the spectra were obtained, the number of accumulation cycles was 3. The degree of dye destruction was assessedaccording to the results of spectrophotometric studies.

Results and discussion

Figure 3 shows the absorption spectra of the initial and ozone-treated solution of the dye "Lanasyn Black M-DL" which were recorded during the research. The change in spectra reflects the depth of destruction of the dye. Spectrum 1, corresponding to an untreated dye solution, clearly shows an absorption maximum at a wavelength of 575 *nm*; however, even at an ozone dose of 0.04 *mg / mg of the dye*, the height of the maximum peak of the dye spectrum is compared with the height of an additional peak in the region of 400 *nm* which indicates chemical changes in the dye solution. With an increase in the specific dose of ozone, the changes in the spectra in comparison with the initial one increase, the height of the peak at a wavelength of 575 *nm* decreases and with an ozone dose of more than 0.4 *mg / mg of the dye*, the absorption maximum shifts to a shorter wavelength region (400 *nm*) which may indicate complete destruction of the original dye.

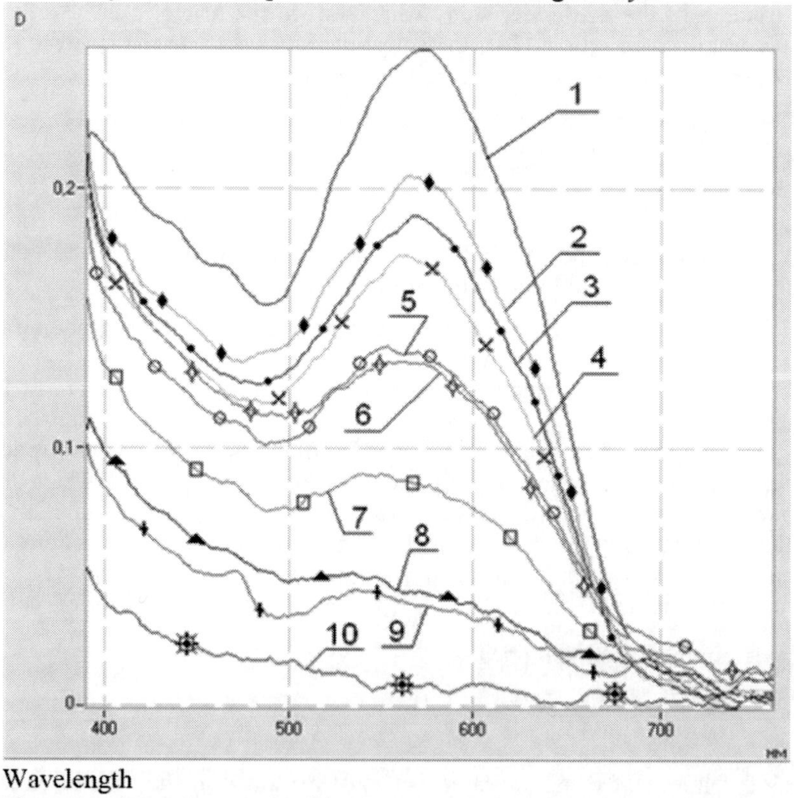

1 - initial solution;
2 - ozone dose 0.04 mg / mg;
3 - ozone dose 0.06 mg / mg;
4 - ozone dose 0.1 mg / mg;
5 - ozone dose 0.17 mg / mg;
6 - ozone dose 0.22 mg / mg;
7 - ozone dose 0.27 mg / mg;
8 - ozone dose 0.33 mg / mg;
9 - ozone dose 0.4 mg / mg;
10 - ozone dose 0.52 mg / mg

Wavelength

Figure 3 - Absorption spectra of the "Lanasyn Black M-DL" dye solution treated with different doses of ozone.

Based on the obtained spectra using the above-described method, the dependence of the color intensity of the "Lanasyn Black M-DL" dye solution on the specific dose of ozone was built and it is shown in Figure 4. Analysis of this dependence shows that ozone destroys the dye at any specific dose. At the same time, to reduce the color by 81% which is equivalent to a decrease in the color intensity in the degree of dilution from 1: 100 to 1:20 a dose of 0.23 *mg* ozone per 1 *mg* of the dye was sufficient.

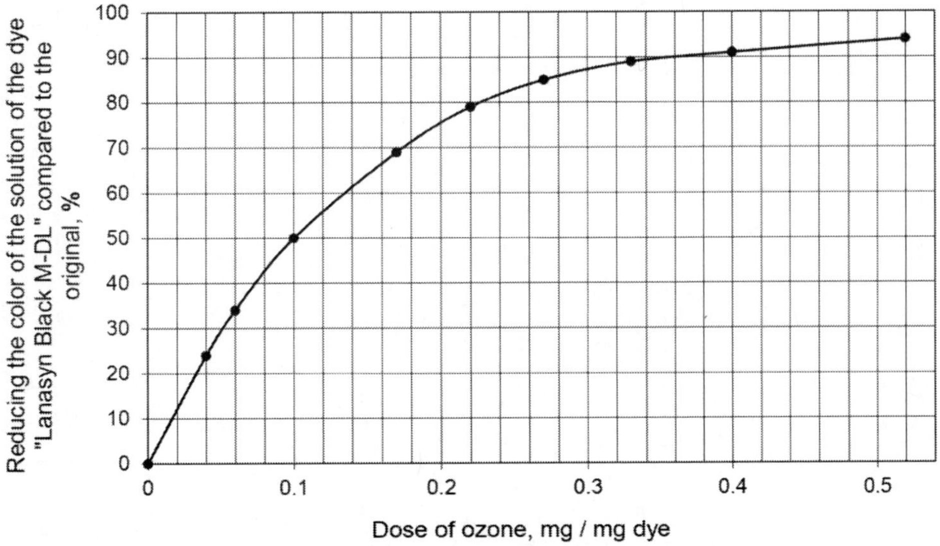

Figure 4. - The effect of the specific dose of ozone on the decrease in the color intensity of the "Lanasyn Black M-DL" dye solution.

We also investigated the effectiveness of using ozone to reduce the color intensity of the Methylene Blue dye solution. This dye does not belong to the class of azo dyes but according to the structure of the chromophore is a thiazine acid dye with weak oxidizing properties. The dye "Methylene blue" according to the literature data is relatively resistant to oxidants but it is relatively easy to recover and has a relatively low molecular weight [5] [6], therefore, according to the efficiency of its destruction using ozone and it is possible to more accurately estimate the specific doses of ozone required for the destruction of various classes of dyes.

By analogy with the previous dye, during the research, the spectra of the initial and treated with various doses of ozone solutions of the dye "Methylene blue" were recorded and it is shown in Figure 5. Changes in the spectra of the dye solution treated with different doses of ozone show a proportional decrease in absorption levels with increasing ozone dose, however, the forms the spectrum remain practically unchanged. This observation can be explained by the fact that the decomposition products of the dye absorb exclusively in the ultraviolet region, therefore, ozone also acts selectively on the molecules of this dye, i.e. on certain connections. However, the mechanism of action of ozone on dye chromophores is practically not studied. It can also be noted that, at ozone dose of more than 1.25 *mg / mg of the dye* spectral analysis in the visible region did not reveal noticeable absorption.

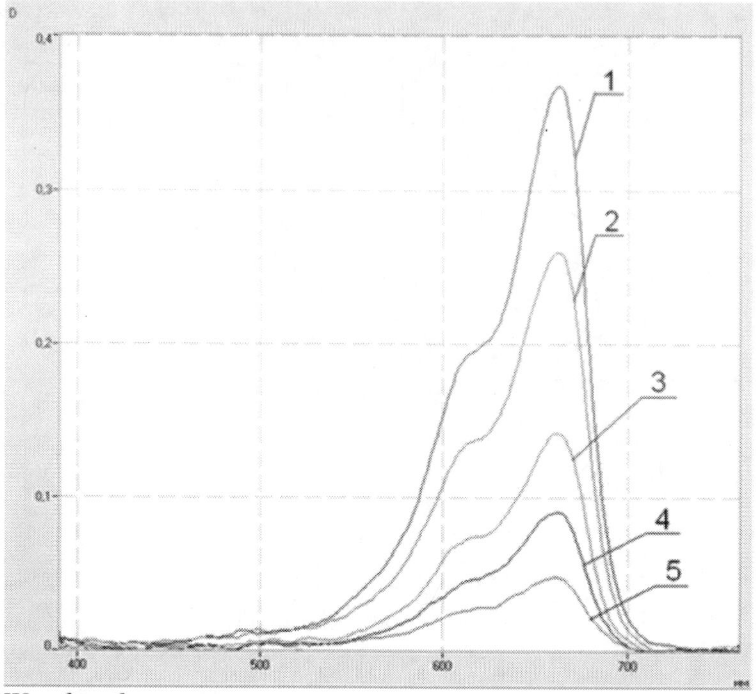

1 - initial solution;
2 - ozone dose 0.25 mg / mg;
3 - ozone dose 0.52 mg / mg;
4 - ozone dose 0.75 mg / mg;
5 - ozone dose 1 mg / mg;

Wavelength

Figure 5 - Absorption spectra of the Methylene Blue dye solution treated with various doses of ozone.

On the basis of the obtained spectra, the dependence of the efficiency of ozone application for reducing the color intensity of the solution of the dye "Methylene blue" with a concentration of 2 *mg / l* was plotted which corresponds to the color intensity at a dilution ratio of 1: 200 and is shown in Figure 6.

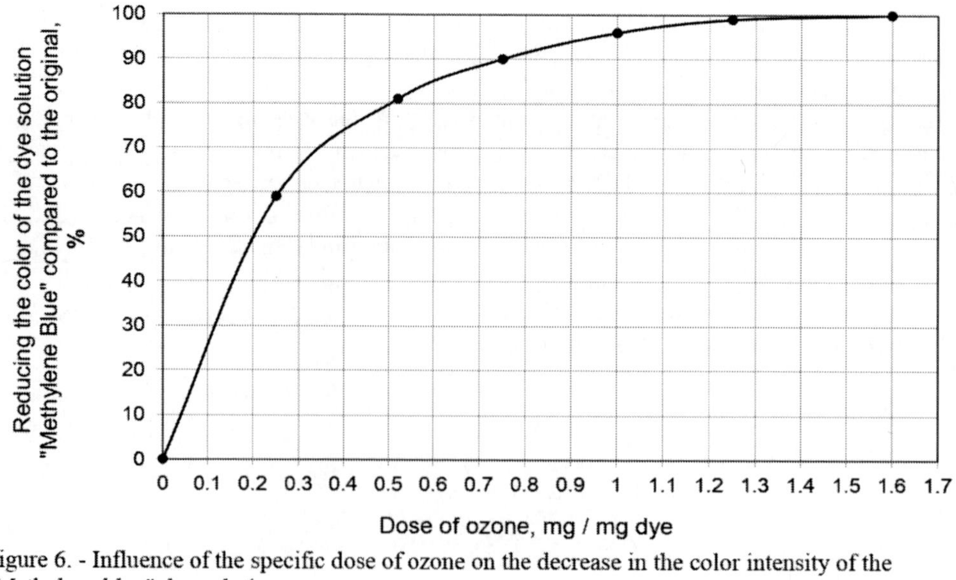

Figure 6. - Influence of the specific dose of ozone on the decrease in the color intensity of the "Methylene blue" dye solution.

Ozone also effectively reduces the color of the solution of this dye despite the fact that "Methylene Blue" dye is more resistant to oxidation. Thus, at a specific ozone dose of 0.75 *mg /mg of the dye*, the color intensity of the ozonated model solution was already achieved at a dilution ratio of 1:20 which meets the requirements for the permissible color of wastewater discharged into the municipal drainage network. At an ozone dose of more than 1.5 *mg / mg of the dye* the decrease in the color intensity of the "Methylene Blue" dye solution was 100%; i.e. solution became completely colorless.

Conclusion

According to the results of the studies carried out to reduce the color of solutions of dyes "Lanasyn Black M-DL" and "Methylene blue" by treatment with ozone, the following conclusions can be drawn:

1. Ozone is an effective reagent for reducing the color intensity of wastewater caused by synthetic dyes. In the studies performed, ozone acted equally effectively on dyes with different chromoform systems (on azo dye and on a dye with a heterocyclic thiazine chromophore).
2. Doses of ozone required to reduce the color intensity by 5 ... 10 times according to the degree of dilution of dye solutions are 0.25 ... 0.75 *mg / mg of the dye* which in model solutions in terms of the concentration of dyes was *1.5 ... 3, 75 mg ozone per liter of solution*. An even stronger decrease in color occurred at higher doses of ozone, up to complete discoloration.
3. The effect of color reduction under the influence of ozone is irreversible which indicates the destruction of the chromophore part of the dye molecule.
4. In view of the high reactivity of ozone which makes it possible to effectively destroy carbocyclic and heterocyclic compounds, it is possible not only to effectively reduce the color of wastewater but also to destroy the decay products of dyes which is especially important for the protection of the surrounding aquatic environment.

Bibliography

1. Vasiliev G.V., LaskovYu.M., Vasilyeva E.G. Water management and waste water treatment of textile industry enterprises. - M .: Light industry, 1976 .-- 224 p.
2. Wang Z. and others. Textile Dyeing Wastewater Treatment // Advances in Treating Textile Effluent. , 2011.
3. Yaseen D. A., Scholz M. Textile dye wastewater characteristics and constituents of synthetic effluents: a critical review // Int. J. Environ. Sci. Technol. 2019.
4. Krasnoborodko, I.G. Destructive purification of waste water from dyes / I.G. Krasnoborodko. - L: Chemistry, 1988 .-- 192 p.
5. Holkar C. R. et al. A critical review on textile wastewater treatments: Possible approaches // J. Environ. Manage. 2016.
6. Chichibabin, A.E. Basic principles of organic chemistry / A.E. Chichibabin, M: Goskhimizdat, 1958 .-- 767 p.
7. Krichevsky, G.E., Chemical technology of textile materials / G.E. Krichevsky, M.V. Korchagin, A.V. Senakhov. - M: "Light industry and consumer services", 1985. - 640 p.
8. Henze, M. Wastewater Treatment: Per. from English. / M. Henze, P. Armoes, J. La-Cour-Jansen, E. Arvan - M .: Mir, 2004 .-- 480 p.

9. Alekseev E.V. Ecological aspects of wastewater treatment containing biologically stable organic substances // Water and ecology: problems and solutions. 2015. No. 4. P. 68–78.
10. Soares O. S. G. P. et al. Ozonation of textile effluents and dye solutions under continuous operation: Influence of operating parameters // J. Hazard. Mater. 2006.
11. Tehrani-Bagha A. R., Mahmoodi N. M., Menger F. M. Degradation of a persistent organic dye from colored textile wastewater by ozonation // Desalination. 2010.
12. Wu J., Doan H., Upreti S. Decolorization of aqueous textile reactive dye by ozone // Chem. Eng. J. 2008.
13. Turhan K., Turgut Z. Decolorization of direct dye in textile wastewater by ozonization in a semi-batch bubble column reactor // Desalination. 2009.
14. Kozhinov, V.F. Ozonation of water / V.F. Kozhinov, I.V. Kozhinov. - M .: Stroyizdat, 1974 .-- 159 p.
15. Orlov, V.A. Ozonation of water / V.A. Orlov. - M .: Stroyizdat, 1984 .-- 88 p.
16. Selcuk H. Decolorization and detoxification of textile wastewater by ozonation and coagulation processes // Dye. Pigment. 2005.
17. Snider E. H., Porter J. J. Ozone destruction of selected dyes in wastewater // Am. Dyest. Report. 1974.
18. Razumovsky, S.D. Ozone and its reactions with organic compounds / S.D. Razumovsky, G.E. Zaikov. - Moscow: Nauka, 1974 .-- 324 p.
19. Rawat D., Mishra V., Sharma R. S. Detoxification of azo dyes in the context of environmental processes // Chemosphere. 2016.
20. Neylond, O. Ya. Organic chemistry / O. Ya. Nylon. - M: Higher school, 1990 .-- 751 p.
21. Babu B. R. et al. Cotton textile processing: Waste generation and effluent treatment // J. Cotton Sci. 2007.
22. Brief chemical encyclopedia: in 5 volumes / IL. Knunyants [and others]; under total. ed. I.L. Knunyants. - M: "Soviet Encyclopedia", 1964. –T. 3.- 1112 p.
23. Turhan K. et al., Decolorization of textile basic dye in aqueous solution by ozone, Dye. Pigment. 2012.
24. Belov S.G., Naumchik G.O. Development of an instrumental method for determining the color intensity of waste water from textile enterprises // Water supply and sanitary engineering. 2017. No. 3. P. 53–65.

Translocation of heavy metals and methods of their detoxification in podzolized chernozem

Olga Chernikova[1*], *Yuri* Mazhaysky[1], *Svetlana* Buryak[2], *Tatyana* Seregina[1]

[1]Academy of law management of the federal penal service of Russia, 390036 Ryazan, Sennaya str., 1, Russian Federation
[2]Ryazan State Agrotechnological University named after P.A. Kostychev, 390044 Ryazan, Kostycheva str., 1, Russian Federation

Abstract. At present, the fact that all unforeseen negative consequences of anthropogenic human activity significantly affect the biochemical regime of the natural environment is obvious. Therefore, agriculture should be aimed at reducing the release of chemicals from the biological cycle. Field observations were conducted to study the effect of detoxification techniques on the accumulation of heavy metals in soil. Six experiment variants were developed, using detoxicants with various combinations of them. The results of this research indicate that the organic matter of manure binds soluble metal salts into organometallic complexes, and fertilizer phosphates convert them into sparingly soluble compounds. Thus, they improve (heal) the soil contaminated by heavy metals.
Keywords: heavy metals, detoxification methods, translocation, chernozem.

Introduction

The increase in anthropogenic pressure on the soil cover leads to degradation processes, among which chemical degradation of soil and, in particular, their contamination by heavy metals has become widespread in recent decades.

Heavy metals enter the biosphere due to technogenic dispersion from emissions of high-temperature technological processes (ferrous and non-ferrous metallurgy, thermal power plants, etc.). One of the most important components of the environment is the soil, which takes on the effects of industrial emissions and waste, residual pesticides and other toxicants, making a difference as a buffer and detoxifier. The source of soil contamination can serve as sewage sludge used as fertilizer. In addition, secondary contamination due to the removal of heavy metals from dumps of mines or metallurgical enterprises by water or air flows [3,7].

Technological emissions from stationary and mobile sources of environmental contamination enter the atmosphere, and then, falling to the earth's surface, accumulate in the upper horizons of the soil, are again included in natural and man-made migration cycles.

* Corresponding author: chernikova_olga@inbox.ru

The intensification of agricultural production, the dynamic development of industry, transport and energy lead to increased anthropogenic pressure on ecosystems. Contamination of the soil cover by heavy metals leads to a weakening of its resistance to technogenesis, the development of degradation processes, a decrease in fertility, and, consequently, a decrease in the productivity of both natural and agricultural landscapes, which significantly affects the volume and quality of productive resources. Agricultural products grown on contaminated soil are often toxic due to the increased content of heavy metals [1,2, 4-8].

The soil with a heavy particle-size distribution exhibits its buffering properties. The danger of plant contamination by heavy metals on such soil is much less, which is associated with a greater retention capacity of the silt fraction.

With the organic matter of the soil, metals can form organometallic complexes that are not capable of overcoming cell membranes at the soil – root contact. The most stable compounds in the soil form heavy metals with humic compounds. The stability of complexes of heavy metals with fulvic acids increases with increasing of pH.

Chernozem is a strong accumulator of heavy metals; in this regard, the urgent problem is the development of agromeliorative practices for the rehabilitation of technologically contaminated chernozems and the production of safe crop products.

The aim of our work was to study the influence of detoxification techniques for podzolized chernozem, contaminated by heavy metals, on their accumulation in soil.

Methods

Developing agrochemical methods for the decontamination of contaminated soil in a stationary lysimetric experiment (Fig. 1), a preliminary stage was carried out in which the total content of Zn, Cu, Pb, Cd in podzolized chernozem, its hydrolytic acidity (Ha) in each lysimeter were studied, and acidity was neutralized .

According to the regional gradation of soil contamination levels, compiled on the basis of the geochemical background, the content of Cu - 90 mg / kg, Zn - 110 mg / kg, Pb -40 mg / kg, and Cd - 0.6 mg / kg in the soil represents increased contamination.

Modeling of the increased complex level of soil contamination was carried out by adding to the soil. In this case, chemically pure salts were used: $Zn(CH_3COO)_2 \times 2H_2O$; $CuSO_4 \times 5H_2O$; $Pb(CH_3COO)_2$; $CdSO_4$.

For this, a soil layer of 20 cm depth was selected from the lysimeter. The calculated dose of Cu, Zn, Pb, and Cd salts was thoroughly mixed with this soil and placed in the same lysimeter.

The following fertilizer systems were studied: organic (cattle manure), organo-mineral and mineral, where double superphosphate was used periodically and annually at higher doses. For podzolized heavy loamy chernozem, the manure rate of 100 t / ha was adopted (Table 1). The annual norms of mineral fertilizers, depending on the crop, are adopted according to the recommendations for our zone.

Dates and methods of fertilizing in the rotation link were performed in the following way. In autumn, after harvesting the rotation crop, cattle manure (cattle) was introduced at the rate of 100 t / ha, with a humidity of 85%.

This organic rate was buried to a depth of 25 cm. The calculated norms of mineral fertilizers were evenly distributed manually on the surface of the lysimeter, and then the soil was dug to a depth of 12-15 cm. The surface of the soil was leveled by a rake. A mixture of herbs (fescue, timothy, clover) was sown as a leveling culture.

Urea (N - 46%), double superphosphate (P2O5 - 44%), potassium sulfate (K2O - 48%) were used in the experiment.

The gross determination of Cu, Zn, Pb and Cd in manure and mineral fertilizers was carried out by the atomic absorption method.

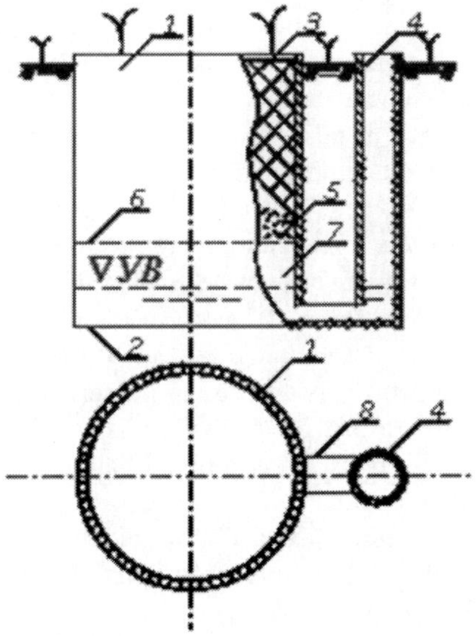

Figure 1. Diagram of a water balance lysimeter (legend: 1 - lysimeter case, 2 - lysimeter bottom, 3 - soil monolith, 4 - pocket for water extraction, 5 - gravel packing, 6 - strainer, 7 - groundwater level, 8 - communication channel).

Analytical research in determining the gross and mobile forms of heavy metals in the soil was carried out in the ACTC NIMRM. When determining the total content of elements, the soil sample was decomposed with a mixture of acids: HCl, HNO3, HF, HClO4 with heating. Mobile forms were recovered using an ammonium acetate buffer - pH of 4.8. The heavy metal content was determined using an inductively coupled plasma mass spectrometer Elan - 6100, and an atomic emission spectrometer Optima - 4300.

Table 1. The scheme of establishment and performance of field lysimetric experiment. Station on podzolized heavy loamy chernozem.

Varian t №№	Names of variants, fertilizer application systems in the rotation link	Abbreviations in the tables of variant names
1	Without fertilizers	W/f
2	Cattle manure 100 t / ha - periodic application	H100
3	Cattle manure 100 t / ha - periodic application, N60 (N90) P60K60 (K120) - annually depending on the crop (kg / ha)	H100 N1P1K1
4	P2 - periodic application of phosphorus, once every 2 years at a dose of 120 kg / ha, annual use of N60 (N90) K60 (K120)	P2N1K1
5	P4 - periodic application of phosphorus, once every 4 years at a dose of 240 kg / ha, annual use of N60 (N90) K60 (K120)	P4N1K1
6	P2 (e) - annual application of an increased dose of phosphorus (120 kg / ha) and optimal doses of N60 (N90) K60 (K120)	P2(e)N1K1

Statistical processing of the obtained results was carried out by generally accepted methods (assessment of significance according to Fisher's and Student's criteria) using the Microsoft Excel software package.

Results and Discussion

The total soil contamination is characterized by the gross content of heavy metals. Changes in the content of chemical elements in the soil characterize steady changes in soil properties. The negative effect of high concentrations of heavy metals on the soil depends on their mobility and solubility. While the elements are firmly connected with the constituent parts of the soil, they are inaccessible to plants, their vertical movement in soil is weakly expressed. Therefore, the content of mobile forms of heavy metals in the soil characterizes the sanitary-hygienic situation and determines the need for amelioration detoxification measures.

The problem of mobility of heavy metals in soil is very complex and multifactorial. Depending on the extractant used, a different amount of the mobile form of the heavy metal, which with a certain conditionality can be considered available for plants, is recovered.

The availability of a mobile form of heavy metal for plants depends largely on the properties of the soil and the specific characteristics of the plants. Moreover, the behavior of each element in the soil has its own specific laws inherent only to it.

For the extraction of mobile forms of heavy metals, various chemical compounds that have unequal extracting power are used. A number of authors proposed a scheme of various fractions of mobile metals in the soil:

1) the total stock of mobile forms (extracted by acids);
2) active mobile form (extracted by buffer solutions);
3) exchange (extracted by neutral salts);
4) water soluble.

According to the researchers [9–15], the majority of mobile compounds of heavy metals are confined to the humus, humus-eluvial horizons, where the roots of plants, in which biochemical processes are actively occurring and which contain a lot of organic substances, are located. Under the influence of contamination by heavy metals of the soil, organic complexes, which have high mobility, are formed in it.

Nazarova L.K. et al. [16] note that the systematic long-term use of ballast and concentrated fertilizers, including phosphorite meal on sod-podzolic heavy loamy soil did not significantly affect the gross content of heavy metals and toxic elements in the soil, but led to the almost doubled increase in their mobility in it. Alloway B.J. et al. [17-19] indicated that complete mineral fertilizer, creating optimal conditions for the growth and development of plants, significantly reduced the accumulation of mobile forms of lead and cadmium in the soil.

There is evidence of a close relation between the total content of Zn, Pb, Cd and their mobile forms, extracted by acetate-ammonium buffer (AAB) pH 4.8. Therefore, we determined the mobile forms, their mobile parts in our research of the rehabilitation measures of contaminated soil by heavy metals. AAB pH 4.8 was used (Table 2) for the extraction of mobile forms of heavy metals.

Mobile forms of metals are the most aggressive part of the gross content of HM, which can enter directly into the root system. These mobile forms are involved in migration flows, including intra-profile ones. Therefore, the liquid phase of soil is a direct source of HM for higher plants and soil biota.

The research was carried out in annual soil samples, which were selected after crop accounting. According to years of the research, variability was observed in the quantities of mobile forms of heavy metals. The average metal content in the acetate-ammonium buffer indicated that they were extracted significantly more in the variant without fertilizers. The maximum degree of mobility was observed in Cd (73.3%). All fertilizer systems used for the rehabilitation of contaminated by HM chernozem, on average, reduced the accumulation of mobile compounds of Zn, Cu, Pb, Cd. Thus, the degree of their mobility also decreased. The organo-mineral fertilizer system reduced the mobility of zinc from 42.8% to 21.2%, copper from 15.7% to 8.3%, lead from 53.5% to 25.5% and cadmium from 73.3% to 51.6%.

Table 2. The influence of fertilizer systems on the study of the rehabilitation of contaminated by heavy metals podzolized chernozem.

Experiment variant	Element	The amount of element extracted from the soil $AAB_{pH=4,8}$, mg / kg				
		1 year	2 years	3 years	Average, mg / kg	%*
W/f	Zn	26,3	45,3	69,7	47,1	42,8
	Cu	32,3	10,1	19,5	14,1	15,7
	Pb	31,2	10,7	22,2	21,4	53,5
	Cd	0,47	0,29	0,57	0,44	73,3
H100	Zn	17,3	36,2	26,1	26,5	24,1
	Cu	11,1	11,4	4,1	8,9	9,9
	Pb	20,4	12,9	9,8	14,4	36,0
	Cd	0,49	0,25	0,30	0,35	58,3
H100N1P1K1	Zn	4,2	26,2	39,4	23,3	21,2
	Cu	13,2	3,6	5,7	7,5	8,3
	Pb	10,0	11,0	9,7	10,2	25,5
	Cd	0,50	0,17	0,27	0,31	51,6
P2N1K1	Zn	60,0	37,9	36,8	44,9	40,8
	Cu	31,1	6,7	7,8	15,2	16,9
	Pb	12,5	7,1	11,2	10,3	25,8
	Cd	0,41	0,26	0,32	0,33	55,0
P4N1K1	Zn	67,0	63,5	47,4	59,3	53,9
	Cu	14,0	7,5	3,8	8,4	9,3
	Pb	7,2	10,2	8,5	8,6	21,5
	Cd	0,39	0,32	0,30	0,34	56,7
P2(e)N1K1	Zn	7,8	64,0	44,1	38,6	35,1
	Cu	0,7	13,0	6,4	6,7	7,4
	Pb	2,5	20,7	12,2	11,8	29,5
	Cd	0,13	0,25	0,25	0,21	35,0

The note: * data are given on the degree of mobility of HM (the ratio of mg / kg in AAB to the gross content in the soil, expressed in %)

Almost close changes in this indicator were in the variant of the annual introduction of P2 (e) N1K1. Especially in this variant, the accumulation of mobile Cd decreased. Another ecotoxicant Pb minimized the degree of mobility from the periodic use of phosphorus fertilizers (P4) against the background of N1K1. It decreased from 53.5% (W/f variant) to 21.5% in the P4N1K1 variant.

Conclusions

So, the studied application of different fertilizer systems on podzolized chernozem significantly remove and reduce the aggressiveness of pollutants. The results of this research indicate that the organic matter of manure binds soluble metal salts into organometallic complexes, and fertilizer phosphates convert them into sparingly soluble compounds. Thus, they improve (heal) the soil contaminated by heavy metals.

References

1. A. Zafarzadeh, H. Rahimzadeh, A. H. Mahvi. Health Risk Assessment of Heavy Metals in Vegetables in an endemic esophageal cancer region in Iran, Health Scope, **7(3)**: e12340 (2018) doi: 10.5812/jhealthscope.12340.
2. X.G. Zhao, B. Li, M. Gao, Y. Li. Detection and analysis of heavy metals in vegetables in Xinzhu vegetable base, Advances in Engineering Research, 2nd ICSD, **94**, 252-254 (2016).
3. S. Kamran, A. Shafaqat, S. Hameed, S. Afzal, F. Samar, B. S. Muhammad, A. S. Bharwana, M. T. Hafiz. Heavy metal contamination and what are the impacts on living organisms, Greener Journal of Environmental Management and Public Safety, **2 (4)**, p 172-179 (2013).
4. G. Sandeep, K.R. Vijayalatha, T. Anitha. Heavy metal and its impact in vegetable crops, International Journal of Chemical Studiez, **7(1)**, 1612-1621 (2019).
5. M. Abou Auda, I. Abu Zinada, E. El Shakh Ali. Accumulation of heavy metals in crop plants from Gaza Strip, Palestine and study of the physiological parameters of spinach plants, Journal of the Association of Arab Universities for Basic and Applied Sciences, **10**, 21–27(2011).
6. I.I. Ekpe, S. E. Okere, L.C. Agim, C.M. Ahukaemere, S. C. Ihemtuge, C. Okoye, M.D. Onuora, M.O. Nwaigwe. Effect of Organic Wastes on Soil Heavy Metal Concentration and Growth Characteristics of Cucumber (Cucumis sativus L.) in an Ultisol, FUTOJNLS, **2 (2)**, 335 – 345 (2016).
7. A. O. Igwegbe, C. H. Agukwe, C. A. Negbenebor. A survey of heavy metal (lead, cadmium and copper) contents of selected fruit and vegetable crops from Borno State of Nigeria, International Journal of Engineering and Science, **2(1)**, 01-05 (2013).
8. R. Lacatusu, A.R. Lacatusu. Vegetable and fruits quality within heavy metals polluted areas in Romania, Carpth. J. of Earth and Environmental Sciences, **3(2)**, 115-129 (2008).
9. Harmanescu M., Alda L.M., Bordean D.M., Gogoasa I., Gergen I., Heavy metals health risk assessment for population via consumption of vegetables grown in old mining area; a case study: Banat County, Romania, Chemistry Central Journal, **5:64**, (2011) http://journal.chemistrycentral.com/content/5/1/64.
10. I. Gergen. Analysis of agri-food products, Eurostampa Publishing House, Timişoara, (2004).
11. I. Fatykhov, N. Busorgina, B. Borisov, Ch. Islamova. Chemical composition of sod-strongly-podzolic light-loamy soil during long-term agricultural USE. Vestnik of Kazan State Agrarian University, 82-86 (2019). doi:10.12737/-article_5db961e997b831.34073353.
12. N. Gomonova, I. Skvortsova, G. Zenova. Effect of the long-term application of different fertilization systems on soddy-podzolic soil. Eurasian Soil Science, **40**, 456-462 (2007). doi:10.1134/S1064229307040126.
13. I. Jaskulska, D. Jaskulski, M. Kobierski,. Effect of liming on the change of some agrochemical soil properties in a long-term fertilization experiment. Plant, Soil and Environment, **60**, 146-150 (2014). doi: 10.17221/850/2013-PSE.
14. A. Łukowski, J. Wiater. Influence of Mineral Fertilization on Lead, Cadmium, and Chromium Fraction Contents in Soil. Polish Journal of Environmental Studies, **20**, 951-960 (2011).

15. R. Mouhamad. Effect of Mineral Fertilizers Application on Accumulation of Heavy Metals in Soil and Tomato Plant, **6**, 394-404 (2019).
16. L.K. Nazarova, I.K. Dilmukhametova,V.S. Yegorov, N.A. Kirpichnikov, E. Morachevskaya, M.M. Karpukhin. Influence of long-term use of mineral fertilizers and liming on the state and balance of lead in agrocenosis on soddy-podzolic heavy loamy soil in the moscow region, **2**, 18-23 (2019).
17. S. Czarnecki, R.-A. Duering. Influence of long-term mineral fertilization on metal contents and properties of soil samples taken from different locations in Hesse, Germany. SOIL Discussions, **1**, 239-265 (2014). doi:10.5194/soild-1-239-2014.
18. B.J. Alloway Sources of Heavy Metals and Metalloids in Soil. In: Alloway B. (eds) Heavy Metals in Soil. Environmental Pollution, **22** (2013).
19. N. Mărin, N. Vrînceanu, N. Lupaşcu, M. Dumitru. Heavy metals from the soil and mineral fertilization Scientific Papers. Series A. Agronomy, **LXI (1)**, 101-108 (2018).

Energy saving in wastewater treatment plants

Natalia Ciobanu*
Technical University of Moldova, 39 Dacia avenue, Chisinau, Republic of Moldova

Abstract. This paper uses data from the Chisinau Wastewater Treatment Plant (WWTP) (2018 year) to analyze the potential for energy recovery from wastewater treatment plant via anaerobic digestion with biogas utilization with electricity generation. These energy recovery strategies could help offset the electricity consumption of the wastewater treatment plants and represent possible areas for sustainable energy policy implementation. We estimate that anaerobic digestion could save approximately 14,444,918 kWh annually in Chisinau WWTP. Anaerobic digestion is widely considered as an environmentally friendly technology for sewerage sludge. This study aims to highlight the potential as well as to provide a starting point for further studies regarding the treatment as sewerage sludge using anaerobic digestion in Republic of Moldova and recovery energy that could further reduce electricity cost and reduction of sludge cake.
Keywords: wastewater treatment, energy, sludge, anaerobic digestion, biogas.

1. Introduction

Wastewater collected from municipalities and communities must ultimately be returned to receiving waters or to the land or reused [1]. Chisinau WWTP has been constructed in successive phases and is operated by "*Apa Canal Chisinau*" (ACC). The wastewater flow rate was approximately 435 thousand m^3/day (yearly average) with average chemical oxygen demand (COD), 5 day biological oxygen demand (BOD_5), and total suspended solids (TSS) concentrations of 900 mg/l, 300 mg/l, and 500 mg/l respectively. Chisinau WWTP has been largely oversized and is currently running at only one third of its capacity (142 thousand m^3/day). The wastewater treatment process features the conventional steps of a medium-load activated sludge plant modified with contact stabilization tank for the aeration of returned activated sludge to the head of secondary treatment. The lack of an appropriate sludge treatment line combined with extremely poor conditions of the works and pieces of equipment and the absence of online sensors and of control systems impede the optimal operation of the plant. The current practices in sludge management by dewatering and drying in geotubes must be changed since they cause significant odor problems in the city especially in nearby settlements.

This is a study of analyzing scenario in terms of their technological and economical feasibilities focusing on the sludge management processes to increase the sustainability of the WWTP operation in Chisinau by using biogas recovered from waste activated sludge (WAS). The formation of biogas composed of mainly CH_4 and CO_2 in anaerobic digester

*Corresponding author: natalia.ciobanu@fua.utm.md , ciobnata@hotmail.com

processing WAS as the substrate is proposed for rendering organic load produced in the secondary treatment and reducing solids content of WAS.

2 Methods

2.1 Data collection

The existing Chisinau Wastewater Treatment Plant dates back from the 70s and was built in several stages although it has never worked at full capacity according to the best of our knowledge. About 35-40 % of the works are being used currently. The treatment of wastewater in Chisinau WWTP takes place in several stages as follows:

a) *The Preliminary Treatment* consists of two physical operations: screening and sand removal. Screening (six pieces, 10.5 or 16 mm for spacing) removes large solids, which are retained by the screens. The main reasons for screening are to protect the pumps and pipe works, downstream treatment units and tanks. The sand is removed by sedimentation through grit chamber (with four sections).

b) *The Primary Treatment* consists of primary settling process. Primary settling tanks (4 cylindrical tanks with a volume of 4500 m^3 each) allow solids to settle gravimetrically with the settling velocity of 0.7 m/s minimum. Primary sludge is transferred directly to geotubes to be dewatered.

c) *The Secondary Treatment* is a conventional activated sludge process (aeration and secondary settling) modified by contact stabilization of returned activated sludge (RAS) which is 70 – 80% to maintain the bacterial population in the reactor. There are currently 6 aeration tanks in operation with (2 – 22500 m^3, 4 – 12500 m^3) with 10-15 days solids retention time (SRT). The secondary settling tanks (volume 4500 – 9000 m^3 each, four work out of eight) remove mixed liquor suspended solids (MLSS) produced in the aeration tanks by gravitational force. The clarified wastewater is discharged to the receiving water body (Bac River) and sludge is dewatered in geotubes.

d) *The Sludge Management* is simply dewatering of primary and secondary sludge in geotubes followed by landfills. The mixture of primary and secondary sludge used to be stored directly on sludge platforms (32 ha) for dehydration, which generated serious odor problems. The geotubes were implemented in 2009 as a quick and easy attempt to mitigate odor issues generated by the sludge platforms and it proved to be quite efficient in odor removal. Sludge is pumped into the geotubes and are added the polymers, which makes the solids bind together and water separate. Clear effluent water simply drains from the geotube through the small pores in the specially engineered textile. The decanted water is returned to the head of plant. After the final cycle of filling and dewatering, the solids remain in the bag and continue to density due to desiccation as residual water vapor escapes through the fabric. Volume reduction can be as high as 90 percent. When full, the geotubes (appr.600 kg each) and sludge cake are deposited at a landfill. Thus, capital cost of sludge management can be estimated as 391,308.8 €/year (160 Geotubes) [2]. After dewatering process, geotubes are opened and the sludge cake is transferred by trucks to a landfill site. The geotubes generated approximately 500,768.30 m^3/year of sludge in 2018.

Nowadays, the Chisinau WWTP is powered with electricity 100% from the grid. The main power consumption apparatus in the plant are pumps and blowers. All other auxiliary power needs such as lighting and potable water pumps are also supplied from the grid. This data are essential for predicting the energy requirements for an alternative power system before doing an upgrade. The average yearly was calculated as 19,450,795 kWh and cost including all taxes was 1,623,436.073 Euro/year or 0.083 Euro per kWh. For a 24 hour period the power consumption for the Chisinau WWTP was estimated 53,310 kWh. The electricity consumption of the individual units is shown in *Table 2.1*, this data indicate that the blowers consumed the highest amount of electricity.

Table 2.1. Evaluation of equipment electrical consumption by current operation

No	Unit/Equipment name	Electricity consumption, kWh/day	Percentage of energy consumption, %
1.	Blower	39,500	74
2.	Main Station Pump	13,810	26
3.	Return sludge pump		

2.2 Two case scenarios

There are two scenarios investigated for the Chisinau WWTP focused on sludge production and its management including current operation to determine the best alternative with the maximum energy savings. The scenario is designed to employ anaerobic digestion in sludge management and mass and energy balance calculations and analyses are described in this section.

CASE 1: Current wastewater treatment operation. The current operation of the Chisinau WWTP is described in Section 2.1 in detail. The mass flow of solids (kg/day) is calculated using the average yearly flow rate (141,400 m^3/day) and the concentration of TSS (74,845.85 kg/d) in untreated wastewater (Figure 2.1). The measured values were used when available in mass balance calculations. There is 1,372 m^3/day of sludge production mainly composed of primary settlings (>95% wet volume), is discharged to geotubes for the dewatering process. The removal efficiency of TSS and BOD_5 is determined as 95 and 96% respectively for the year of 2018 (see Table 2.2). The energy consumption for the treatment of 1 m^3 wastewater is 0.380 kWh/m^3, which is similar to the energy used to treat 1 m^3 of wastewater (no industrial input) by secondary treatment; such as, 0.485 kWh/m^3 in Slovakia [3], 0.420 kWh/m^3 in Sweden [4], and 0.304 kWh/m^3 in Japan [5]. The average energy consumption of using an activated sludge system for the treatment of wastewater has been reported as 0.34 kWh/m^3 in U.S.A. [6].

Table 2.2. Treatment performances and the quality of the treated effluent (average values for the period from 01.01.2018 until 12.31.2018)

Parameters	**Influent WWTP**	**Maximum allowable concentration in WWTP [7]**	**Effluent WWTP**	**Maximum allowable concentration at the discharge in the river Bac [8, 9]**	**Minimum removal rate %**
COD, mg/l	967.79	500	83.98	125.0	91.32
BOD_5, mg/l	299.20	225	13.41	25.0	95.51
TSS, mg/l	529.32	350	26.97	35.0	94.90
pH	7.653	6.5-8.5	7.615	6.5-8.5	-
T, °C	18.641	8 - 30	19.534	30	-
TKN, mg/l	63.031	-	28.139	10.0	55.0
NH_4, mg/l	41.044	30.0	21.547	2.0	47.5
P, mg/l	12.768	5.0	3.491	2.0	73.0
N, mg/l	63.400	-	30.822	15.0	51.0

In *Figure 2.1* is shown a schematic diagram of activated sludge process with wasting from the sludge return line on current operation plant.

https://doi.org/10.1051/e3sconf/202021201003

Table 2.3. Monthly average of sludge in Chisinau WWTP (2018)

Months	Sludge flow, m3/day
January	1,402.23
February	1,207.34
March	1,019.52
April	976.15
May	1,226.69
June	951.18
July	689.95
August	948.47
September	985.33
October	1,487.14
November	2,565.25
December	2,786.29
2018 average	**1,340.32**

As mentioned in p.2.1, in the Chisinau WWTP, dewatering sludge is accumulated on the sludge platforms (32 ha), where during the last years thousands cubic meters of sludge have been collected (only in 2018 about 500,768 m^3/day, see the table 2.2). This sludge is one of the sources of unpleasant odor that has taken over the city of Chisinau in the last 3-4 years. In 2009, geotubes were purchased as a temporary solution, but it turned into a long-term problem, for which Local Authorities did not find a solution regarding its use. The main issues of the Chisinau WWTP are following: (1) The current wastewater treatment performances of the plant do not comply with proposed discharge limits for all parameters; (2) The current energy consumption from the grid can be replaced with other sources; (3) The lack of modern sludge treatment facilities formed in the wastewater treatment process; (4) The lack of a strategy for using dewatering sludge.

Despite producing a massive amount of sewerage sludge annually, the utilization of sewerage sludge in Moldova did not get as much attention or progress as seen in other regions. The current state of sewerage sludge treatment sees it primary being used for landfill.

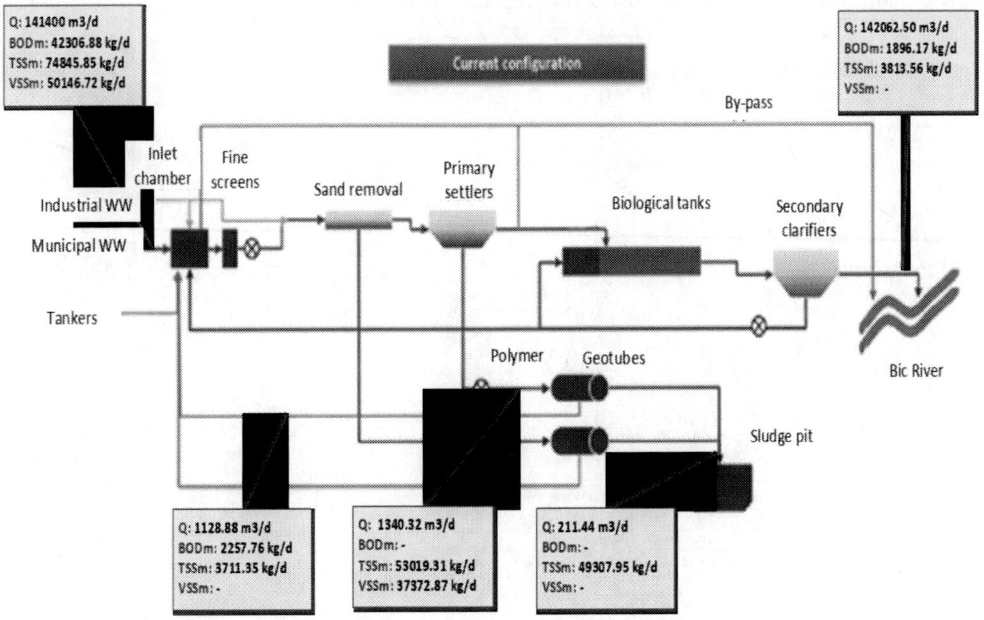

Fig. 2.1. Mass balance based on current operation of Chisinau WWTP

CASE 2: Wastewater treatment operation with anaerobic digestion of sewage sludge. Anaerobic digestion (AD) is one of the most widely used stabilization technologies. During AD, a combustible gaseous mixture composed principally of CO_2 and CH_4 is produced by

biological activity [10], investing sludge with commercial value as a bioenergy source. AD is a complex biochemical process that involves the action of microorganisms forming syntrophic consortia. It follows four fundamental steps: hydrolysis, acidogenesis, acetogenesis and methanogenesis. The *hydrolysis* step degrades both insoluble organic material and high molecular weight compounds such as lipids, polysaccharides, proteins and nucleic acids, into soluble organic substances. The components formed during hydrolysis are further split during *acidogenesis*, the second step. The third stage in AD is *acetogenesis*, where the higher organic acids and alcohols produced by acidogenesis are further digested by acetogens to produce mainly acetic acid as well as CO_2 and H_2. The final stage of *methanogenesis* produces methane by two groups of methanogenic bacteria: the first group split acetate into methane and carbon dioxide and the second group uses hydrogen as electron donor and carbon dioxide as acceptor to produce methane [1,11]. Previous studies by the Biosystems and Agricultural Engineering Department at Michigan State University demonstrated that many things can be done at the AD in order to save energy and one of them is by producing biogas [12,13,14,15,16].
The goal of this paper is to reduce the amount of sludge, that needs to be disposed and the most widely employed method for sludge treatment is anaerobic digestion. The treatment of wastewater sludge, from primary and secondary treatment steps, consists of two main phases: in the 1st step, all incoming flows of sludge are combined, and the mixture is heated to a mild temperature to accelerate biological conversion, which the mixture is allowed to undergo digestion. In further processes, the settled sludge is dewatered for, to separate as much water as possible to decrease the volume of sludge.
The simulation of the AD was based on the real data (2018 year) collected from Chisinau WWTP. A good approach to estimating solids is to prepare a material mass balance for the entire Chisinau WWTP. A material balance is prepared for the key components of *flow*, *BOD* and *TSS*. In addition, I have been developing a solids mass balance with AD, and all the calculations are done using Microsoft Excel, as shown in Figure 2. 2.

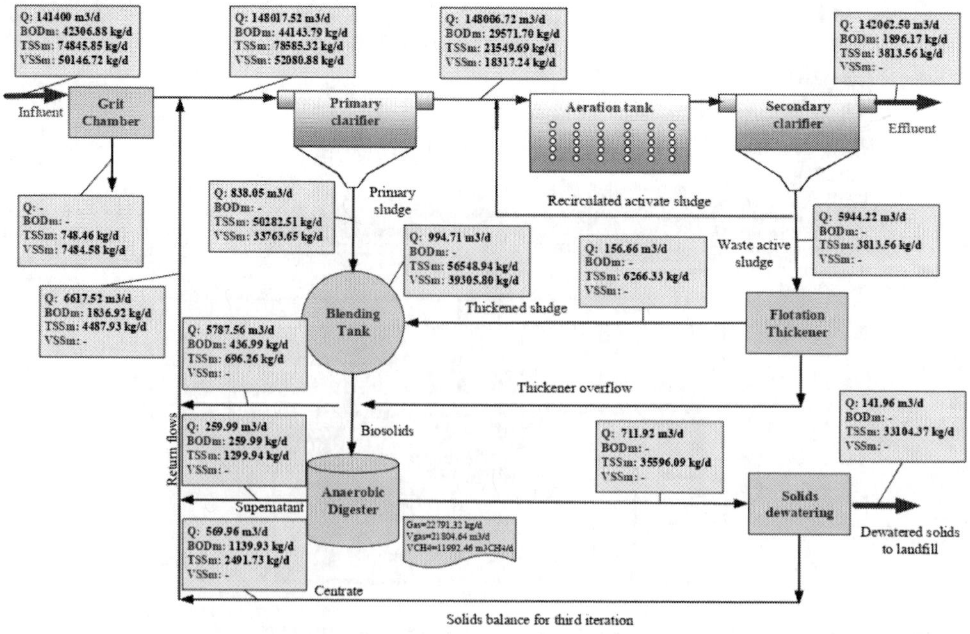

Fig. 2.2. Mass balance based on actual condition with anaerobic digester

As a result of the third iteration of mass balance, we can see the daily production of sludge 994.7 m^3/day and biogas 21,804.46 m^3/day Figure 2.3 and gas production 11,992.46 Figure 2.4.

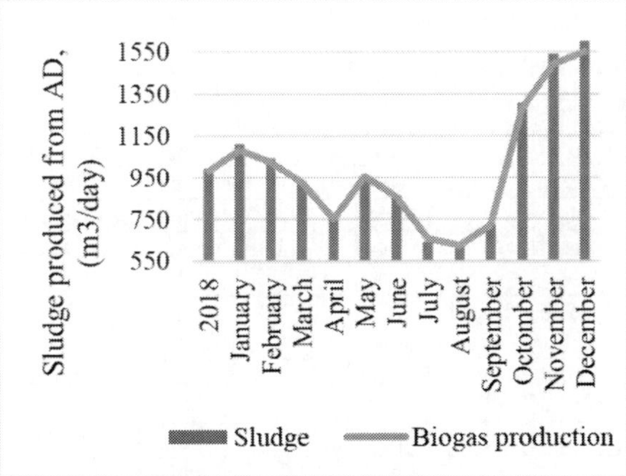

Fig. 2.3 Annual monthly profile of Sludge and Biogas Generation

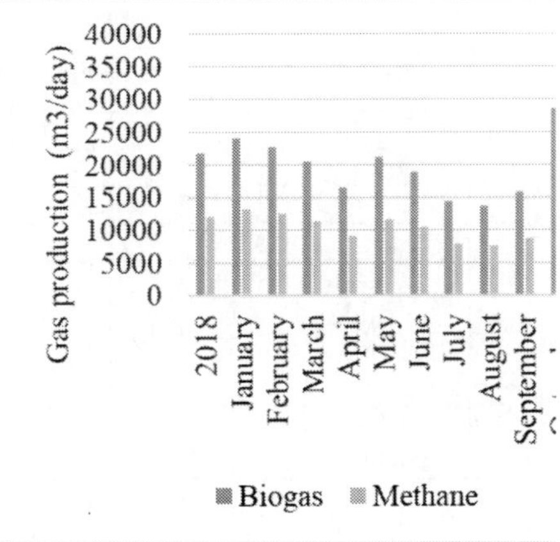

Fig. 2.4 Daily gas and methane (%55 of biogas is methane)

The current state regarding sewage sludge in Moldova is reviewed, with activities around other regions including for comparison. Challenges and tasks faced when treated sewerage sludge using anaerobic digestion in Moldova are discussed, with anaerobic co-digestion which would be include food waste, suggested as a possible approach. This study aims to provide a greater understanding as well as a starting point for further study regarding the future aspect of utilizing sewage sludge in Moldova. Sludge digestion would be very beneficial for the following reasons [1,11,17,18]:

- Reduces the emissions of greenhouse gases (GHG), has a green image and is one of the best available technologies for sludge treatment;
- Reduces the volume of sludge, which consequently reduces the associated costs of downstream sludge treatment. That solution is financially relevant on a long term basis for large WWTPs;
- Production of energy. During anaerobic fermentation, the bacteria themselves are separating the energetic content of the sludge (transformed into methane gas) from the water. This is the main advantage of sludge digestion. No energy is consumed to recover the energy of the sludge itself, except for a very small percentage required for the heating of the digester.

3. Results and discussion

Most of the WWTPs are designed to meet a demand for the effluent quality without consideration of energy requirements. According to the European Benchmarking Cooperation [19], the average electricity consumption for wastewater treatment was 33.4 kWh/PE. The WWTP [20] shows the energy consumption in Europe at wastewater treatment plants by means of activated sludge at the level of 0.15 - 0.7 kWh/m^3 often representing over 50% of the total site energy demand.

Energy efficiency of WWTP is of increasing interest, not only due to economic but also due to environmental aspects. Hence, the optimization of energy consumption and generation in WWTPs is an important topic. Examples for energy self-sufficient WWTPs in the world are described in [21].

A mass and energy balance were conducted to evaluate the system performance. The operational parameters for current WWT plant and anaerobic digester are listed in table 2.4 and 2.5. The total of sludge produced from Chisinau WWTP is 1,371.97 m^3/day and the total electricity consumption per day is 53,310 kWh. The AD system generated 22,792 kg/day, approximately 39,575 kWh/day.

Table 2.4. Main results of the statistical analysis at Chisinau WWTP. Analysis section

No		Unit	Value	Unit	Value
1.	Wastewater flow	m^3/d	141,400	m^3/year	51,594,337
2.	Total sludge at WWTP	m^3/d	1,371.97	m^3/year	500,768.30
	a. Primary sludge	m^3/d	1331,56	m^3/year	486,018.31
	b. Waste activated sludge	m^3/d	40.41	m^3/year	14,749.99
3.	Total amount sludge to dewatering	m^3/d	1,340.32	m^3/year	489,216.8
4.	Total Electricity consumption	kWh/d	53,310	kWh/year	19,450,795
	a. pumping	kWh/d	13,810	kWh/year	5,036,254
	b. treatment	kWh/d	39,500	kWh/year	14,414,541
5.	Electricity consumption to remove **1** kg CBO_5	kWh/1kgCBO_5 per day	1.69	kWh/1kgCBO_5 per year	1.69
6.	Electricity consumption to treat **1** m^3 water	kWh/d	0.38	kWh/year	0.38
7.	Expenses for electricity	lei MD/month	2,917,619.25	lei MD/year	35,011,431
		€uro/month	135,286.339	€uro/year	1,623,436.073
8.	Geotubes	Pieces/year	160	€uro/year	391,308.8
9.	Polymer	kg/year	19,825	lei MD/year	1,467,087.75
				€uro/year	73,911.68

Table 2.5. Sludge management data in Chisinau WWTP modified by the addition of Flotation Thickener, Anaerobic Digestion, and Sludge dewatering processes (data from year 2018)

No		Unit	Value	Unit	Value
1.	Total amount of sludge produced from WWTP	m^3/d	994.70	m^3/year	363,065.50
	a) primary clarifier	m^3/d	838.04	m^3/year	305,884.60
	b) secondary clarifier	m^3/d	156.66	m^3/year	57,180.90
2.	Total amount of sludge sent to dewatering	m^3/d	711.92	m^3/year	259,851.49
3.	Volume of mass of water from sludge dewatering process	m^3/d	6,617.52	m^3/year	2,415,393.80
4.	Amount of sludge cake	m^3/d	141.96	m^3/year	51,814.30
5.	Amount of Gas	kg/d	22,791.32	kg/year	8,318,831.80
6.	Amount of Biogas production	m^3/d	21,804.64	m^3/year	7,958,629.30
7.	Amount of Methan	m^3 CH_4/d	11,992.46	m^3 CH_4/year	4,377,247.90
8.	Amount of Electricity (3.3 kWh/1m^3)	kWh/d	39,575.12	kWh/year	14,444,918.07

Energy recovery at WWTPs represents an important policy lever for sustainability. It can be done through biogas production. The biogas produced in a digester via anaerobic digestion has the energy potential of 6.5 kWh/m^3 (65% methane content) and is the main

energy source in WWTP. It was estimated that WWTPs with sludge digestion consume about 40% less net energy than wastewater treatment plant without AD digestion [22].
The Chisinau WWTP in 2018 year was consumed approximately of 19,450,795 kWh of electricity for treatment wastewater, which is about 45-50% of the overall operations and maintenances costs, then the power generated through biogas from anaerobic digestion could be helpful in reducing total energy consumption, as well as serving as an alternative power generation source.

4. Conclusions

Chisinau WWTP can substantially reduce grid electricity consumption, especially by utilizing biogas from anaerobic digestion. With increasing stricter discharge requirements, and aging/old infrastructure, as well as rising energy prices and concerns about climate change, Chisinau WWTP face many challenges that could significantly increase energy and costs. In additional to reduced energy consumption, anaerobic digestion at WWTP could reduce greenhouse gas emissions by decreasing the amount of grid electricity required for operations.

5. Acknowledgments.

This study is done within the FULBRIGHT PROGRAM at Michigan State University, Anaerobic Digestion Research and Education Center.

6. References.

[1] Metcalf, E. (2003) *Wastewater Engineering Treatment and Reuse*, 4th ed., McGraw Hill, New York.
[2] Feasibility Study, 2011. *Chisinau Water Supply and Sewage Treatment*. Contract No:C21156/ECWC-2010-01-01.

[3] Bodik, I., Kubaska, M., 2013. *Energy and sustainability of operation of a wastewater treatment plant. Environment Protection Engineering*, 39(2): 15-24. DOI:10.5277/-EPE130202.
[4] Yang, L., Zeng, S., Chen, J., He, M. and Yang, W., 2010. *Operational energy performance assessment system of municipal wastewater treatment plants.* Water Science and Technology, 62(6), pp.1361-1370. https://doi.org/10.2166/wst.2010.394
[5] Mizuta, K. and Shimada, M., 2010. *Benchmarking energy consumption in municipal wastewater treatment plants in Japan.* Water Science and Technology, 62(10), pp.2256-2262. https://doi.org/10.2166/wst.2010.510
[6] Arpke, A. and Hutzler, N., 2006. *Domestic water use in the United States: A life- cycle approach.* Journal of Industrial Ecology, 10(1- 2), pp.169-184). https://doi.org/10.1162/-108819806775545312
[7] HG nr. 950 din 25.11.13 *Regulamentul privind cerințele de colectare, epurare si deversare ape uzate în sistemul de canalizare şi/sau în corpurile de apă pentru localitățile urbane şi rurale.* Publicat: 13.03.2020 în *Monitorul Oficial al Republicii* Moldova Nr. 75-83 art. 219, în vigoare după modificare: 13.03.2020.
[8] HG nr.802 din 09.10.2013 *Regulament privind condițiile de deversare a apelor uzate in corpurile de apa* modificata prin HG nr. 862 din 08.10.2014. Publicat: 31.10.2014 în *Monitorul Oficial al Republicii* Moldova Nr. 325-332 art. 946, în vigoare după modificare: 31.10.2014.
[9] Council Directive of 21 May 1991 *concerning urban wastewater treatment (91/271/EEC),* Published: Official Journal of the European Communities L 135, 03.05.1991, p.40 https://eur-lex.europa.eu/legal-content/EN/TXT/PDF/?uri=CELEX:-31991L0271&from=EN
[10] Henze M., Harremoës P., La Cour Jansen J. and Arvin E., *Wastewater Treatment. Biological and Chemical Processes*, 2nd ed., Springer-Verlag, Berlin, Heidelberg, 1997.
[11] Turovskiy I. S., Mathai P. K.. *Wastewater sludge processing*, 355 p.
[12] Alvarez R. E. A., Bustamante M., Kirk D., Chavarria J. A. M., Baudrit D., Pereira J. F. A., Montere W. R., Reinhold D., Liao W., *Technical and economic feasibility of a solar-bio-powered waste utilization and treatment system in Central America*, Journal of Environmental Management 184 (2016), p.371-379. https://doi.org/10.1016/-j.jenvman.2016.10.007
[13] Bustamante M., Liao W., *A self-sustaining high-strength wastewater treatment system using solar-bio-hibrid power generation*, Bioresource Technology 234 (2017) p.415-423. DOI:10.1016/j.biortech.2017.03.065
[14] Chen R., Rojas-Downing M., Zhong Y., Saffron C., Liao W. *Life Cycle and Economic Assessment of Anaerobic Co-digestion of Dairy Manure and Food Waste*, Industrial Biotechnology, Vol.11, No.2, April 2015, pag.127-139.
[15] Liu Z., Stromberg D., Liu X., Liao W., Liu Y.. *A new multiple-stage electrocoagulation process on anaerobic digestion effluent to simultaneously reclaim water and clean up biogas,* Journal of Hazardous Materials 285 (2015) p.486-490.
[16] Valenti F., Porto S., Dale B., Liao W., *Spatial analysis of feedstock supply and logistics to establish regional biogas power generation: A case study in the region of Sicily*, Renewable and Sustainable Energy Reviews 97 (2018) p.50-63. https://doi.org/10.1016/-j.rser.2018.08.022
[17] Appels L., Baeyens J., Degreve J., Dewil R. *Principles and potential of the anaerobic digestion of waste-activated sludge*, Progress in Energy and Combustion Science 34 (2008), p.755-781. DOI: 10.1016/j.pecs.2008.06.002
[18] de Oliveira S. Jr., *Exergy,* Green Energy and Technology, DOI: 10.1007/978-1-4471-4165-5_9, Springer-Verlag London 2013, pag.281-303.

[19] *EU Reference Scenario 2016 Energy, Transport and GHG Emissions Trends to 2050*. Available online: https://ec.europa.eu/energy/sites/ener/files/documents/20160713%20-draft_publication_REF2016_v13.pdf .
[20] *WssTP. Water and Energy*: Strategic Vision and Research Needs. The Water Supply and Sanitation Technology Platform. Available online: http://www.danishwaterforum.dk/-activities/WssTP_Water_and_Energy_Publication%2009_2011.pdf .
[21] Nowak, O., Keil, S. & Fimml, C., *Examples of energy self-sufficient municipal nutrient removal plants*. Water Science and Technology 64 (1), 1–6., IWA Publishing 2011.
[22] Neczaj E., Grosser A., *Circular Economy in Wastewater Treatment Plant – Challenges and Barriers*, presented at the 3rd EWaS International Conference on „Insights on the Water-Energy-Food Nexus", Lefkada Island, Greece, 27-30 June 2018.
[23] *Indicatorii financiari si de producţie ai activităţii întreprinderilor de alimentare cu apă şi de canalizare* - membre ale Asociaţiei "Moldova Apă-Canal" anul 2018. http://www.amac.md/public/files/documente/indicatorii_financiari_productie_intreprinderilor_apa_canal_membre_2018.pdfhttp://www.amac.md/public/files/documente/indicatorii_financiari_productie_intreprinderilor_apa_canal_membre_2018.pdf

Determination of Contours of Flooded Areas Due to Possible Accident of Zhinvali (Georgia) Earth Dam and Calculation of the Hydrodynamic Parameters of a Destructive Wave at a Dam

Givi Gavardashvili[1,*], Tamriko Supatashvili[1], Edward Kukhalashvili[1], George Natroshvili[1], Irma Qufarashvili[1], Inga Iremashvili[1] and Konstantin Bziava[1]
[1]Tsotne Mirtskhulava Water Management Institute of Georgian Technical University, 0179, 60-b I. Chavchavazde Ave., Tbilisi, Georgia

Abstract. Using theoretical and field studies, computer software programs (VOLNA-4, and MIKE-21,) and GIS technology, and taking into account the possible accident (destruction) of the Earth Dam with a height of 102 meters, the contours of the riverbed of flooded territories in riverbeds for rivers Aragvi and Mtkvari have been defined in the given article considering the configuration of the water-catchment basins of rivers.
By means of basic differential equations of hydraulics, geometrical dimensions of Zhinvali Earth Dam, hydrological basic parameters of the reservoir, and computer software, the basic hydrodynamic parameters of the destructive wave at the Zhinvali Earth Dam are calculated taking into account the time of flow movement.
Keywords: earth dam, accident, flood, risk, flooded territories.

1. Introduction

The reason for the disruption of hydraulic facilities, including Zhinvali Earth Dam (Georgia) may be natural calamities (earthquake, storm, mudflow, etc.), technogenic factors (corrosion and collapse of the building structures, violation of water intake regimes, etc.), as well as sabotage and terroristic acts and use of arms destructing dams during the war.

In case of disruption of a hazardous hydrodynamic plant, a 102-meter-high Zhinvali Earth Dam (Fig. 1) in our case, breakdown tsunami-type waves are formed potentially. The strength of the waves depends on the amount of water and wave velocity. Therefore, the hydrodynamic objects with great amount of water in their reservoirs and with a great difference between the heights of their head and tail races (high head), are hazardous in this respect.

* Corresponding author: givi_gava@yahoo.com

Fig. 1. General view of Zhinvali Earth Dam and reservoir

A breakdown wave and great water mass can overwhelm everything on their way: buildings, premises and agricultural plots, and may result in victims and great material losses [2].

The main function of Zhinvali Earth Dam is to protect the environment, economic objects and human life from the floods formed in the bed of the Aragvi River. However, in terms of incorrect exploitation of the dam and under the action of extreme natural phenomena (earthquake, catastrophic floods, etc.), Zhinvali Earth Dam itself may be a reason for a big catastrophe.

Following the category of a safety risk of a hydraulic facility (technological, environmental or social) and by considering a safety declaration of a facility owner, the terms of a dam operation must be observed by means of a safe operation and checks of the terms of no-failure operation of the plant units and all auxiliary facilities, buildings and equipment.

By considering the EU Directive on the assessment and management of flood risks in Georgia (January 1, 2012, http://www.slovakaid.sk/), a harmonization [2] and implementation mechanism must be incorporated in the legislation of the country what implies as follows:

- Developing a roadmap for harmonization and implementation of the EU Directive on the assessment and management of flood risks in the legislation of Georgia;

• Developing a draft responsibility and actions Directive in the field of flood forecast and early warning for relevant ministries, what means:

➢ The development of methodologies in the area of the flood risks and hydrological monitoring;

➢ Developing the methodologies to draft the area of floods risks;

➢ Developing the map of the zones of flood risks in the basins in the GIS system for the whole country;

➢ Introducing modern hydrological models to predict floods;

➢ Providing automated hydrological posts in the water catch basins of the rivers;

➢ Training young experts in flood prediction issues, etc.

2. THE METHODOLOGY TO CALCULATE HYDRODYNAMIC PARAMETERS OF TSUNAMI-LIKE WAVE

With the aim to provide a computer imitation of possible technogenic catastrophes of Zhinvali Earth Dam, modified software VOLNA-4 and MIKE-4 algorithms were used and field studies of Zhinvali Dam were accomplished on February 10-27 (Fig. 2 and 3) what allows calculating the hydrodynamic parameters of the dam breakthrough wave in case of possible dam destruction, in particular: wave velocity, covered distance and more importantly, the sizes of the flooded area in relation to time.

The initial data were divided into two parts by us: the first is the permanent values and another is variable values. Permanent values consider the parameters not dependant on some or other condition; as for the variable values, they depend on flood, degree of dam destruction, etc.

The construction height of Zhinvali Earth Dam is 102 m and its effective height (water flooding height) is 96 m. The dam width varies within 415 m; the capacity of Zhinvali water reservoir is 520 mln. m^3, while the water surface area is 733 mln. m^3 [4].

In our case, in case of destruction of Zhinvali Earth Dam, destructing tsunami-like waves are formed potentially. The strength of the waves depends on the amount and velocity of water. Therefore, the hydrodynamic objects with great amount of water in their reservoirs and with a great difference between the heights of their head and tail races (high head), are hazardous in this respect.

A breakdown wave and great water mass can overwhelm everything on their way: buildings, premises and agricultural plots, and may result in victims and great material damage.

Fig. 2. Swimming on a motorboat in Zhinvali reservoir during the field studies

Fig. 3. General view of colloid and sandy material deposited in Zhinvali reservoir

Figure 4 shows a longitudinal profile of a tsunami-type wave formed in case of distruction of Zhinvali Earth Dam.

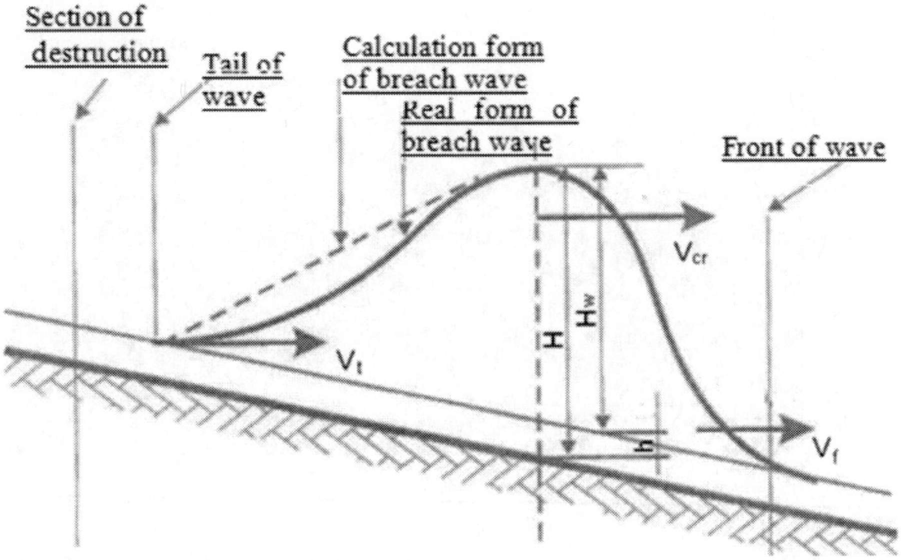

Fig. 4. Longitudinal profile of a tsunami-like wave

The volume of water in the reservoir (W_0) was calculated with the following dependence [2,5,5,10]:

$$W = \frac{H_B S_B}{3} \qquad \text{(mln. m}^3\text{)} \qquad (1)$$

Where H_B is the water depth at the height of normal dam flooding (m); S_B is the area of the water surface in the reservoir (mln. m^3);

The width of river is taken from the topographic maps. As for the number of points, it must not exceed 3 points on the each side of the river axis, i.e. total of 6 points, and must cover the whole water catch area. In order to calculate the area of the flooded territory, the number of sections from the dam must not exceed 8, with the distance between them to be plotted on a topographic map in advance.

The wave velocity (V) during the flood is calculated by the following formula in the tailrace of the facility [10]:

$$V = V_0 (H_1 / H_0)^{2/3}, \quad \text{(m/sec)} \tag{2}$$

The degree of the dam destruction (E_p) was established with the following expression [64]:

$$E_p = \frac{F_B}{F_0}, \tag{3}$$

Where F_B is the area of the bank breach (m^2); F_0 is the surface area (m^2).

Besides the above-mentioned, the algorithm considers: the height of the river bank edge (m), number of sections along the river, distance between the sections (km), width of the riverbed (m), water current velocity in the riverbed (m/sec); width of the bed of the Nogha River (m), values of the river levels (m), etc.

Zhinvali water reservoir, with its Dam representing a structure with a fill clay is located in village Zhinvali, Dusheti Region. The construction height of the Dam is 102 m and its effective height (water flooding height) is 96 m. The Dam width varies within 415 m; the capacity of Zhinvali water reservoir is 520 mln. m^3, while the water surface area is 733 mln. m^3 (see Fig.5).

Fig. 5. View of Zhinvali Earth Dam from the tailrace

3. IDENTIFYING THE CONTOURS OF THE AREAS FLOODED AS A RESULT OF A POSSIBLE ACCIDENT OF ZHINVALI EARTH DAM

The number of sections from the Aragvi River to the Mtkvari River taken to predict the catastrophe of Zhinvali dam is 8 (Fig. 6).

The sections were taken in the following settled areas located at different distances from the dam, in particular, 1) village Misaktsieli – 30,0 km; 2) Avchala settlement - 35,0 km; 3) Dighomi (Shalikashvili) Bridge - 44,5 km; 4) Tamar Mepe (Chelyuskinites) Bridge - 48,0 km; 5) Ortachala Bridge - 54,2 km; 6) new settlement of Rustavi - 74,0 km; 7) center of Rustavi - 77,0 km; and 8) end point of Rustavi - 80,0 km.

By using the theoretical and field studies, computer software (VOLNA – 4 and MIKE-21) and GIS technology [2-8,10,12], by considering a possible accident of a 102-meter-high Zhinvali Earthen Dam (Georgia) with 0.75 degree, the contours of the beds of the flooded areas in the Aragvi and Mtkvari Riverbeds were identified by considering the configuration of the water catch basins of the Rivers. The results of the calculation are given Table #1.

Detailed data about Zhinvali Dam are given in the Tables below (Tables 2 and 3) and the sizes of the flooded areas in the riverbed and in the adjacent areas are given in Figures 7-15.

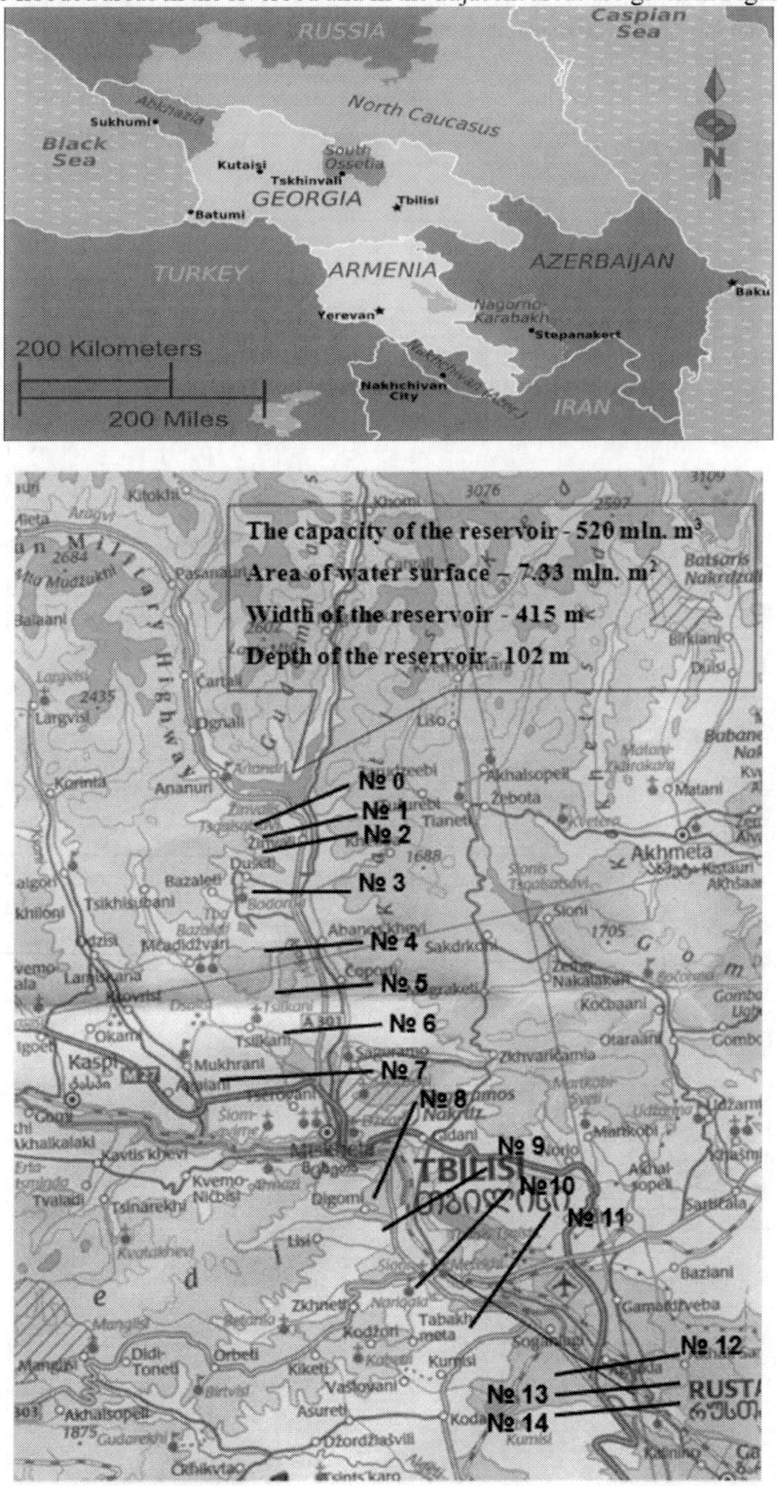

Fig. 6. Plan of location of design sections

Table 1. Description: Zhinvali Dam (With Destruction Degree $E_3 = 0{,}75$)
Time: 29.02.2020, 14:02:40

Initial data of the hydrological system :	Unit of meas.	Number
1. Water reservoir capacity at standard flooding level (SFL)	mln. m^3	520
2. Reservoir depth at SFL	m	96
3. Water surface area at SFL	mln. m^3	7,33
4. dam width at SFL	m	415
5. River depth in the Dam tailrace	m	1
6. River width in the Dam tailrace	m	25
7. River velocity in Dam tailrace	m/sec	1
8. Reservoir depth at the moment of the dam accident	m	96
9. Dam destruction degree	m	0,75
10. Height of the riverbed bank breach	m	24
11. Water standard flooding level in the reservoir	m	816
12. Number of lateral profiles in the riverbed	pcs.	8

Table 2. Calculated numerical values

Cross sections	Unit of meas.	Section #1	Section #2	Section #3	Section #4	Section #5	Section #6	Section #7	Section #8
1	2	3	4	5	6	7	8	9	10
Distance of *a* i^{th} section from the Dam	km	30	35	44,5	48	54,2	74	77	80
Specific current:									
Water flooding level	m	480	425	398	393	375	327	322,7	313
Depth	m	1	3	2	2	1	2	1	1
Width	m	70	60	80	82	80	150	100	93
Current velocity	m/sec	1	1	1	1	1	1	1	1
Left bank									
Height of the river bank breach	m	7	2	3	5	4	2,5	3	0,5
Width of the Nogha River bed	m	50	10	50	20	20	5	5	40
Level #1	m	490	435	402,5	400	383	330	327,5	315
Distance from the river axis to level #1	m	137	50	440	71	180	100	60	225
Level #2	m	520	440	405	405	385	340	330	320
Distance from the river axis to level #2	m	687	70	670	371	280	125	1310	1295
Level #3	m	680	490	410	415	388	350	332,5	322,5
Distance from the river axis to level #3	m	2437	340	970	800	720	265	1610	1345
Right bank									
Height of the river bank breach	m	15	8	3	5	4	1	1	5
Width of the Nogha River bed	m	30	25	50	20	100	300	300	50
Level #1	m	520	435	402,5	410	380	330	325	340
Distance from the river axis to level #1	m	912	60	840	90	510	525	750	440
Level #2	m	600	445	405	415	385	340	337,5	350
Distance from the river axis to level #2	m	1137	70	1680	200	660	600	900	570
Level #3	m	680	490	410	425	395	343	345	400
Distance from the river axis to level #3	m	1637	540	2000	230	940	1300	1650	840

https://doi.org/10.1051/e3sconf/202021201004

Table 3. Hydrological values of the flooded area

Dam destruction parameters	Unit of meas.	Section #0	Section #1	Section #2	Section #3	Section #4	Section #5	Section #6	Section #7	Section #8
Distance of the section from the hydrological station	km	0	30	35	44,5	48	54,2	74	77	80
Water peak discharge in the section	thous. m^3/sec	137	24,4	21,9	18,12	17,2	15,59	12,32	11,84	11,5
Time:										
Of wave front descending	min	0	47,1	57,8	76,9	90,9	108,5	175,5	186,7	201
Of wave descending	min	0	80,2	99,1	143,4	158	180,7	288,7	301,9	314,8
Wave tail descending	min	149	649	732	890,4	949	1052	1382	1432	1482
Of flooding	min	149	602	674	813,5	858	943,6	1207	1245	1281
Maximum flow velocity	m/sec	18,7	11,2	14,1	5,03	9,53	6,77	6,86	4,91	5,26
Wave height	m	56,6	18,6	29,7	6,15	14,8	8,29	7,88	4,88	5,41
Maximum flooding depth	m	57,6	19,6	32,7	8,15	16,8	9,29	9,88	5,88	6,41
Maximum flooding level	m	778	499	455	404,2	408	383,3	334,9	327,6	318,4
Maximum flow height:										
On the left bank of the river	m	152	294	862	591,6	614	211,7	112,2	97,83	955,8
On the right bank of the river	m	152	185	524	1394	84,8	608,8	561,6	780,9	103

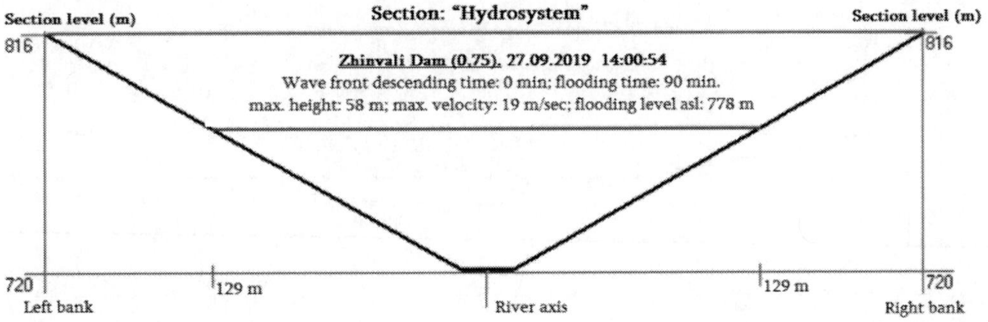

Fig. 7. Section of Hydrosystem

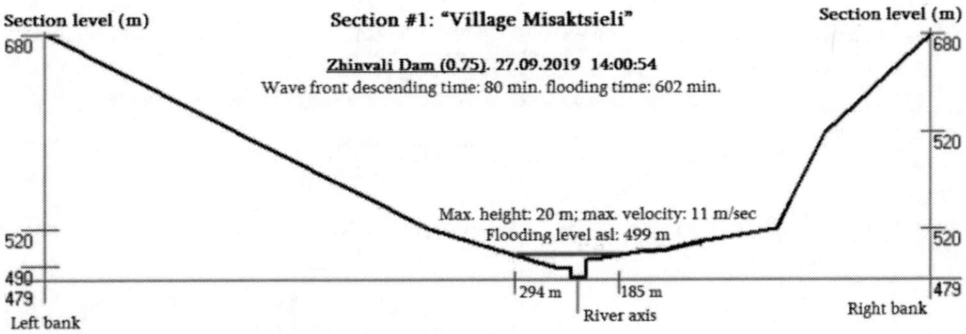

Fig. 8. Section #1, Village Micaktsieli

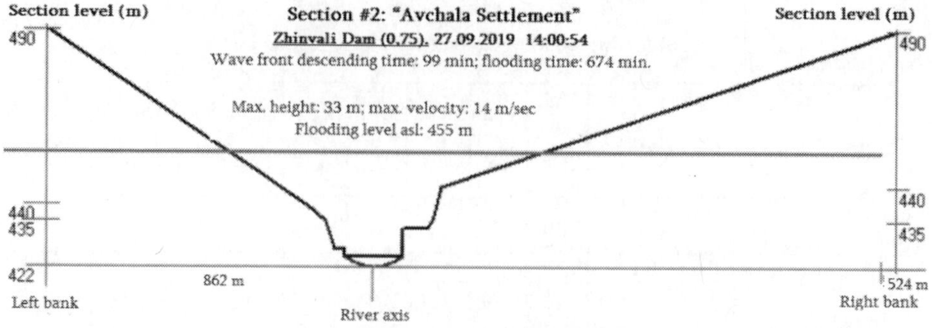

Fig. 9. Section #2, Village Avchala Settlement

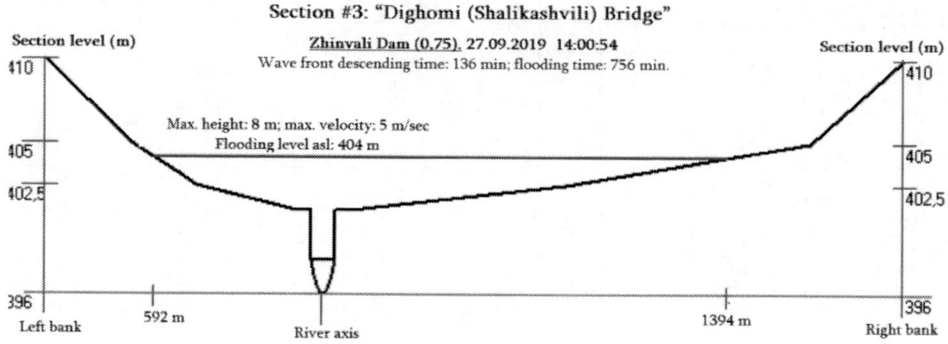

Fig. 10. Section #3, Digomi (Shalikashvili Bridge)

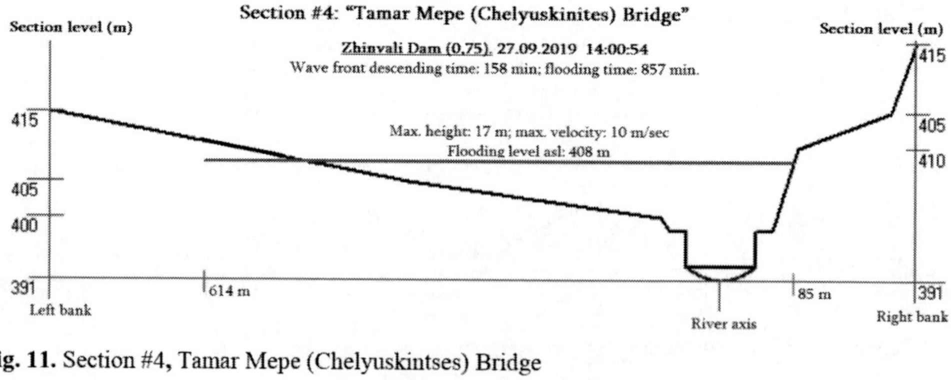

Fig. 11. Section #4, Tamar Mepe (Chelyuskintses) Bridge

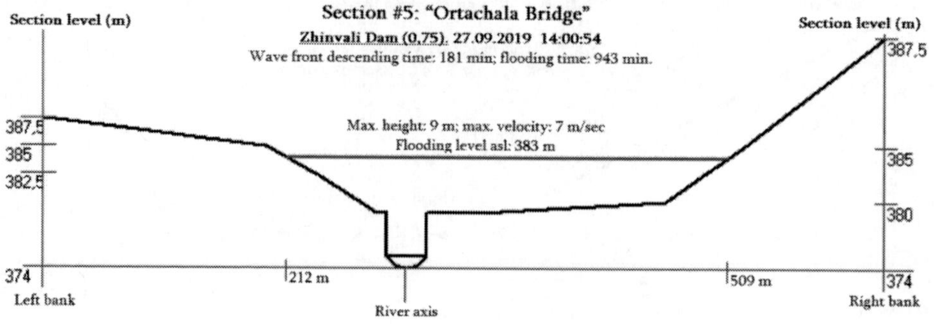

Fig. 12. Section #5, Tamar Ortachala Bridge

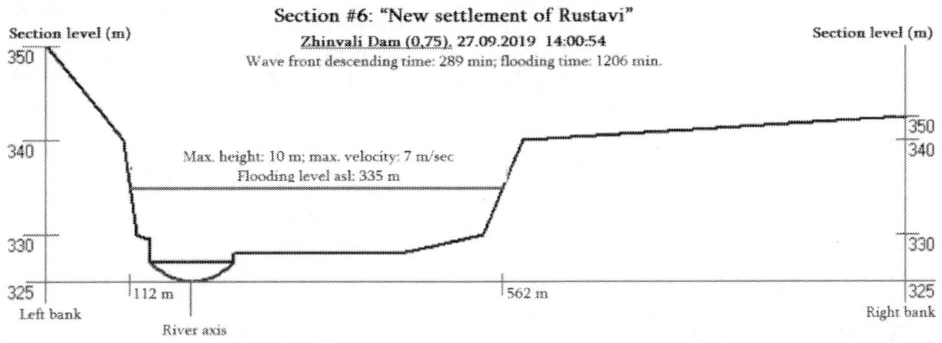

Fig. 13. Section #6, New settlement of Rustavi

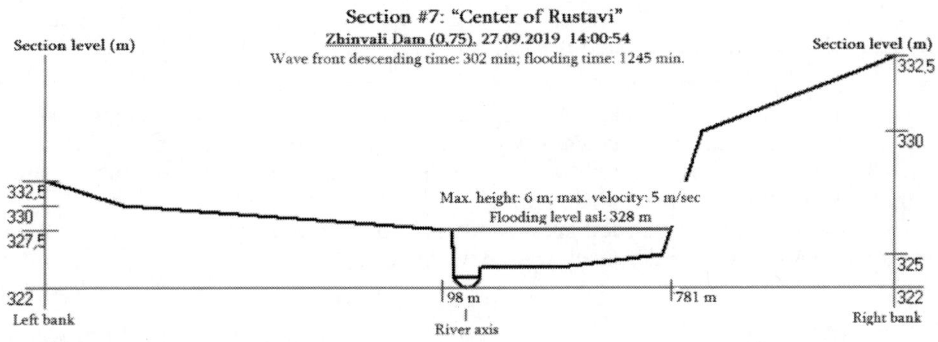

Fig. 14. Section #7, Center of Rustavi

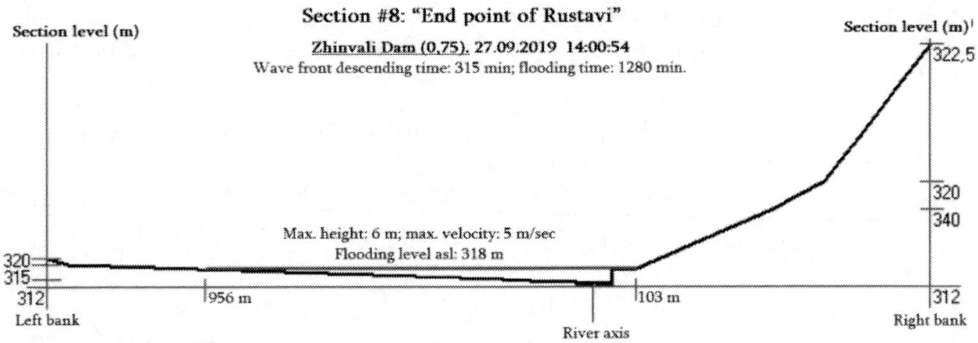

Fig. 15. Section #8, End point of Rustavi

As for the hydrodynamic data of a tsunami-type wave formed following the accident with 0.75 destruction degree of Zhinvali Dam by considering the time factor, the geometrical data of the flooded area are given in Table #4.

Table 4. Geometric data of the flooded area

#	Section	Tsunami wave motion time (min)	Width of the flooded area from the center of symmetry of the riverbed (m)		Height of tsunami wave (m)	Velocity of tsunami wave (m/sec)
			Right	Left		
1	Village Misaktsieli	47,1	322	319	21	11
2	Avchala settlement	57,74	**The canyon gets flooded on both sides**			
3	Dighomi (Shalikashvili) Bridge	76,34	1603	649	9	5
4	Tamar Mepe (Chelyuskinites) Bridge	90,1	715	88	18	10
5	Ortachala Bridge	107,0	629	239	10	7
6	New settlement of Rustavi	172,6	570	115	11	7
7	Center of Rustavi	183,3	786	293	6	5
8	End point of Rustavi	197,3	110	1055	7	5

Thus, based on the field, theoretical and experimental studies, the principal hydrodynamic properties of Tsunami wave formed as a result of possible destruction of Zhinvali Earth Dam were identified by considering the relevant time factor and the geometry of the flooded area was determined as well.

The risk zones of the flooded areas were specified and the norms of behavior for the population in the state of emergency were developed [2,9].

4. Conclusions

Under the financial support of Grant Project # FR17_615 "Assessment of safety risks of vulnerable infrastructure during the formation of expected catastrophes", based on the complex studies accomplished during the field scientific expeditions for Zhinvali Earth Dam in 2019-2020, we may draw the following conclusions:

1. As a result of the studies accomplished in Zhinvali reservoir, the statistical material was processed, which allows specifying the volume of the solid drift deposited in the headrace of the Dam and minimum useful water capacity;
2. The samples of colloid fraction were taken from the reservoir area and a chemical analysis of the suspension was provided with the aim to use it for commercial purposes;
3. Based on the scientific studies in the tail and head races of Zhinvali Earth Dam, a longitudinal profile of a tsunami-type wave formed as a result of the possible dam destruction was determined by calculating relevant hydrodynamic parameters;
4. For the case of a 0,75% destruction degree of 102-meter-high Zhinvali Earth Dam, by using a mathematical model (VOLNA – 4 and MIKE-21) and GIS-technology, the contours of the riverbed of flooded territories in riverbeds for rivers Aragvi and Mtkvari have been defined by considering the configuration of the water catch basins of the rivers.
5. Based on the obtained data, the risk zones of the flooded areas were specified and the norms of behavior for the population in the state of emergency were developed.

Acknowledgment

The research was financial supported by Shota Rustaveli National Science Foundation of Georgia, Grant Project # FR17_615 "Theoretical research of vulnerable infrastructure security risk during formation of predictable disasters".

References

1. B.M. Ayyub, *Risk Analysis Engineering and Economics*. A CRC Press Company. Boca Raton, London, New York, Washington, D.C., 571 p. (2003)
2. *Determination of the risk zones of flooded areas and evaluation of the rules of population behavior in an emergency situation in a case of potential accident of Zhinvali earth dam*. Grant Project Leader: Givi Gavardashvili - Doctor of Technical Sciences, Professor. Tbilisi, Public House „Sachino", 63 p. (2019)
3. G.V. Gavardashvili - Forecasting the Security of the Local People in Village Nakra of Mestia Region (Georgia) Against Floods and Mudflows. // Environments, ITS, www. itspoa. com/ jurnal/envi, UK, pp. 13-24 (2018)
4. G. Gavardashvili, E. Kukhalashvili, T. Supatashvili, G. Natroshvili, I. Qufarashvili, K. Bziava. *The Calculation of Maximal and Average Speed of Debris Flow Formed as a Result of Outstretched Water Wave on the Land Dam of Zhinvali. International Conference on Construction and Environmental Engineering*. WASET, Barcelona, Spain, pp. 1029- 1032 (2019)
5. G. Gavardashvili, E. Kukhalashvili, T. Supatashvili, G. Natroshvili, I. Qufarashvili, K. Bziava. *The Ecological Awareness Level Arising of the Population Living in the Flood Risk Zone in case of Possible Accident of Zhinvali Land Dam and Designing Precautionary Measures*. XIII International Conference on Environmental, Biological, Ecological Sciences and Engineering. WASET, Rome, Italy, pp. 755- 758 (2019)
6. Ts.E. Mirtskhulava. *Hazards and Risk (at some water and other systems. Types, Analysis, Assessment)*. 2 Books. Tbilisi, "Metsniereba", 807 p. (in Russian) (2003)
7. Mike 21 Flow Model. *Hints and recommendations in applications with significant flooding and drying* [interaktyvus] [žiūrėta 2012-12-10]. Prieiga per internetą: <http://www.dhigroup. com/ upload/dhisoftwarearchive/papersanddocs/hydro-dynamics/MIKE21SignificantFlodry Guidelines.
8. O.G. Natishvili, G.V. Gavardashvili. *Dynamics of Gully-Formation by Considering the Wave Motion of Flow*. American Scientific Research Journal for Engineering, Technology and Sciences (ASRJEST), Vol 55, #1, pp. 17-26. http://asrjtsjournal.org/index.php/American Scientific_Journal/ issue/view/81 (2019)
9. A. Prangishvili, N. Bochorisvili, O. Lanchava. *Life Safety - Emergensy Management and Sivil Defense* (in Georgian). Tbilisi, 638 p. (2011)
10. S.K. Shoigu. *Operative prediction of the engineering situation in emergency situations* (Book 2), Moscow, 176 p. (1998)
11. D.V. Shterenlikht. *Hydraulic*. Energoatomizdat, Moscow, 494 p. (1984)
12. V. Urbanavičius. *Kaunas's Impact on Hydrodynamics of the Neman River*. Alexander Stulginski University. Kaunas, Lithuania, 54 p. (2013)

Biological method in monitoring small rivers of the Oka River basin

Tatyana Guseva[1,*], *Iyri* Mazhayskiy [2]

[1]FSBEI HE Ryazan Medical University named after I.P. Pavlov, Ministry of Health of Russia, Ryazan, Russia
[2]Meshcherskiy branch of the National Scientific Research Institute of Hydraulic Engineering and Amelioration named after A.N. Kostyakov, Ryazan, Russia

Abstract. The Oka basin is characterized by serious environmental problems associated with the depletion of water resources. Communities of aquatic organisms are the first to respond to negative changes in the aquatic environment, such communities are used as indicators in biological monitoring. The aim of the work was to assess the ecological state of the small river of the Oka basin by bioindication. The quality of the water in the reservoir was assessed by chemical indicators - biogenic and technogenic elements. The saprobity index of the reservoir was determined by the species diversity of periphyton. The results revealed technogenic and organic pollution of the small river. According to the results of biological monitoring, the value of the saprobity index is 2.85, which corresponds to class 4 of the reservoir quality, and the water is assessed as polluted. Monitoring the degree of pollution of the small river by the biodiversity of the periphyton community allows you to determine the degree of water pollution quickly.
Keywords: small river, biogenic elements, heavy metals, saprobity, periphyton.

The Oka basin is characterized by intensive development of the territory, and as a result, the accumulation of environmental problems associated with the depletion of water resources. A significant negative effect on water bodies is exerted by runoff from agrolandscapes containing both biogenic elements and heavy metals [4, 5, 7]. Anthropogenic impacts on water bodies are multifactorial in nature, which makes it difficult to assess the quality of the aquatic environment only by abiotic parameters. Currently, environmental researchers have universally recognized and legally supported the simultaneous use of chemical analysis, bioindication, and biotesting methods for an objective assessment of the ecological state of water bodies [15, 16]. The community of hydrobionts, which can act as biological indicators, is the first to respond to all negative changes in the parameters of the aquatic environment. Biological monitoring of water bodies has a number of significant

* Corresponding author: guseva.tm@yandex.ru

advantages, since indicator organisms respond to relatively weak anthropogenic pressures that are recorded without the use of time-consuming and expensive methods [2, 3, 8, 10, 11, 12, 13]. In world practice, a large number of monitoring methods, including various characteristics of groups of aquatic organisms, are used. Biological control methods are various: determining the species diversity of aquatic organisms from microorganisms to ichthyofauna, assessing the saprobity of a reservoir, using phytoplankton, zooplankton, benthos, periphyton and higher aquatic vegetation as indicators. Bio-indicator organisms should be ubiquitous in the researched area and have a well-defined reaction to ongoing changes in the environment. Based on many years of research, certain types of hydrobionts - bioindicators, capable of living in water of varying degrees of pollution are identified. In practice, a chemical analysis of surface water is usually carried out. But, having determined the composition of aquatic organisms, it is possible to obtain the most important environmental information: not only to assess the state of the environment, but also to predict the development of the aquatic ecosystem as a whole [1, 6, 9, 17, 18, 19, 20].

The purpose of work: to assess the ecological status of the small river of the Oka basin by bioindication.

To achieve the goal, the following tasks:

-to assess the water quality of the small river of the Oka basin by chemical characteristics;

-to conduct biological monitoring of the reservoir (to assess the species diversity of aquatic organisms, to determine their belonging to the saprobity zone by the importance of indicator species).

The research was carried out at an ecological testing ground located on the territory of the Meshchera lowland, including landscapes on the left bank of the Oka River basin. Ecopolygon is a natural model that includes the water-collecting area of a small river, where ameliorated landscapes are located on. This water body receives runoffs from irrigated, drained territories, arable land, pastures and also summer cottages, that is, almost all types of use of agrolandscapes that can have a negative impact on the aquatic ecosystem are presented here. Over a long period of time, the chemical indicators of the aquatic environment and the determination of biogenic and technogenic pollution were monitored at the ecological polygon. Chemical analysis of water was carried out according to generally accepted methods: ammonium nitrogen, nitrates, nitrites - by photometry [GOST 33045-2014], heavy metals - by adsorption spectrometry [GOST R 57162-2016]. The periphyton organisms as the most typical communities represented by attached organisms are the most indicative to assess the water quality of small reservoirs using the biological method. The samples were collected with a scraper. The periphyton was microscoped, noting the frequency of occurrence of indicator organisms. The saprobity index was calculated according to the Pantle-Bucca method in the modification of Sladchek V. [GOST 17.1.3.07-82]. The indicator significance and the saprobity zone were determined for each type of periphyton according to the lists of organisms "Unified methods for the water quality research".

The number of individuals found in the sample was evaluated using a 5-point system: 1 point - single specimens, 2 points - few (2 - 4 individuals), 3 points - average number (5 - 7 individuals), 4 points - many (8 - 11 individuals), 5 points - in the mass (from 12 individuals). The saprobity index (Ind S) was calculated by the formula:

$$Ind\, S = \Sigma\, (Sh) \,/\, \Sigma\, h \quad (1)$$

where: S is the indicator significance of individual species, h is the relative number of individuals of the species.

The object of the research is a small river directly connected with the water system of the Oka river. Its main characteristics: length - 6.4 km, water-collecting area - 30.0 km^2, average width - 1.0 m, average depth - 0.4 m, maximum water flow rate - 0.3 m^3 / s. The water of the small river belongs to the calcium hydrocarbonate class, transparent, without color and odor, pH is 6.5-7.0, hardness is 1.2-2.5 mg-equ / l, the amount of dissolved O_2 is 6.9 - 7.4 mg / l. The studied water body for a considerable period of time experiences anthropogenic stress as a result of the functioning of the irrigation system, artificial drainage, the presence of summer cottages, and private lands on the model landscape. A water body is a collector that receives both surface and subsurface runoff from the landscape, that affects the hydrochemical characteristics of the reservoir. The results of long-term monitoring of biogenic and technogenic pollution of a water body are presented in table 1.

Table 1. Hydrochemical characteristics of the small river (long-term average indicators, mg / l).

Indicator	Value (long-term average)	Maximum allowable concentration (MAC_f)*
NH_4^+	1,2	0,5
$N0_3^-$	1,5	40,0
NO_2^-	0,01	0,08
Cu	0,003	0,001
Zn	0,017	0,01
Pb	0,012	0,006
Cd	0,001	0,005

* MAC_f - maximum allowable concentration for water in commercial fishery reservoirs (in accordance with the order of the Ministry of Agriculture of Russia of October 12, 2018 N 454).

To assess the negative impact of pollutants specifically on aquatic organisms, the long-term average values of the considered parameters were compared with the MAC_f. The content of nitrogen compounds in water did not exceed the permissible values, with the exception of $NH4_4^+$, which is explained by the action of runoffs coming mainly from the lands allocated for summer cottages. The concentration of Cd is in the range of MAC_f, and the content of Pb, Cu, and Zn in some cases exceeded this indicator. The results of chemical assessment of water indicate the existing technogenic pressure, as well as organic pollution, which worsens the ecological condition of the reservoir and creates unfavorable conditions for the aquatic ecosystem.

One of the informative indicators of the anthropogenic load on aquatic ecosystems is the qualitative composition of hydrobionts, undergoing significant changes under the influence of pollutants. In order to determine the species diversity of the biota of the researched water body, the periphyton samples were taken to determine the saprobity index of the small river (Table 2).

Table 2. Dominant species of periphyton of the small river.

Organisms	Saprobity Zone, S	Species frequency, h	Sh
	Seaweed		
Scenedesmus acuminatus	2	1	2
Spirogyra sygmoide	2	2	4
Closterium acerosum	3	4	12
Cyclotella menengiata	3	2	6
Navicula viridua	3	3	9
Navicula cryptocephala	3	2	6
Navicula rhynchocephala	3	5	15
Nitzschia sigmoidea	2	3	6
Navicula hungarica var. capitata	2	3	6
		$\sum h = \mathbf{25}$	$\sum Sh = \mathbf{66}$

Protozoa			
Euglena viridis	4	4	16
Chilidonella cuculata	3	3	9
Colpoda cuculus	3	2	6
		∑h = **9**	∑Sh= **31**
		∑h = **34**	∑Sh= **97**

The results of microscopy of periphyton representatives: seeweed - 9 species, protozoa - 3 species. Comparison of tabular data with a list of saprobity indicator organisms has shown that most of the hydrobionts found in a water body belong to α, β - mesosaprobes, and polysaprobes are also found. Based on the results of biological monitoring, the saprobity index was calculated: *Ind S* = 97/34 = 2.85. This value corresponds to class 4 of the quality of the reservoir, and water is assessed as polluted. The depletion of the species composition of hydrobionts of the researched reservoir is an adequate indicator of its pollution.

Thus, controlling the degree of pollution of a small river by the biodiversity of the periphyton community helps to quickly establish its state and determine the degree of pollution. The biological method, and in particular the saprobiological analysis, by the ratio of indicator organisms allows you to assess the ecological situation of the aquatic ecosystem quickly and without additional costs.

References

1. E. Yu. Afonina, A.V. Afonin Assessment of the status of hydrobiocenoses of small watercourses in the Argun river basin, Theor. and appl. ecol., **3**, 57 (2017)
2. T.Ya. Ashikhmina, N.M. Alalykina, L.I. Domracheva *Biological monitoring of natural-technogenic systems* (Publishing House of Komi Scientific Center Ural Branch of the Russian Academy of Sciences, Syktyvkar, 2011)
3. E.A. Bronovitskaya, V.V. Petrash, T. Ya. Ashikhmina, L.V. Kondakova Comparative toxicity assessment of surface water and soil based on the results of biotesting by two independent laboratories, Theor. and appl. ecol., **4**, 59 (2015)
4. R.G. Dzhamalov, A.M. Nikanorov, O.S. Reshetnyak, T.I. Safronova Waters of the Oka basin: chemical composition and sources of pollution, Wat. and Ecol.: Probl. and Sol., **3**, 114 (2017)
5. *Report on the state and environmental protection in the Ryazan region in 2018.* (Ryazan, 2019).
6. I.E. Dubovik, M.Yu. Sharipova Land and water epiphytic cyanoprokaryotes and seeweed and the possibility of their use in assessing the state of the environment, Theor. and appl. ecol., **1**, 51 (2016)
7. Yu.A. Mazhayskiy, T.M. Guseva Monitoring of heavy metals in the ecosystem of the small river of the Oka basin, Theor. and appl. ecol., **2**, 54 (2017)
8. G.G. Matishov, S.V. Kreneva, V.M. Muraveyko, I.A. Shparkovskiy, G.V. Ilyin *Biotesting and prediction of variability of water ecosystems in conditions of anthropogenic pollution* (KNC RAN, Apatity , 2003)
9. E.V. Makhanova Diagnosis of the ecological state of a reservoir: a comparison of the results of chemical analysis, biotesting and bioindication, Wat. and Ecol.: Probl. and Sol., **2 (78)**, 102 (2019)
10. O.P. Melekhova, E.I. Sarapultseva *Biological environmental control: bioindication and biotesting* (Academy, Moscow, 2010)
11. E.E. Minchionok, N.A. Pakhomova Assessment of the state of urban aquatic ecosystems by hydrobiological indicators, Theor. and appl. ecol., **3**, 48 (2016)

12. A.S. Olkova Modern trends in the development of the methodology of biotesting of aquatic environments, Theor. and appl. ecol., **3**, 19 (2018)
13. A.S. Olkova Actual directions of development of the methodology of biotesting of aquatic environments, Wat. and Ecol.: Probl. and Sol., **2 (74)**, 40 (2018)
14. A.S. Olkova, E.V. Makhanova Selection of biotests for environmental studies of waters polluted with mineral forms of nitrogen, Wat. and Ecol.: Probl. and Sol., **4 (76)**,70 (2018)
15. Electronic fund of legal and scientific and technical documentation (2020). *RD 52.24.309-2016. Organization and conduct of regime monitoring of the state and pollution of surface water of land* [online] Available at: http://docs.cntd.ru/document/495872993
16. The electronic fund of legal and scientific and technical documentation (2020). *Water Code of the Russian Federation (as amended on April 24, 2020)* [online] Available at: http://docs.cntd.ru/document/901982862
17. W. Brack, S. Ait-Aissa, R.M. Burgess Effect-directed analysis supporting monitoring of aquatic environments. An in-depth overview, Scien. of The Total Environ., **544**, 1073 (2016)
18. A. Cattaneo, G. Methot, B. Pinel-Alloul, T. Niyonsenga & L. Lapierre Epiphyte size and taxonomy as biological indicators of ecological and toxicological factors in Lake St.-Francois (Quebec), Environ. Pollut., **87**, 357 (1995)
19. J.M. O'Connell, E.D. Reavie, J.P. Smol Assemblent of water quality using epiphytic diatom assemblages on Cladophora from the St. Lawrence river (Canada), Diat. Resear., **12**, 55 (1997)
20. A.G. Solimini, R. Ptacnik, A.C. Cardoso Towards holistic assessment of the functioning of ecosystems under the Water Framework Directive, TrAC Trend. in Analyt. Chem., **28**, 143 (2009)

MONITORING OF THE PETROGENICALLY POLLUTED TERRITORIES

S.R. Hajiyeva, E.M. Gadirova*
Baku State University, Baku, Azerbaijan
*elmina2010@mail.ru

Abstract. Environmental monitoring has been carried out in the regions of Azerbaijan, chemical analysis of crude oils taken from Siyazan and May 28 has been counducted, and the degree of oil contamination of these areas has been determined. At the same time, the offshore waters of Mardakan and Gunashli regions have been taken and organic toxic substances have been identified in its content with the help of chromatographic analysis.
Key words: PAH, chemical analysis, gas chromatography, monitoring , water samples.

Introduction:

Aromatic compounds enter the biosphere in various ways and their sources are industrial enterprises, transport, and household waste. The particular attention given to aromatic compounds is largely due to their carcinogenic properties. Polycyclic aromatic hydrocarbons (PAHs) enter the atmosphere as a result of emissions and wastes from coke plants, some chemical plants, exhaust from internal combustion engines, and combustion products of various fuels. Groundwater is often contaminated with PAHs due to various wastewater sludge.

Aromatic hydrocarbons in petroleum range from 5 to 55%. These are the most toxic components of oil, and at a concentration of only 1% in water, they kill aquatic lower plants. Oil, containing 38% of aromatic hydrocarbons, significantly inhibits the growth of higher plants, and with an increase in the aromaticity of oil, their herbicidal activity increases. The content of all groups of polycyclic aromatic hydrocarbons gradually decreases with the gradual transformation of oil in the soil.

Polycyclic aromatic hydrocarbons (PAH) formed during fuel combustion are a multicomponent mixture, the individual analysis of each component of which is difficult without preliminary preparation of the sample for analysis. After preliminary chromatography of the sample by the method of column chromatography in a thin layer or by the method of differential sublimation, it is possible to analyze individual PAH using UV spectroscopy. The method allows to determine the concentration of various PAH in air up to 10-5%, or 0.1 μg / ml in solution [1-3].

Organic compounds such as polycyclic aromatic hydrocarbons (PAH) are a global environmental concern as they cause inflammation and skin cancer. As you know, there are two types of anthropogenic sources of hydrocarbons: petrogenic and pyrogenic sources.

Petrogenic sources include crude oil and petroleum-derived hydrocarbon compounds. Pyrogenic sources of hydrocarbon compounds are formed as a result of incomplete combustion of organic substances such as oil, wood, coal, etc. [4,5]. About 6.1 million tons of oil products are thrown into the ocean every year, most of which are of anthropogenic origin. Before entering the open sea, these oil products pass through the entire coastal zone.

The Caspian Sea is a very sensitive ecosystem. Over the past decades, under the influence of anthropogenic and biochemical factors, the state of ecosystems in general has deteriorated sharply, and especially in the northeastern part of the sea. Observations of recent years show that the waters of the Caspian Sea, especially along the coast of the National Park, are also polluted by oil and sewage [4,5]. Industrialization and urbanization in the Caspian region has been developing rapidly over the past several decades and the associated increase in hydrocarbons is a concern in the region. Offshore production and accidental oil spills, industrial waste, wastewater, discharges flowing down from river water are considered the main source of anthropogenic hydrocarbons in the marine environment [6-8]. Industry is believed to be the main source of oil pollution in the Caspian Sea. The total amount of industrial waste discharged into the Caspian Sea averages 2342.0 million m3 per year. Such waters contain 122.5 thousand tons of oil, 1.1 thousand tons of phenols, 9.9 thousand tons of organic chemistry products. The total content of hydrocarbons in the North-Western part of the South Caspian was small - 32-54.2 μg / g. In this area, in the vicinity of oil fields, the concentration of phenols was 0.002-0.003 μg / g [9,10]. Pollution of water and bottom sediments is noted throughout the Absheron Peninsula and in the Baku Bay. The main volume of pollution (90% of the total) enters the Caspiaan Sea with river runoff [11,12].

In the reviewed article, we also analyzed wastewater taken from the territories of Mardakan and Hovsan. It was important for us to calculate the amount of PAHs in these polluted waters. After purification, these waters are discharged into the Caspian Sea, and even in small quantities these harmful substances are dangerous for the flora and fauna of the sea and the environment. As is already known, PAHs are very dangerous for the environment, for living organisms in aquatic ecosystems, and therefore, the identification of hazardous substances and the application of methods for their destruction is very important [13,14].

Experimental part:

Crude oil were extracted from two regions of Azerbaijan – Siyazan , 28 May and their chromatographic analysis was carried out. At the same time, water samples has been taken for analysis from the two sites: Mardakan and Hovsan coasts at depths of 0.5 m from the sea level with the "12 L Niskin" sampler. Quality analysis of crude oils were carried out with A flame-ionizing GC equipped with ZB-1 (Phenomenex, USA) -FID 6890 (Agilent, USA) gas chromatographic detector. In the analysis, helium was used as a gas-bearing agent.

Chromatographic analysis of oil from two regions of Azerbaijan (Siyazan and May 28) was carried out. The determination of the chemical analysis of these oils, as well as the determination of organic toxic substances in offshore waters from Mardakan and Hovsan fields was implemented with the support of the Caspian Ecology Laboratory(Azerbaijan).

Explantion of the results:

The chemical analysis of crude oil gives us a reason to say that the sulfur content in our local oil ranges from 0.2 to 0.3%. This is not a big indicator, and this is one of the superior features of our oil. However, the little amount of sulfur content of the oil has environmental problems [15, 16].Car engines working with such oil fractions produce sulfuric gases into the environment by contaminating the atmosphere, causing greenhouse effect[17]. Therefore, even the least sulfur content is not satisfactory. There are various methods to reduce sulfur content in the oil.The gas chromatographic analysis of crude oil from the three regions of Azerbaijan is shown below (Figure 1-2).

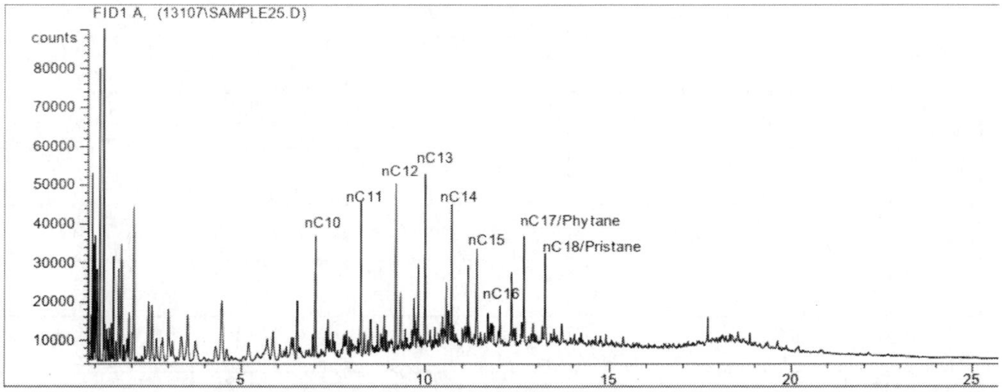

Figure 1. Gas chromatographic analysis of 28 May oil

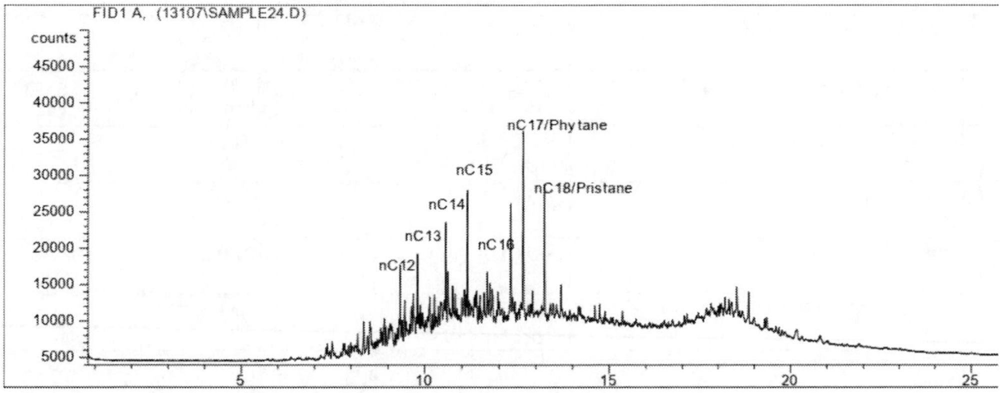

Figure 2. Gas chromatographic analysis of Siyazan oil

Although the composition of crude oil mainly consists of alkanes and naphthenes, mainly composed of C_{10}-C_{16} alkanes, aromatic hydrocarbons are formed as a result of transformation processes in the content of oil. The presence of salt over norm in the analysis of seawater and the nearness of the pH to the alkaline environment suggests that their composition is contaminated with organic and inorganic compounds[18-20]. At the same time, the content of water contaminated with oil and oil products has been determined by us, and analysis has been carried out to establish the toxic components from organic origin. For this purpose, water samples from the two parts of the Absheron peninsula: from the Mardakan and the Caspian Sea near Hovsan were taken. As a result of this water analyzes, the content of these organic substances has been clarified and chromatographic analysis is done, and the results are given in table 1 and figure 3-4.

https://doi.org/10.1051/e3sconf/202021201006

Table 1. Organic content of water samples taken from Mardakan and Hovsan areas

Polysilic aromatic hydrocarbons ,µg/l	**Mardakan**	**Hovsan**
Naphthalene	0,09	0,19
Achenthylene	<0.01	<0.01
Acenaften	0,01	0,01
Fluoren	0,04	0,07
Fenantren	0,09	0,17
Anthracene	0,01	0,01
Fluoranten	0,01	0,01
Piren	0,01	0,01
Benz (a) anthracene	<0.01	0,00
The chrezen	0,02	0,01
Benz (b + j + k) fluorantene	0,01	0,03
Benz (a) pyrene	0,01	0,01
Inden (1,2,3-cd) pyrene	<0.01	<0.01
Benz (ghi) perilen	<0.01	<0.01
Dibenz (ah) antracen	<0.01	<0.01
Σ 16 Individual PAC	0,29	0,53

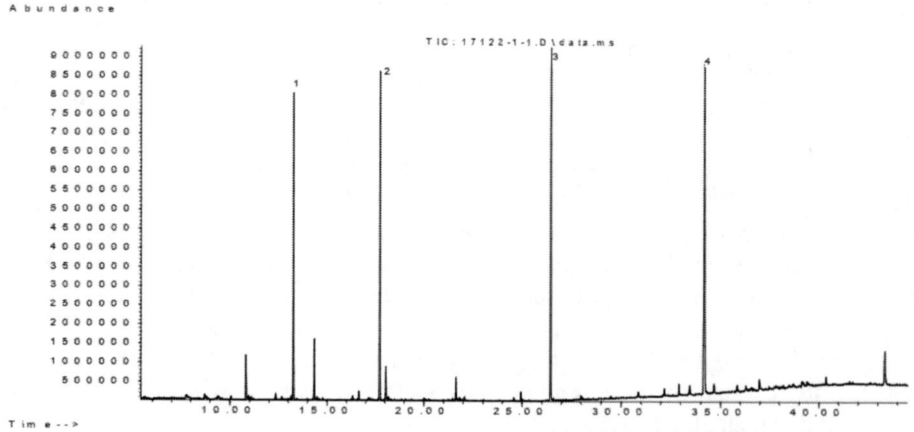

1.

Figure 3. Chromatographic analysis of water samples taken from Mardakan field

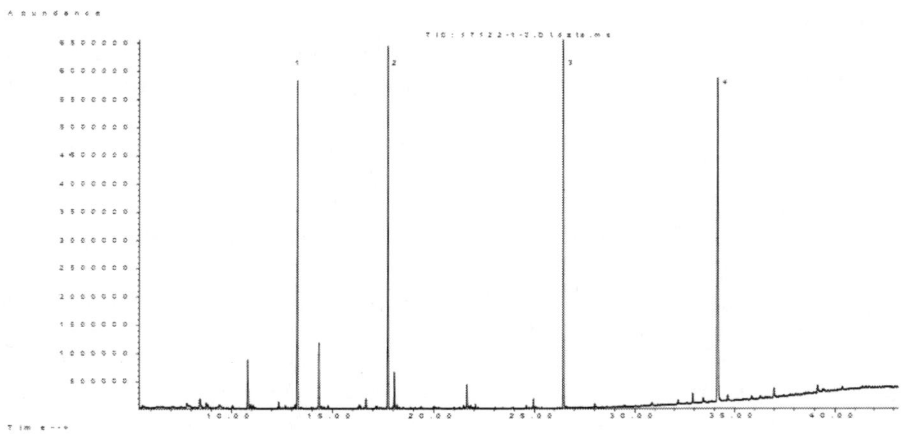

Figure 4. Chromatographic analysis of water samples from Hovsan field

Result and Conclusion:

According to the results of monitoring, pollution of oil products was observed in the southern part of the Caspian Sea. In this case, pollution of the soil, in particular contamination of water basins with oil and petroleum products is unwelcome, because the phenolic compounds have a high permissible concentration in the food products related to water (such as fish) and soil content (vegetable crops), which are ecologically useless, and ultimately lead to unwanted malignant tumors for the human body. More pollution of the Hovsan area was observed. The purpose of environmental monitoring was to expose these qualities. The continuation of these monitoring in the future will be the main subject of our research.

References:

1. A. S. Stepanovsky.General ecology: Textbook. Ed. // M.UNITI, 2000, 510 p.

2.Korshenko A., Gul A.G. Pollution of the Caspian Sea.// Hdb. Env. Chem. Vol. 5. Part P. Springer-Verlag, 2005.p.105-142.

3. Novikov Yu.V. Ecology of the environment and human: Moscow. 2005, p. 347.

4. Hajiyeva S.R., Gadirova E.M., etc. Methods for cleaning water contaminated with oil // Azerbaijan Chemistry Journal Baku, 2014, No. 1, p. 35-38.

5. Caspian Sea. State of the Environment // Report of the Interim Secretariat of the Framework Convention for the Protection of the Marine Environment of the Caspian Sea and the Bureau for Management and Coordination of the CASPECO Project. 2011, p.28.

6. Ostroumov S.A. Problems of ecological safety of water supply sources. 2006, No. 5, p. 17-20.

7. Neff J. M. Bioaccumulation in Marine Organisms. Amsterdam, Elsevier, 2002, 460 p.

8.Yunker M. B., Macdonald R. W., Vingarzan R. et al. PAH in the Fraser River basin: a critical appraisal of PAH ratios as indicators of PAH source and composition //Organic Geochemistry. 2002, vol. 33, p. 489-515.

9. Bogdanovski G.A. Chemical ecology // Moscow: Moscow State University. 1994. 237 p.

10. Problems of ecological safety of water sources. Ecological systems and appliances.2006,N 5. p.17-20.

11. Kostianoy Ed. A.,Kosarev A.,Korshenko A.,Gul A.G.The Caspian Sea Environment Vol. 5 Water Pollution Pollution of the Caspian Sea. Hdb. Env. Chem.V. 5. Part P, Springer-Verlag,2005, p.109–142.

12. Abdul Kasymov.Ecology of the Caspian Sea plankton, Baku,2004,p. 19–20.

13. Korshenko A.,Gul A.G. Pollution of the Caspian Sea. Hdb. Env. Chem.V. 5. Part P, Springer-Verlag,2005,p.109–142.

14.S.R.Hajiyeva,E.M.Gadirova. Advances in Biology&Earth Sciences. Baku, 2018, vol.3.No.3. p.248–256.

15. S.N. Gosling, N.W. Arnell. A global assessment of the impact of climate change on water scarcity. Clim. Change 134,2016,p.371-385.

16. S. Mohammadi, A. Kargari, H. Sanaeepur, K. Abbassian, A. Najafi, E. Mofarrah.Phenol removal from industrial waste-waters: a short review. Desalin. Water Treat.2015, 53, p.2215-2234

17. S.R.Hajieva,E.M.Gadirova,N.Sh. Sharifova,etc. Published under licence by IOP Publishing Ltd IOP Conference Series: Materials Science and Engineering, 2019,vol.570, N#1. https://iopscience.iop.org/article/10.1088/1757-899X/570/1/012031

18. Shmal A.G. Environmental safety factors-environmental risks. Publisher: Bronnitsy, MP "ICC BNTV, 2010,192p.

19. KorobkinV.I.,Korobkin L.V.:Peredelsky- Rostov: Ecology, 2003,576 p.

20. Khoruzhaya T.A. Assessment of environmental hazard: "Book service", 2002, 208 p.

Ecological condition of water bodies of the south-west of Belarus in spring 2020

Larysa Kirichenko[1,*], *Aliaksandr* Volchak[1], and *Anna* Golovach[1]

[1]Brest State Technical University, Faculty of Engineering systems and Ecology, Maskouskaya str. 267, 224017, Brest, Republic of Belarus

Abstract. In this work, hydrochemical and hydromorphological characteristics of urban reservoirs of southwestern Belarus in the spring of 2020 were investigated. Hydromorphological characteristics of reservoirs were determined by field methods and GIS mapping methods. Hydrochemical assessment of water quality of reservoirs in the spring period was carried out according to 15 indicators. Multivariate statistical methods were used to analyze surface water quality. Based on the analysis of hydrochemical and hydromorphological indicators, an assessment of the ecological state of the reservoirs of the urboterritories of southwestern Belarus in the spring period of 2020 was given. During the study, patterns of distribution of morphometric characteristics of the reservoirs within the studied territory were established. The analysis of hydrochemical characteristics showed that the water quality of urban reservoirs depends on the types of anthropogenic effects.

Keywords: urboterritory, water morphometry, water pollution level

Introduction

The development of urbanization usually negatively affects the ecological condition of the reservoirs located within the city, which depends on a set of factors, the most important of which is anthropogenic impact. Anthropogenic impact is expressed in the origin of the reservoir, changing its hydromorphological characteristics, the type and sources of pollutants entering the reservoirs of cities, in the degree of recreational load, in unauthorized water intake and waste water discharge, etc. This makes it increasingly difficult to determine the nature of the ecological status of the reservoir.

According to the classification of reservoirs according to the size of their water surface proposed by P.V. Ivanov, very small and small reservoirs (about 90%) with a maximum depth of up to 5 m predominate among the reservoirs of Belarus [1]. Most of them are of anthropogenic or natural-anthropogenic origin. These bodies of water are not included in the State network for monitoring the ecological state of water bodies in Belarus, there are no characteristics of their hydromorphological parameters and ecological state, which does not allow to give a legal assessment of the negative impact on urbanized bodies of

* Corresponding author: lakobrinetch@mail.ru

water [2]. In this regard, the study of the ecological state of urban reservoirs is especially relevant.

Modern climate change has disrupted the hydrological regime not only of rivers, but also of water bodies, including urban ones [3, 4]. This is especially evident in the south-west of Belarus, a vivid confirmation of which is clearly shown by the spring of 2020 characterized by abnormal hydrological conditions: the absence of ice and spring flood on water bodies occurred due to abnormally high average monthly air temperatures and the absence of snow cover in winter. The developed weather conditions affected the ecological state of water bodies. During this period, the greatest transformation of the hydrological regime of water bodies takes place [4–6]. Therefore, the study of the ecological status of urban reservoirs during this period is of particular interest.

The purpose of this work is to study the ecological, hydrochemical and hydromorphological state of the reservoirs of the urbanized territories of southwestern Belarus in the spring in modern conditions.

During the study, the following tasks were solved:

1. Hydromorphological indicators of urboterritory reservoirs have been studied;
2. The assessment of hydrochemical parameters of water quality of urboterritory reservoirs in spring period is given;
3. The state of the investigated reservoirs in the spring period and their main environmental problems are determined.

1 Methods

The object of the study is typical reservoirs of urboterritories of southwestern Belarus. As a result, urban reservoirs were selected, where a representative sample is possible: Kobrin, Zhabinka, Malorita, Brest. All of them are characterized by flat catchments, natural-anthropogenic origin, nutrition due to groundwater and drainage reclamation canals.

When planning selection points, urban reservoirs are divided into four groups depending on the nature of anthropogenic impact: a) reservoirs in the area of private residential development (characterized by relatively low population density within the city limits); b) water bodies of the park area with an expressed recreational load; c) quarry reservoirs and d) reservoirs exposed to the expressed technogenic effects.

Hydromorphological characteristics of reservoirs were determined by field methods and GIS mapping methods according to the following morphometric parameters: maximum length (L), maximum width (B), water mirror area (A), bankline length (L_l). Based on these data, the bankline elongation indicator ($L^* = \frac{L}{B}$) and the degree of bankline development ($S = \frac{L_l}{2\pi\sqrt{\frac{A}{\pi}}}$) were calculated [7–10].

Water sampling, preservation and transportation to determine hydrochemical indicators were carried out in accordance with the register of methods of chemical analysis of surface and wastewater of the Republic of Belarus. Water samples were taken from the near-surface part of the reservoir from a depth of 0,3-0,5 m. The samples were transported to the laboratory and stored in a refrigerator at 4°C for 24 hours. Water samples were analyzed within 24 hours after sampling. Water analysis by hydrochemical indicators was carried out by standard methods (Table 1).

The assessment of the environmental and hydrochemical status of urban reservoirs was carried out in accordance with the regulatory documents in force in Belarus [11–15].

Table 1. Water quality monitoring methods.

Parameters	Unit	Analytical method	Minimum detection limit
pH	-	potentiometric method	0,01
Hardness total	mmol/dm^3	titrimetric method	0,1
Na^+, Ca^{2+}, Mg^{2+}	mg/dm^3	drip electrophoresis method	0,5
Cl^-, SO_4^{2-}	mg/dm^3	drip electrophoresis method	0,5
CO_3^{2-}, HCO_3^-	mg/dm^3	acid-base titration method	0,01
PO_4^{3-}	mg/dm^3	spectrophotometric method with ammonium molybdate	0,005
Chemical oxygen demand COD	mgO_2/dm^3	dichromatic method	4
Dissolved oxygen DO	mgO_2/dm^3	slack method, iodometric titration	1
Biochemical oxygen demand BOD_5	mgO_2/dm^3	slack method, iodometric titration	1
Iron total $Fe_{(total)}$	mg/dm^3	spectrophotometry with sulfosalicylic acid	0,1
Total mineralization	mg/dm^3	design method	-

2 Results and discussion

Morphometric characteristics of the investigated reservoirs of southwestern Belarus are shown in Table 2. On the basis of these indicators, the typification of urban reservoirs of southwestern Belarus was carried out and characteristic features of urbanized reservoirs were identified.

According to the accepted classification, the investigated reservoirs belong to very small 43% (area 0,01-0,1km^2), small 28,5% (0,1-1,0 km^2) reservoirs and to reservoirs with an area of water mirror <0,01 km^2 that were not included in the classification 28,5%. These bodies of water are not included in the national monitoring system of water bodies, since their water surface area is less than 1 km^2 [16]. However, based on the data obtained, 2/3 of the urban reservoirs are very small and water bodies with an area of up to 0,01 km^2.

As a result of statistical data processing, based on the bankline elongation indicator L^*, the typification of urban reservoirs by type of pits is established:

a) bodies of water close to the oval shape ($L^* = 3 - 5$) - 1;

b) water bodies close to the rounded shape ($L^* = 1{,}5 - 3{,}0$) - 2;

c) rounded water bodies ($L^* = 1 - 1{,}5$) - 9;

d) water bodies close to square shape – 1 (Vychulki pond).

The distribution of reservoirs according to the degree of development of the bankline (minimum value S = 1,0 – characterizes an absolutely round lake) in the western part of the Polesskaya lowland indicates that lakes with a small bankline rugged (slightly cut, S not more than 1,5) prevail in the studied territory. These are rounded reservoirs with a smooth bankline, characterized by the predominance of residual and technogenic-disturbed basins developed within the reclamation drained swampy Polesskaya lowland.

Thus, rounded water bodies prevail within the city. However, the main criterion for forming the shape of the urban reservoir is the degree of change (straightening, strengthening by embankments) of the bankline. It was established that for urban reservoirs the change in the bankline is significant (> 5%), the bankline is straightened in almost all reservoirs, the banks of park reservoirs are strengthened by embankments.

Table 2. Hydromorphological parameters of some urban water bodies of southwestern Belarus.

Pond	Origin	Square A, km^2	*max* length B, km	*max* width L, km	bankline length L_l, km	Degree of bankline development L^*	Elongation factor S
town of Kobrin							
pond Polesskaya St.	Nature-anthropogenic	0,004	0,084	0,024	0,290	3,50	1,29
Park pond	Nature-anthropogenic	0,02	0,200	0,102	1,545	1,96	3,08
town of Zhabinka							
Mukhina Yama pond	Nature-anthropogenic	0,002	0,064	0,039	0,250	1,64	1,58
Vizzhar reservoir	Nature-anthropogenic	0,238	0,711	0,560	0,936	1,27	0,54
town of Malorita							
quarry Dzerzhinskaya St.	Nature-anthropogenic	0,007	0,144	0,060	0,370	2,4	1,25
Park pond	Nature-anthropogenic	0,016	0,170	0,123	0,512	1,38	1,14
Lake Military	Nature-anthropogenic	0,093	0,452	0,302	1,180	1,50	1,09
Peat-swamp pond	Nature-anthropogenic	0,286	0,784	0,493	2,45	1,59	1,29
town of Brest							
pond Vasnetsov St.	Nature-anthropogenic	0,012	0,156	0,090	0,44	1,73	1,24
quarry Kirpichnaya St.	Nature-anthropogenic	0,060	0,535	0,238	1,66	2,25	1,91
quarry Gershonsky	Nature-anthropogenic	0,201	0,692	0,319	1,86	3,72	1,17
Zerkalka pond	Nature-anthropogenic	0,046	0,286	0,201	0,812	1,42	1,07
Vychulki pond	Nature-anthropogenic	0,234	0,706	0,602	2,120	1,17	2,24
Lower pond May 1st park	Nature-anthropogenic	0,003	0,080	0,053	0,222	1,51	1,14

In natural conditions, water bodies of the region of southwestern Belarus are characterized in composition as calcium-sodium hydrocarbonate [5]. Based on the studies conducted, it was found that in the reservoirs of areas of private development of urboterrritories of southwestern Belarus there is a slight heterogeneity in the ionic composition of water, they can be described as calcium-magnesium hydrogen carbonate-chloride (Figure 1a). In the pond on Polesskaya street in Kobrin Ca^{2+} Na^{+} and HCO_3^- ions dominate. In the pond of Mukhina Yama in the town of Zhabinka the water is calcium hydrocarbonate-chloride. In a quarry on Dzerzhinskaya street in Malorita the water is sodium hydrogen carbonate-chloride. In the water of the pond on Vasnetsov street and pond Zerkalka in Brest established the predominance of Mg^{2+} and HCO_3^- ions. The increased content of Mg^{2+} and Cl^- in the water of reservoirs is associated with the ingress of unorganized stormwater runoff from household areas and roads [17–20].

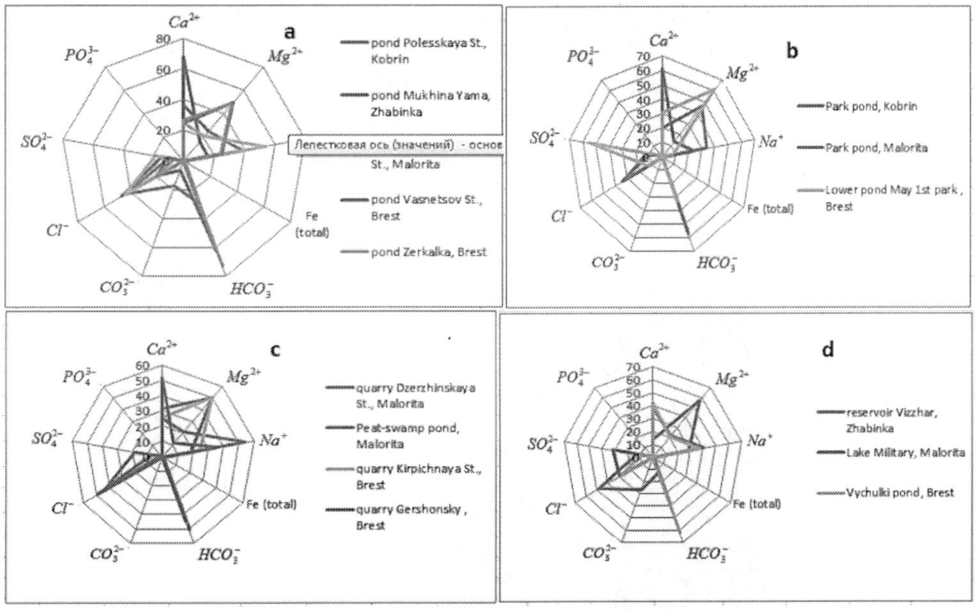

Fig. 1. The composition of water of the urban reservoirs in the spring period of 2020: a) of private residential development; b) in park ponds; c) quarry reservoirs and d) reservoirs with expressed anthropogenic influence.

In the park ponds of Kobrin and Brest, increased concentrations of Mg^{2+}, SO_4^{2-} and Cl^- ions (Figure 1b) are connected with the ingress of perennial unorganized storm drains from the adjacent green territory of the parks (annual fertilization of vegetation) and high recreational load (in these parks there is an annual high attendance of people). In the Park pond of Malorita, the least anthropogenic impact was established, the water in the reservoir belongs to hydrocarbonate-calcium with a noticeable presence of chlorides (Figure 1b). The recorded increased content of Cl^- indicates a slight anthropogenic impact. This is confirmed by the low recreational load on the reservoir (park attendance is low).

The water composition of quarry reservoirs of urboterritories is heterogeneous (Figure 1c) and depends on the previously carried out type of developments, flowing reclamation channels, unorganized storm drains from adjacent roads. In the water of the quarry on the Kirpichnaya street in Brest and the Peat-swamp pond magnesium ions, HCO_3^- and Cl^- ions dominate. The water of the Gershonsky quarry is characterized as calcium-sodium hydrogen carbonate-chloride. The high concentration of Cl^- is due to the fact that the reservoir is formed as the result of the development of a clay quarry, in addition, surface stormwater flows from roads and the adjacent territory of the private sector enter the reservoirs.

The ionic composition of water of reservoirs of the fourth considered group directly depends on the type of the technogenic influence. In the Vizzhar reservoir, water is calcium-sodium chloride. In the Military Lake of Malorita, magnesium chloride-sulfate water. The water in the pond Vyshulki, Brest, calcium-sodium hydrocarbonate. The increased concentrations of Mg^{2+}, SO_4^{2-} and Cl^- ions determine the water characteristic of these reservoirs according to the ion composition and confirm the technogenic effect on them.

In the water bodies under study, pH, dissolved oxygen, the content of calcium ions, hydrocarbonate ions, chlorides, sulphates, phosphates, meets the standards for water bodies within settlements and recreational zones. In terms of the total mineralization of water in the water bodies under study, they belong to weakly mineralized (Figure 2), which is characteristic of the region under study [3, 4]. However, in the pond Mukhina Yama of

Zhabinka and the Lower Pond of May 1st park, Brest, the mineralization of water is increased and approaches the maximum permissible concentrations (MPC 1000 mg/dm^3) for the surface waters of settlements and recreational zones. This is due to the increased content of calcium and magnesium salts in the water (Figure 1b).

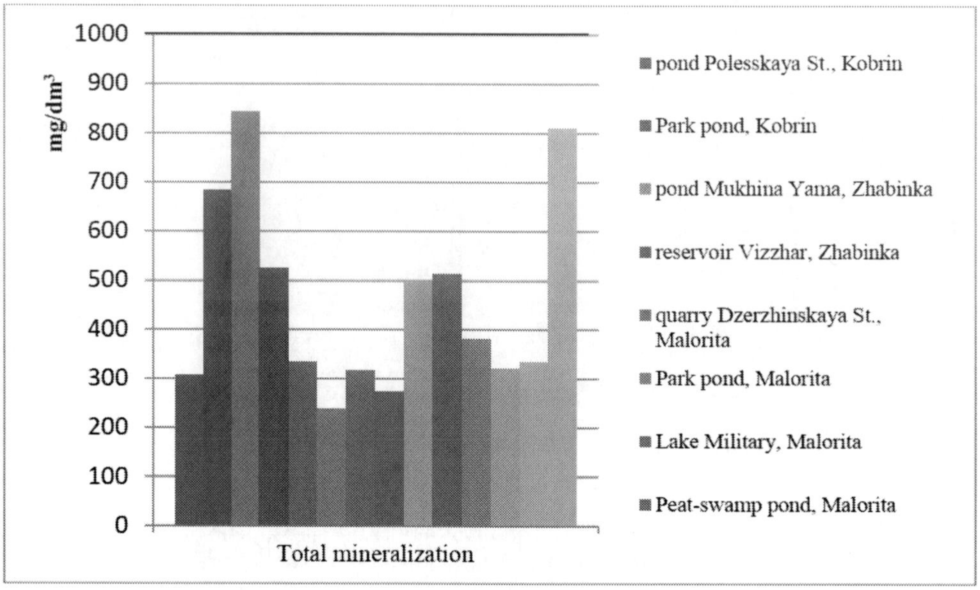

Fig. 2. Indicator of total mineralization of reservoirs of southwestern Belarus in the spring 2020.

To determine the water pollution of urban reservoirs, indicators were considered whose content exceeded the maximum permissible concentration (MPC) with the most stringent (minimum) values from the combined lists of information documents of the Republic of Belarus on the quality of surface water within settlements, recreational zones, and MPC drinking water. The characteristics of the level of contamination with substances are given by the multiplicity of the excess MPC [22, 23]. Based on the results of the study, the level of pollution of water bodies in terms of the multiplicity of excess MPC is set as low and medium. The content of the pollutants for the spring period of 2020 in the reservoirs of the areas of private residential development by the multiplicity of MPC is shown in Figure 3.

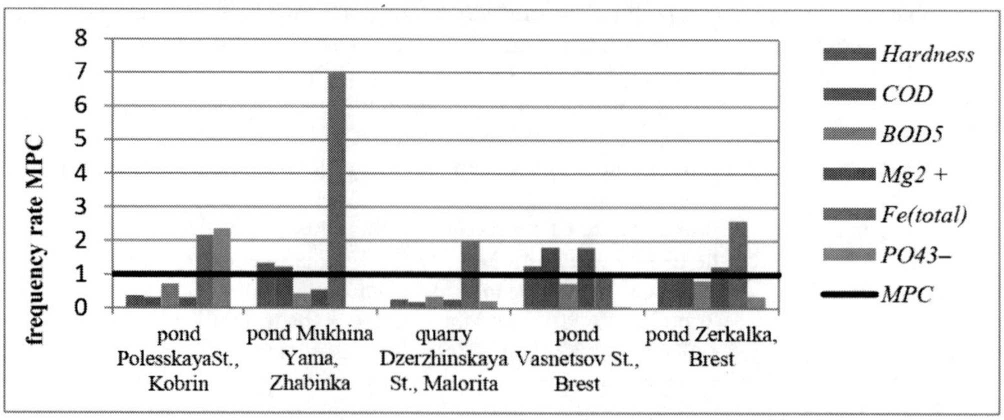

Fig. 3. The content of pollutants in the reservoirs of the private residential areas of the urboterritories of southwestern Belarus, spring 2020.

Based on the results of the study, the increase in the level of contamination with iron and phosphates with substances was established for the pond on Polesskaya street, Kobrin, for the pond Mukhina Yama in Zhabinki - for hardness, COD and general iron. In the ponds of Brest, a similar trend is observed: increased COD, hardness, Mg^{2+}, Fe total. In the quarry on Dzerzhinsky street the water is clean, the recorded 2 times excess of MPC Fe is generally typical for the region under study.

Water contamination by the magnitude of the MPC excess ratio was recorded by hardness, Mg^{2+} and iron common for most park reservoirs (Figure 4). There is a slight excess of MPC in the water of the park pond of Malorita according to COD, BOD_5, general iron.

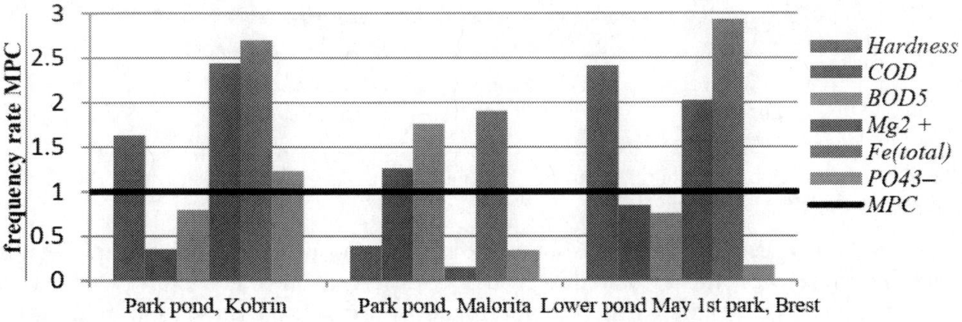

Fig. 4. The content of pollutants in the park ponds of the urboterritories of southwestern Belarus, spring 2020.

The content of pollutants for the spring period of 2020 in quarrying reservoirs of urbanized territories by multiplicity of MPC is shown in Figure 4. In all reservoirs, there is a tendency to increase the total iron content. In a quarry on Dzerzhinskaya street in Malorita, the content of substances in other indicators does not exceed the MPC, the water in the reservoir is clean. In the water of Peat-swamp, an excess of COD and BOD_5 standards was found. This indicates the contamination of water with readily oxidizable organic and inorganic substances [24–26]. In the watered quarries of Brest, the water is hard, an excess of MPC of 1,1-1,5 times is observed. The water hardness in these reservoirs is directly proportional to the concentration of Mg^{2+} ions (Figure 5). The increase in Mg^{2+} concentrations is due to the abnormal discharge of surface stormwater from the adjacent area.

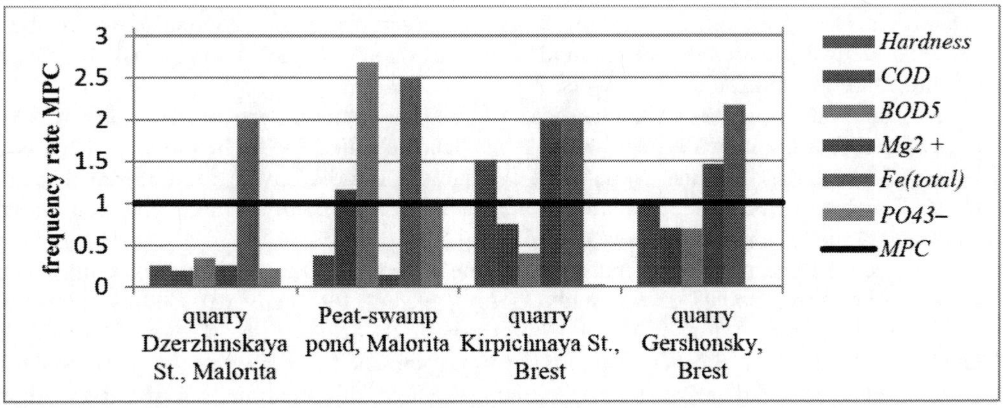

Fig. 5. The content of pollutants in the quarry reservoirs of the urboterritories of southwestern Belarus, spring 2020.

The waters of the water bodies with the expressed anthropogenic influence are characterized by exceeding MPC in several indicators (Figure 6). As in other reservoirs, there is a tendency to increase the total iron content by 2-5,5 times.

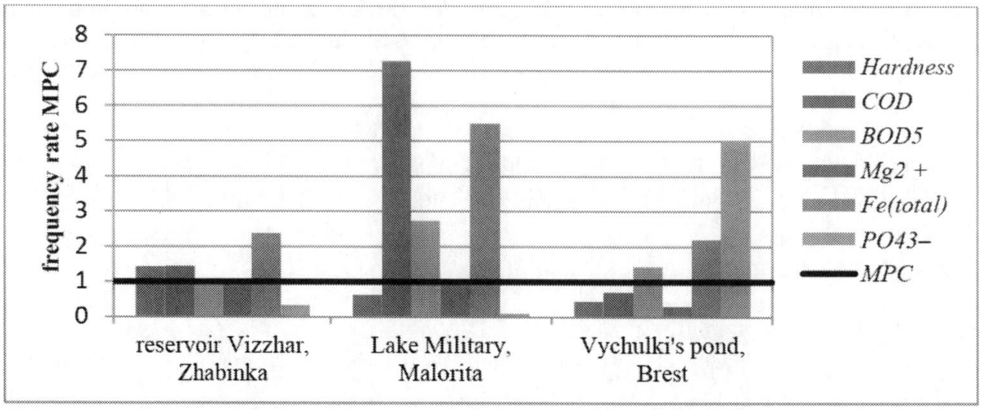

Fig. 6. The content of the pollutants in the water bodies of the urboterritories with the expressed anthropogenic influence, spring 2020.

In the reservoir of Vizzhar, Zhabinki, the multiplicity of excess MPC of pollutants is insignificant (1,1-1,5 times), but this excess is observed on 5 indicators: COD, BOD_5, content of ions Mg^{2+}, hardness and general iron. This is due to the presence of under-treated industrial wastewater in the water of JSC Zhabinkovsky Sugar Plant. The water of Lake Military of Malorita is contaminated with slightly oxidizable organic substances, according to BOD_5 the MPC indicator is 2,7 times exceeded, and for COD – 7,3 times. This indicates the eutrophication of this reservoir [27–29]. In the pond of Vyshulki in the town of Brest, a steady excess of MPC phosphates by 5 times was established. This pollution is due to the ingress of unorganized surface wastewater and storm water from the adjacent territory of the JSC Greenhouse Plant Berestier into the reservoir.

Conclusion

The conducted hydroecological studies have led to the following conclusions:

1. Based on the analysis of hydromorphological indicators, the typification of the urban reservoirs by water surface area was carried out. The predominance of very small and very small reservoirs of natural-anthropogenic origin in the considered urboterritory has been established. The predominance of the urbanized reservoirs of a rounded shape has been revealed. However, it has been found that the shape of the reservoir is completely dependent on the anthropogenic effects.

2. The waters of the urbanized reservoirs of southwestern Belarus are weakly mineralized, but the increased salt content has been revealed for the reservoirs of the Park Pond of Kobrin, the Mukhina Yama pond of Zhabinki and the Lower Pond of May 1st park, Brest. These reservoirs are characterized by a high content of calcium and magnesium compounds and, as a result, increased hardness.

3. Based on hydrochemical indicators in the spring of 2020, the level of pollution of most of the water bodies under study is low, except for the water bodies subject to technological impacts - for these water bodies the level of pollution is set as average. The level of pollution at the Vizzhar in Zhabinki depends on the discharged industrial wastewater of JSC Zhabinkovsky Sugar Plant; the level of contamination of Lake Military in Malorita depends on the discharged wastewater of the Bakery of RPO Maloritsky

District and Maloritsky Canning and Vegetable Drying Plant; the pond Vychulki of Brest - from the surface wastewater and storm water entering the reservoir of JSC Greenhouse Plant Berestier.

References

1. Vlasov, B.P. (2001). Nature–economic classification of lakes of Belarus. *Selected scientific dressings of the Belarusian State University*: at 7 Vol., V. 7.: 314–332 (in Russian). https://elib.bsu.by/handle/123456789/5030
2. Kirichenko, L.A. Problems of monitoring small water bodies of urbanized territories. *Land reclamation and rural construction. Search for youth: Collection of scientific works of students, undergraduates, graduate students and applicants*. (Belarusian State Agricultural Academy): 67–68 (2019) (in Russian).
3. Koronkevich, N.I., Barabanova, E.A., Zaitseva, I.S., Semenov, V.A., Korytny, L.M., Babkin, V.I., Dolgov, S.V., Chepalyga, A.L., Georgiadi, A.G., Shaporenko, S.I., Kichigina, N.V., Naprasnikov, A.T., Malinin, V.N., Fedorov, V.N., Shmakin, A.B., Taratunin, A.A., Borshch, S.V., Dobrovolsky, S.G., Istomina, M.N., Volchak, A.A., and other. Extreme hydrological situations. (Media-Press, Moscow (2010) (in Russian).
4. Volchak, A. A., Meshyk, A. P., Sheshka, M. M., & Bulskaya, I. V. (2016). Floods on the Territory of Polesie. In *Procedia Engineering* (Vol. 162, pp. 91–97). Elsevier Ltd. https://doi.org/10.1016/j.proeng.2016.11.020
5. Volchak, A.A., Bulskaya, I.V. (2017). Water Resources of Belarus under Changing Climate Conditions: Current Status and Prognosis. *Environmental Processes* 4, 125–136. https://doi.org/10.1007/s40710-017-0231-1
6. Szumińska, Danuta & Czapiewski, Sebastian & Goszczyński, Jacek. (2020). Changes in Hydromorphological Conditions in an Endorheic Lake Influenced by Climate and Increasing Water Consumption, and Potential Effects on Water Quality. *Water*. 12. 1348. 10.3390/w12051348.
7. Malozemova, O.V. (2012). Morphometric characteristics of lakes in various landscapes of the east of the Leningrad region. *News of the Russian State Pedagogical University named after A.I. Herzen*, (144), 112-121 (in Russian).
8. Gorodnichev, R.M., Levina, S.N., Ushnitskaya, L.A., Davydova, P.V., Pestryakova, L.A. (2020). Morphometric parameters of different lake types in the north of Yakutia. *Advances in current natural sciences*, 1: 18–25 (in Russian). http://www.natural-sciences.ru/en/article/view?id=37316
9. Ryanzhin, S.V., Myakisheva, N.V. & Zhumangaliev, Z.M. (2015). Morphometric and hydrochemical characteristics of Kazakhstan lakes. *Water Resour* **42,** 658–669. https://doi.org/10.1134/S0097807815050139
10. Chetverova, A. , Skorospekhova, T. , Morgenstern, A. , Alekseeva, N. , Spiridonov, I. and Fedorova, I. (2018). Hydrological and Hydrochemical Characteristics of Lakes in the Lena River Delta (Northeast-Siberia, Russia) , *Polarforschung,* Bremerhaven, Alfred Wegener Institute for Polar and Marine Research, 87 (2), pp. 111-124 . doi: 10.2312/polarforschung.87.2.111
11. Sanitary rules and norms 10-124 RB 99. Drinking water. Hygienic requirements for water quality of centralized drinking water supply systems. Quality control (1999) (in Russian).
12. Sanitary rules and norms 2.1.2.12-33-2005 "Hygienic requirements for the protection of surface waters from pollution," approved by Decree No. 198 of the Chief State Sanitary Doctor of the Republic of Belarus of November 28, 2005 (in Russian).

13. TAP 17.06-17-2018 (33140) Technical code of established practice. Environmental protection and management. Hydrosphere. Comprehensive environmental risk assessment and calculation of permissible recreational loads on water bodies in recreation areas of Belarus. Ministry of Natural Resources, Minsk. p. 24 (2018) (in Russian).
14. Sanitary norms and rules "Requirements for the maintenance of surface water bodies during their recreational use"; Hygienic standard "Permissible values of water safety indicators of surface water bodies for recreational use," approved by Decree of the Ministry of Health of the Republic of Belarus dated December 5, 2016 No. 122 (in Russian).
15. Hygienic standards 2.1.5.10-21-2003 "Maximum permissible concentrations (MPC) of chemicals in the water of water bodies of drinking and cultural and domestic water use," approved by Decree of the Chief State Sanitary Doctor of the Republic of Belarus dated December 12, 2003 No. 163(in Russian).
16. Kirichenko, L.A. (2019). State of ecological status of reservoirs of the Western Bug River basin. *Bulletin of the Brest State Technical University: Water construction, heat power engineering, geoecology*, **2** (115): 78 – 81. (in Russian).
17. Bulskaya, Ina & Volchek, Aliaksandr. (2014). Inorganic constituents in surface runoff from urbanised areas in winter: The case study of the city of Brest, Belarus. Oceanologia. 56. 373-383. 10.5697/oc.56-2.373.
18. Bulskaya, I., Volchek, A. (2015). Pollution of surface runoff from the territory of Brest, Belarus. *Water Supply* 1 April 2015, 15 (2): 256–262. https://doi.org/-10.2166/ws.2014.106
19. Rashid I., Romshoo SA. (2013). Impact of anthropogenic activities on water quality of Lidder River in Kashmir Himalayas. *Environ Monit Assess*, 185(6):4705-4719. doi:10.1007/s10661-012-2898-0
20. Subba Rao, N., Sunitha, B., Adimalla, N. *et al.* (2020). Quality criteria for groundwater use from a rural part of Wanaparthy District, Telangana State, India, through ionic spatial distribution (ISD), entropy water quality index (EWQI) and principal component analysis (PCA). *Environ Geochem Health*, **42,** 579–599. https://doi.org/10.1007/s10653-019-00393-5
21. Leslie DL, Lyons WB. (2018). Variations in Dissolved Nitrate, Chloride, and Sulfate in Precipitation, Reservoir, and Tap Waters, Columbus, Ohio. *Int J Environ Res Public Health*. 15(8):1752. Published 2018 Aug 15. doi:10.3390/ijerph15081752
22. Hydrochemical Institute (2002). RD 52.24.643–2002. Metodicheskiye ukazaniya. Metod kompleksnoy otsenki stepeni zagryaznennosti poverkhnostnykh vod po gidrokhimicheskim pokazatelyam [Regulatory Document RD 52.24.643–2002. Methodical guidelies. Method of comprehensive assessment of the pollution rate of surface waters using hydrochemical indicators]. Saint Petersburg: Gidrometeoizdat, 50 p. (in Russian).
23. Dvurechenskaya, S. Ya., Bulycheva, T. M. (2017). Opredeleniye kachestva vody vodokhranilishcha po integralnym pokazatelyam v periody raznoy vodnosti [Determination of the water quality of the reservoir by an integral indicators in different periods of water content]. *Water and Ecology: Problems and Solutions*, No. 1 (69), p. 44–53 (in Russian).
24. Prokop, Pawel & Wiejaczka, Łukasz & Syiemlieh, Hiambok & Kozłowski, Rafał. (2019). Response of Water Chemistry to Long-Term Human Activities in the Nested Catchments System of Subtropical Northeast India. *Water*. 11. 988. 10.3390/w11050988.

25. Justus B, Meredith B. (2014). Water quality of potential reference lakes in the Arkansas Valley and Ouachita Mountain ecoregions, Arkansas. *Environ Monit Assess*. 186 (6): 3785- 3800. doi:10.1007/s10661-014-3657-1
26. Alobaidy, A., Abid, H. and Maulood, B. (2010). Application of Water Quality Index for Assessment of Dokan Lake Ecosystem, Kurdistan Region, Iraq. Journal of *Water Resource and Protection*, Vol. 2 (9), 792-798. doi: 10.4236/jwarp.2010.29093.
27. Collins Robert, Jenkins Alan. (1996). The impact of agricultural land use on stream chemistry in the Middle Hills of the Himalayas, Nepal. *Journal of Hydrology*, Vol. 185, Issues 1–4, 71-86. ISSN 0022-1694, https://doi.org/10.1016/0022-1694(95)03008-5.
28. Bhatt, M.P., Hartmann, J., & Acevedo, M.F. (2018). Seasonal variations of biogeochemical matter export along the Langtang-Narayani river system in central Himalaya. *Geochimica et Cosmochimica Acta, 238*, 208-234.
29. Wang, R., Xu, T., Yu, L. *et al.* (2013). Effects of land use types on surface water quality across an anthropogenic disturbance gradient in the upper reach of the Hun River, Northeast China. *Environ Monit Assess*. **185,** 4141–4151. https://doi.org/10.1007/s10661-012-2856-x

Features of technogenic morphogenesis on the territory of south-western Belarus

Evgeniy Kukharik[1*], *Mikalai* Hrachanik[2], *Maxim* Bogdasarov[2], and *Aleksey* Matveyev[1]

[1]Institute for Nature Management of the NAS of Belarus, 220114 Minsk, 10, F. Skoriny str., Belarus
[2]Brest State University named after A.S. Pushkin, 224665 Brest, 21, blvd. Kosmonavtov, Belarus

Abstract. The classification of modern technogenic relief with the identification of its types, subtypes and forms is substantiated for the territory of southwestern Belarus. Information about the types of impact on the geological environment in the process of technomorph formation is given. Based on the generalization of field research materials, topographic maps and published literature, a diagram of the technogenic relief of the region under study was constructed which shows the largest artificial forms of its various types. Using information from technical documentation, special measurements and calculations, the averaged parameters of technomorphs were determined and their main morphological features were characterized.
Keywords: technogenic geological processes, technogenic relief, technogenesis, geomorphology.

Introduction

A directed effect on the earth's surface of the territory of southwestern Belarus occurs as a result of human economic activity and it leads to its significant transformation and the emergence of technomorphs. The transformation of the relief is associated with the construction of residential and industrial buildings, highways and railways, hydraulic structures, drainage reclamation, mining, land use in agriculture and other activities.

Purposeful study and mapping of technogenic geological processes and the artificial morphogenesis (technomorphs) formed by them are currently being actively pursued in various regions of the world as evidenced by the solid published literature [1–7]. The study of the processes of technogenic morphogenesis in the territory of south-western Belarus was carried out by A.V. Matveyev, S.F. Savchik, M.F. Hrachanik and others. Much attention in the works of these authors was paid to the creation of a classification of technogenic geological processes and the resulting technomorphs. So, A.V. Matveyev et al. [8, 9] proposed a rubrication of the relief of the south of Belarus in which technogenic formations were identified in the rank of a group. Accumulative and depleted forms, channels of technogenic watercourses, basins of artificial reservoirs and technogenic-mediated taluses, landslides, etc. were distinguished in the group. M.F. Hrachanik [10], in relation to the territory of the eastern part of the Podlaska-Brest depression, compiled a classification with

* Corresponding author: shzhk@mail.ru

the allocation of an anthropogenic class including a technogenic group which is divided into several types and subtypes.

Methods. Based on the generalization of the materials of the mentioned authors, a classification of the technogenic relief supplemented by the results of their own research was developed reflecting its diversity, genesis and the most general features of morphology. The allocated classification units will be discussed in detail below. The territorial distribution of the technogenic relief is shown in a specially drawn up diagram (figure).

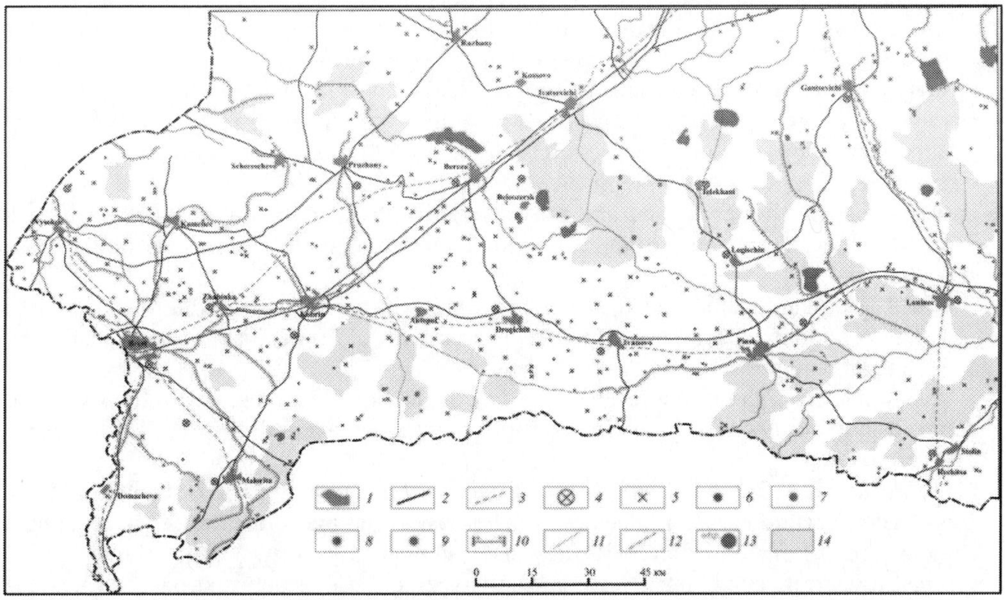

Legend: 1 – complexes of forms of residential spaces; 2 – large road technomorphs; 3 – large railway technomorphs; 4 – solid waste landfills; 5 – mini-landfills for solid waste; 6 – open pit for sand extraction; 7 – quarries for the extraction of clay; 8 – quarries for the extraction of peat; 9 – quarries for the extraction of chalk; 10 – straightened river sections; 11 – large reclamation canals; 12 – large hydraulic structures (canals); 13 – bowls of reservoirs; 14 – large reclamation systems

Fig. Scheme of technogenic relief of the territory of south-western Belarus.

The main source of factual material was the data of field observations, medium-scale topographic maps of the territory of southwestern Belarus, scientific publications. Cartographic measurements were made within the squares of a kilometer grid having a side length of 2 km on a scale and an actual area of 4 km^2 in order to determine the length of the formed road and irrigation technomorphs. All obtained values were entered into the table. Further, the obtained materials were generalized and reduced to a single scheme at a scale of 1 : 500,000. For this, a transparent cartographic base with a grid of squares with an area of 100 km^2 was compiled. The territory of the region under study was divided into 322 squares for which the volumes of technogenically displaced soils were calculated. Additionally, using online cartography tools (Google Earth) and Landsat-8 satellite images the location of solid waste landfills was determined and applied to a cartographic base accompanied by calculations of accumulation volumes.

Results and Discussion. *Mining type*. Among the negative forms of relief that arise during the extraction of minerals the most common are quarries which are divided into 2 groups: industrial and intraeconomic. There were 266 open-pit mines (of which 63 were industrial, and 203 were on-farm) on the territory of south-western Belarus, as of October 22, 2018 according to the data of the Brest Regional Committee for Natural

Resources and Environmental Protection. Clay, sands, sand and gravel mixtures, gravel, peat, sapropel, chalk are mined by quarrying in the region.

Industrial quarries are wider and deeper due to the greater volume of mining operations. Thus, the largest in southwestern Belarus are the peat quarries (Perelumye (145.27 ha), Zavidnyuvka (254.41 ha), Zubkovo (180.2 ha), etc.). The area of the quarries for the extraction of sand and clay is much smaller and usually does not exceed 10 hectares. The depth of excavations ranges from 1–1.5 to 12 m (Rechitsa sand pit). The area intra economic quarries usually does not exceed 1 hectare, the depth can reach 5 m.

The transformation of the relief of the territory during the development of minerals by quarrying occurs in several stages. At the first stage, an excavation starts at the quarry location based on the parameters specified in the design documentation. The fertile soil layer is removed and stored, later it can be used in the reclamation of the quarry. Thus, already at the stage of laying a quarry positive (dumps of the fertile soil layer) and negative (quarry excavations and other depressions) technomorphs are formed. At the second stage, with the beginning of the development of a mineral, the depth and area of the excavation grows, collapse-talus processes are actively manifested on its walls [11, 12] which are activated under the influence of atmospheric precipitation, melted snow and ground waters and wind.

There are many unauthorized places for mining in the region in addition to industrial and intra economic quarries which are registered with the state. They have a small width and depth and are measured by the first meters.

Accumulative technogenic forms created in the process of mining are embankments and dumps usually located near industrial facilities that process natural raw materials or in areas of its production.

Selitebny type. There are significant differences in the topography of cities and rural areas. By definition E.A. Likhacheva, "the relief of the urban area is a complex combination of natural, technogenic and architectural forms that create specific geomorphological conditions – a special type of polygenetic surface" [13, p. 176]. According to L.L. Rozanov [14, 15], an integral geotechnomorphic surface is formed in cities – a conjugate set of primary (natural) and secondary (technogenically determined) landforms. According to the author of the article, the relief of urbanized territories is part of the forms of the selitebny type since the geological impact on the earth's surface and the emerging technomorphs during the construction of settlements are monotonous.

The general background of urbanized areas is significantly influenced by complexes of one- and multi-storied artificial structures for various purposes. They are complex engineering structures consisting of natural and man-made materials. Usually, their height does not exceed 10 m, however, in large cities of the region, such as Brest, Pinsk, Kobrin, etc., there are multi-storied buildings several tens of meters high. The area of buildings is measured in tens by the first hundred square meters.

Relief planning, filling or cutting of natural forms is carried out in the process of building buildings. Significant masses of rocks are moved at various distances. According to the author's estimates, during the construction of ditches for buildings in the cities of the territory of south-western Belarus about 267 million m^3 of solids were displaced. Urban engineering structures create large static loads on the foundation soils, changing the conditions of their occurrence and physical and mechanical properties. According to data [12], static pressures from multi-storied buildings reach 0.5 MPa and from separate industrial structures – up to 2 MPa as a result of which the base soils are compacted and their moisture content decreases. Deformations occur from the arising stresses in the solid mass which can cause a violation of structural bonds, the development of subsidence phenomena that create a risk of destruction of buildings.

An important part of a settlement is its transport infrastructure which includes embankments and road surfaces, bridges, crossings, sidewalks and other engineering solutions. Highways within settlements are most often two- and four-lane paved which is important to ensure the speed of traffic.

The number of railway lines in cities which are not junction is greater than outside them. The railways form a complicated complex in the settlements where the junction stations are located (the cities of Brest, Zhabinka, Luninets) which ensures the movement of transport in several directions. The construction of embankments and excavations for roads and railways changes the appearance of the relief, creates additional loads on the foundation soils. The risk of destruction of embankments and road surfaces becomes real under the influence of dynamic loads created by transport, unfavorable geological processes (landslide phenomena, activation of erosion processes) can develop.

Movable-motionless technogenic morphogenesis [14] play an important role in the formation of urban relief and its change. These include road, rail, river, air vehicles as well as industrial, agricultural, construction mining machines and mechanisms, military self-propelled units, etc. They have a direct impact on the geological environment in the process of using them for construction and mining, etc.

Hydrological objects located on the territory of cities are subject to significant transformation. Thus, many rivers in the region within the urbanized territories have straightened channels which prevent the development of lateral erosion and meandering processes, high water during floods. Fortified embankments are being built as a result of hydrotechnical measures. The banks are subject to embankments and the accumulation of solid sediments in the riverbed is regulated. Lakes are also affected by the process of technogenic transformation, as drainage areas, the feeding regime of the reservoir and the processes of sedimentation change significantly. Small reservoirs and ponds are being created to regulate the water regime of the territory. Sediments are accumulated in the bowls of reservoirs depending on the mode of their use and processes typical for the bank zones of reservoirs develop.

Places for storage and processing of solid waste industrial waste mounds and waste rock dumps formed as a result of the activities of processing enterprises are formed in the relative proximity to cities and outside them and sometimes in the city limits. The accumulation of significant volumes of waste and dumps creates loads on the underlying rocks. In addition, material is eroded and transported over considerable distances under the influence of wind and atmospheric precipitation. Landfalls and taluses, suffusion can develop within the dump technomorphs.

Fortified settlements and mounds form a special group of residential type objects. The fortified settlements are represented by leveled forms and are the remains of ancient inhabited territories (cities, fortresses, etc.), their cultural layer is often buried under modern sediments. Mounds are hills of various sizes that were created by man in the historical past.

Road type. Ridge and ridge-like embankments of roads and railways are widely developed in the territory of south-western Belarus. Highways are divided into 4 classes – trunk roads, highways with improved coverage, roads with and without coverage. Country and forest roads and trails were not specially considered during the research as they do not make a big contribution to changing the topography of the territory. Highways cover the territory of south-western Belarus with a dense network. The parameters of road technomorphs are determined using technical documentation for their construction, field studies and are presented in Table 1.

Table 1. Average parameters of road technomorphs [by 16, 17]

Road class / number of paths	Subgrade width, m	Subgrade height, m	Volume of 1 running meter of subgrade, m³
Highways			
Motorways	28,5	1	28,5
Highways with improved coverage	27,5	0,7	19,25
Coated roads	12	0,5	6
Unpaved roads	10	0,25	2,5
Railways			
Double Track	6,4	1	6,4
Single Track	2,3	1	2,3

The total length of embankments and excavations of all highways on the territory of the studied region is 11,300 km of which technomorphs are highways – 198 km, highways with improved coverage – 1700 km, with coverage – 6300 km, without coverage – 3050 km. According to the calculations carried out on the territory of south-western Belarus, as a result of road construction, about 84 million m^3 of rocks have been displaced at various distances.

A motor road is a complex engineering structure, the embankment (subgrade) of which is constructed in a shape close to trapezoidal from natural and / or man-made soils, local or imported [16]. Pavement is laid on the upper part of the roadbed it is a multi-layer structure designed to redistribute the pressure on the ground from the action of the load of a moving vehicle. Work is carried out to plan and level the territory when preparing the area for the construction of the road. Embankments are compacted during construction, especially if the underlying soils are weak (peat, sapropel, etc.). The construction of the roadbed and pavement leads to the formation of extended linear technomorphs, which in the conditions of the Belarusian Polesie significantly alter the earth's surface. The higher the class of the road, the higher the degree of relief transformation as for the construction of a multi-lane highway the construction of a larger subgrade is required and large volumes of rocks are moved.

Crossings and bridges are constructed at intersections of roads as well as when a road crosses a river or canal. A subgrade is erected during the construction of crossings and it in height and width exceeds the usual embankments and has a complex configuration. During the construction of bridges, special reinforced concrete or metal structures are used. Embankments are erected adjacent to the two banks of the reservoir.

In some cases, excavation is carried out most often when crossing local elevated sections to build a road, huge volumes of sediments are moved which are later used for the construction of subgrade and road pavements or a cavalier is arranged in which soils that are not used for one reason or another are stored. The greatest transformative impact on natural systems occurs during the construction of roads within areas with weak soils at the base, for example, in swamps. In such cases, soft soil is extracted which leads to the destruction of natural landforms, displacement of cover deposits and their replacement with imported soils. In the areas of road construction, the relief acquires a leveled appearance with extended positive rampart and ridge-shaped technomorphs that stand out against its background.

During the construction of the roadbed of highways, engineering and technical solutions are necessarily applied, aimed at preventing the development of negative manifestations of exogenous processes, however, during the exploitation of the road under the influence of natural factors (precipitation, air temperature, etc.) and man-made (dynamic effects of moving vehicles) impacts, the subgrade can lose stability. This leads to the emergence of the danger of the development of erosion and landslide processes, destruction of the subgrade and disruption of movement or failure of certain sections of the roads.

The railways of south-western Belarus are represented by 6 large branches. The total length of technomorph railways in the region is 770 km of which double-track – 280 km, single-track – 495 km. About 2.9 million m^3 of solids were moved over various distances during the construction of railways.

The subgrade of the region's railways is arranged in the form of a trapezoid. Technomorphs can be arranged both on leveled surfaces and in recesses. Natural and technogenic soils, local or imported are used [17] during the construction of embankments. After the erection of the embankment and before laying the track, the roadbed is covered with a layer of rocky, weakly weathered rocks, which prevent the development of processes of erosion and destruction by the wind and also redistribute the load during the operation of the railway.

The greatest impact on the natural environment occurs when laying paths in areas with damp and damp foundations (in swamps). In addition to the actual construction of excavations and embankments, replacing part of the natural soils with imported ones, additional drainage ditches, drainages, etc. are constructed.

Hydraulic engineering type. The largest hydraulic structures on the territory of south-western Belarus are the Dnieper-Bug and Oginsky canals. Of these two objects, the Dnieper-Bug waterway is navigable. It stretches across the study area from west to east for 243.2 km connecting the rivers Mukhavets and Pina. The guaranteed channel width and depth are 40 and 2 m, respectively. Based on this, the volume of displaced material during the construction of 1 running meter of the canal is 80 m^3 and the total volume of displaced soil is 19.45 million m^3. The Oginsky canal connects the rivers Shchara and Yaselda, however, do not fulfill their functions at the present stage, they are a historical monument.

A significant transformation of the relief occurs during the construction of hydraulic structures. The formed technomorphs are significantly superior in parameters to other types of channels in the region. Cavaliers are constructed from the excavated and unused soils, which reach 5 m in height.

After excavation of the canal to ensure a long service life and to prevent water erosion in the channel and meandering processes, sections of the banks of hydraulic structures are often doned or straightened using engineering and technical methods.

A large number of dams and bunds have been built on the territory of the region which are extended positive forms of relief, they are built mainly from natural soils. Sluice gates and waterworks which are designed to regulate water flow are built of artificial materials and rising significantly above the water surface.

Bowls of reservoirs and ponds also belong to negative hydrotechnical technomorphs. In total, there are 44 reservoirs on the territory of the territory of south-western Belarus which are of 3 types: bulk, lake and channel. Small reservoirs with a water surface area of up to 1–2 km^2 and a total volume of up to 5 million m^3 prevail. Ponds are much smaller in comparison with reservoirs; therefore, their number was not specially counted.

According to V.M. Shirokov et al. [18], the Polesie reservoirs have low and gentle shores composed mainly of sands of various granulometric composition. Reservoirs of Belarus are divided into 5 hydromorphological types: small shallow water; small non-deep; medium complex (valley type); medium-deep and deep (lake and lake-river). The morphometric parameters of the Polesie reservoirs are predetermined by the flat relief, the parameters of the lake basins and the structure of the river valleys.

Small shallow water reservoirs have depths less than 2 m. This type includes the Papernya reservoir on the river Zelvyanka, the average depth of which is 1.13 m, the mirror area is 1.8 km^2 and the total volume is 2.0 million m^3. Small non-deep reservoirs have average depths up to 2–3 m. A reservoir of this type can be the Gornovo-2 reservoir, the average depth of which is 2.7 m, the surface area is 0.38 km^2 and the total volume is 1.04 million m^3. Medium complex (valley type) and medium-deep reservoirs have depths of more than 3 m. The Pogost reservoir belongs to the medium complex type, the average depth is 3.4 m, the surface area is 16.2 km^2 and the total volume is 54.5 million m^3. Deep (lake and lake-river) reservoirs are created on the basis of natural lakes or in river valleys they have a large area, depth (more than 5 m) and volume, a complex configuration. An example of such a reservoir is the Bereza-1 reservoir, created on the basis of the lake Black.

During the construction of reservoirs, the adjacent area is planned, its leveling is carried out, water canals, gateways-regulators, dams, enclosing dams and other hydraulic structures are constructed which significantly alter the natural appearance of the relief. At the final stage of construction, the banks of reservoirs, canals and feeding rivers are embanked and the territory adjacent to the lakes is slightly flooded. Reservoirs activate many exogenous processes, primarily abrasion. Slope processes develop in the coastal zones of reservoirs. Reservoirs affect the groundwater regime which leads to the activation of suffusion and subsidence processes.

Irrigation and drainage type. Open drainage channels of various classes prevail among the destructive irrigation and drainage technomorphs. The averaged parameters of irrigation and drainage technomorphs are shown in Table 2. The set of interconnected drainage canals which are a complicated complex of engineering and technical structures is called a drainage system.

Table 2. Averaged parameters of reclamation canals and canalized river sections [by 19–20]

Type of construction	Width, m	Depth, m	The volume of soils during the excavation of 1 running meter, m^3
Trunk	6	2	12
Open drainage	3	1,2	3,6
Canalized section of the river	10	2	20

According to the drainage method, drainage systems prevail on the territory of south-western Belarus they are represented mainly by open drainage channels which are built in trapezoidal recesses. When laying channels, the excavated soil is used for road construction or is stored in the near-edge parts, forming a cavalier. The depth of the canals is set based on the conditions for ensuring the required drainage rate: the minimum for mineral soils is 1 m, for peat soils – 1.2 m after peat settling, the maximum for shallow canals-dehydrators is 1.4–1.5 m. The distance between the open drainage canals the network is calculated according to various formulas (depending on engineering-geological, geomorphological and other conditions). According to V.S. Anoshko [21], the distance between channels on low-lying peat soils is from 60 to 110 m, on transitional – 50–70 m, on upper – 30–40 m. According to the author's calculations, as a result of irrigation and drainage construction, 115 million m^3 of rocks were moved.

The largest in terms of parameters are the main canals. Their depth can reach several meters, width up to 10 m or more. An example is the Orekhovsky channel. After reconstruction in 1973, its width was 12 m, in the middle of 2018 – 8.2 m. This canal is part of the large reclamation system "Orekhovskaya" located to the north of the agro-town Povitie of the Kobrin region, which, in addition to it includes the Orekhovskoye reservoir, catching and supplying channels and open drainage channels.

The largest number of channels falls on small drainage channels. Their width rarely exceeds 5 m, the initial depth is equal to the minimum design depth (1–1.2 m), however, during operation and the action of natural factors, their depth significantly decreases without additional maintenance measures the channels silt up, overgrow and fail. To ensure observation of the level of groundwater and its composition, observation wells of various depths are being constructed on the territory of the reclamation system.

When laying canals, the extracted soils can be used to fill dams that enclose polder systems or reservoirs, road embankments and can also be stored along the canal in the form of a cavalier which is a shaft up to several meters high. Sometimes the cavaliers are leveled and in such cases have a height 0.15–0.5 m. For the further use of reclaimed land as agricultural land, the earth's surface flattens out and leveled areas of reclamation systems are formed.

Belligerative type. A special type of technomorph for various purposes is emerging in territories that are occupied by military units or set aside for military training grounds,. Barracks and other residential and command premises may correspond in parameters to residential buildings of the residential type but there is a small number of multi-storied buildings among them. The sizes of these buildings are different. Garages and hangars are used for storage of military equipment and can reach significant sizes and occupy large areas. To ensure takeoff and landing of helicopters and airplanes, runways of various configurations with hard surfaces are being built. The warehouses of fuels and lubricants, in addition to the surface part, have large underground reservoirs in which oil products are stored.

A special type of technomorph is formed by various fortifications and long-term defensive points. They are widespread in the vicinity of the cities of Brest and Pinsk and in the Kamenets region between the villages of Stavy and Orlya. From the soils extracted during the construction process, defensive embankments (parapets) are arranged, which have a height of up to 1 m.

During the military exercise, weapons are tested and it is accompanied by blasting operations. The explosion of a projectile leaves an explosive funnel on the earth's surface – the dimensions of which depend on the power of the explosion of the projectile.

Communal type. This type includes all technomorphs that arose in the process of equipping territories for landfills and mini-landfills for solid household and municipal waste as well as during waste storage. Landfills are the largest waste storage areas. There are a total of them on the territory of the studied region 18. An example of a large solid waste accumulator is the landfill located to the south of the village of Vulka-Gorodishchenskaya, Pinsk district, Brest region. Its area is about 78.7 thousand m^2. As waste accumulates with a capacity of several meters, the surface of the landfill is leveled with the help of heavy special equipment, poured over with a layer of soil and compacted. The height of the accumulated waste layer currently reaches 20 m and the volume is approximately 2.66 million m^3. The composition of waste is different: metal, paper, cardboard, plastic, reinforced concrete structures, construction waste, glass, etc. Poorly fixed waste is often carried by the wind over considerable distances.

A moat up to 2 m wide was laid around the storage areas it is intended for the accumulation of liquid waste components. Control wells are installed along the perimeter of the landfill with the help of which the quality control of groundwater is ensured.

Mini-landfills for solid domestic waste occupy a much smaller area – usually no more than 0.1 km^2. There were 196 such objects on the territory of south-western Belarus in total, as of October 1, 2018.

The main area of the mini-landfill is the production zone which can occupy up to 95 % of its entire area [22]. Within its limits, solid waste is stored, in most cases in 1, less often up to 3 layers with each compaction according to technologies used at larger landfills. At the bottom of the stored waste is usually a pit, the soil from which is used for intermediate or final isolation of waste or is placed in cavaliers around the perimeter. The equipped cavalier is a shaft made of local soil up to 3 m high which encloses the landfill area and prevents waste from being transported outside of it. Also on the territory of the landfill outbuildings, bowls of ponds and disinfecting baths, drainage ditches, etc.

There are a large number of unauthorized dumps in the region in addition to landfills and mini-landfills for storage and disposal of solid municipal and domestic waste they are most often located in open pit mining sites or in the forest.

Conclusions. A diverse complex of technogenic landforms arose on the territory of south-western Belarus as a result of intensive human economic activity. Technomorphs are formed as a result of a targeted impact on the geological environment during the construction of various buildings and structures, the movement of huge volumes of soil, the transformation of positive and negative landforms. This led to a significant change in the natural appearance of the earth's surface in the region. A classification of the technogenic relief was made on the base of generalization of field research data and published materials. The highest unit of classification is a group (technogenic). All technomorphs of the region are combined into 7 types: mining, road, residential (selitebny), hydrotechnical, irrigation and drainage, belligerative and communal. Each type includes 2 subtypes of forms – accumulative and destructive. The types of impact on the earth's surface are shown as a result of which technomorphs arise.

Bibliography

1. Osipov V. I. Tekhnogenez i sovremennye zadachi nauk o Zemle [*Technogenesis and Current Tasks in the Earth Sciences*]. Ecology and Industry of Russia, 2016, 20 (3), pp. 4–12. (In Russian). DOI: 10.18412/1816-0395-2016-3-4-12.

2. Szabó J. Anthropogenic Geomorphology: Subject and System. In: Szabó J., Dávid L., Lóczy D. (eds). Anthropogenic Geomorphology. Springer, Dordrecht, 2010, pp. 3–10. DOI: 10.1007/978-90-481-3058-0_1.

3. Peloggia A. U. G. Geological classification and mapping of technogenic (artificial) ground: a comparative analysis. Revista do Instituto Geológico, 2018, 39 (2), pp. 1–15. DOI: 10.5935/0100-929X.20180005.

4. Uraskulov M. R., Bogush I. A., Cherkashin V. I., Gazaliev I. M., Yusupov A. R. Geoecological Problems of Zones of Ore Mining Technogenesis in the North Caucasus. Arid Ecosyst, 2018, 8, pp. 47–52. DOI: 10.1134/S2079096118010110.

5. Mzia K., Tkemaladze S. The Classification Imereti's Landscapes and Its Anthropogenic Transformation. Earth Sciences, 2017, 6 (6), pp. 111–116. DOI: 10.11648/-j.earth.20170606.12.

6. Tsermegas I. Anthropogenic transformation of the relief of the Aegean Islands. Miscellanea Geographica, 2015, 19 (2), pp. 40 – 49. DOI: 10.1515/mgrsd-2015-0009.

7. Nelson F., Outcalt S. I. Anthropogenic geomorphology in northern Alaska. Physical Geography, 1982, 3 (1), pp. 17–48. DOI: 10.1080/02723646.1982.10642216.

8. Matveyev A. V. Istoriya formirovaniya rel'efa Belorussii [*History of the formation of the relief of Belarus*]. Minsk, Navuka i tekhnika Publ., 1990, 144 p. (In Russian).

9. Matveyev A. V. et al. Rel'ef Belorusskogo Poles'ya [*Relief of Belarusian Polesie*]. Minsk, Nauka i tekhnika Publ., 1982, 131 p. (In Russian).

10. Hrachanik M. F., Matveyev A. V., Bogdasarov M. A. Rel'ef territorii Podlyassko-Brestskoy vpadiny [*Relief of the territory of the Podlaska-Brest Depression*]. Brest, BrGU Publ., 2013, 154 p. (In Russian).

11. Galkin A. N. et al. Inzhenernaya geologiya Belarusi (3 t.) [*Engineering geology of Belarus (3 vol.)*]. Vitebsk, VGU Publ., 2016–2018. (In Russian).

12. Trofimov V. T. et al. Inzhenernaya geologiya Rossii. Inzhenernaya geodinamika territorii Rossii [*Engineering geology of Russia. Engineering geodynamics of the territory of Russia*]. Moscow, KDU Publ. House, 2013, 816 p. (In Russian).

13. Likhacheva E. A. Antropogenno-geomorfologicheskie sistemy [*Anthropogenic-geomorphological systems*]. Geomorfologicheskie sistemy: svoystva, ierarkhiya, organizovannost' [*Geomorphological systems: properties, hierarchy, organization*]. Moscow, Media-PRESS Publ., 2010, pp. 176–181. (In Russian).

14. Rozanov L. L. Teoretiko-metodologicheskie osnovy geotekhnomorfologiches-kogo issledovaniya prirodopol'zovaniya. Diss. dokt. geogr. nauk [*Theoretical and methodological foundations of geotechnological and morphological studies of environmental management. Dr. geogr. sci. diss. abstr.*]. Moscow, 1997, 47 p. (In Russian).

15. Rozanov L. L. Geotekhnomorfogenez i organizovannost' geotekhnomorfosistem [*Geotechnomorphogenesis and organization of geotechnomorphosystems*]. Geomorfologiya, 2008, 2, pp. 90 – 96. (In Russian). DOI: 10.15356/0435-4281-2008-2-90-96.

16. Minin A. V. et al. Avtomobil'nye dorogi Belarusi [*Highways of Belarus*]. Minsk, BelEn Publ., 2002, 672 p. (In Russian).

17. TKP 45-3.03-163-2009 (02250) Zheleznye dorogi. Zemlyanoe polotno. Pravila proektirovaniya [*State Standard 45-3.03-163-2009 (02250). Railways. Subgrade. Design rules*]. Minsk, M-vo arkhit-ry i stroit-va Belarusi Publ., 2010, 99 p. (In Russian).

18. Shirokov V. M. et al. Vodokhranilishcha Belorussii: prirodnye osobennosti i vzaimodeystvie s okruzhayushchey sredoy [*Reservoirs of Belarus: natural features and interaction with the environment*]. Minsk, Universitetskoe Publ., 1991, 207 p. (In Russian).

19. Murashko A. I. et al. Melioratsiya [*Land Reclamation*]. Minsk, BelEn Publ., 1984, 567 p. (In Russian).

20. Lagun T. D. Melioratsiya i rekul'tivatsiya zemel' [*Land Reclamation and Recultivation*]. Minsk, IVTs Minfina Publ., 2017, 376 p. (In Russian).

21. Anoshko V. S. Geograficheskie osnovy melioratsii [*Geographical Basics of Land Reclamation*]. Minsk, Vysheyshaya shkola Publ., 1974, 176 p. (In Russian).

22. Sokolov L. I. et al. Sbor i pererabotka tverdykh kommunal'nykh otkhodov [*Collection and processing of municipal solid waste*]. Moscow, Infra-Inzheneriya Publ., 2017, 176 p. (In Russian).

Thermal-hydraulic Assessment of Seven Wire-wrapped rod Bundle

Kefeng LYU[ab], *Xuelei* SHENG[ab] , *Xudan* MA[ab]

[a]Anhui Advanced Technology Research Institute of Green Building, Anhui Jianzhu University, Hefei 230601, PR China

[b]College of Environment and Energy Engineering, Anhui Jianzhu University, Hefei 230601, PR China

Abstract: Lead bismuth eutectic (LBE) is one of the most potential materials for coolant and spallation target for Accelerator Driven Systems (ADS). Thermal-hydraulic behavior of LBE in fuel assembly is a key issue for development of the systems. To get a deeper understanding on the complex thermal-hydraulic features of wire-wrapped rod bundle cooled by upward LBE, an electrically bundle with 7 rods wrapped with helical wire was developed in KYLIN-II thermal-hydraulic forced circulation loop. The flow resistance, thermal entrance characteristic and heat transfer coefficient were investigated. As for the entrance characteristics, during the full heating length (exceeding 140 times the hydraulic diameter), the thermal field did not reach a fully developed and stable condition which is contrary to the ducted flows. The experimental heat transfer coefficient showed that the hexagonal shell has a great influence on the heat transfer coefficient in rod bundle geometry. For this reason the application of empirical correlation should be kept cautious in rod bundle analysis.

Key words: Accelerator Driven Systems; Thermal-hydraulics; Rod bundle; Lead-bismuth eutectic

1. Introduction

Accelerator Driven sub-critical Systems (ADS) are considered as one of the most potential options to incinerate long living fission products by transmutation (C. Rubbia,

1996; W. Gudowski, 2001). With respect to low melting temperature, high boiling temperature, outstanding heat transfer performance, chemical stability, and good neutron economy, Lead-Bismuth Eutectic (LBE) are currently attracting more and more attention word wildly to be both the coolant for sub-critical reactor core and spallation target in transmutation systems (T. Takizuka, 2002; N. Li, 2008). In 2011, the ADS project was launched in China by Chinese Academy of Sciences (CAS), and a 10MWth experimental reactor cooled by LBE named CLEAR-I (Y. Wu and Y. Bai) was designed by FDS team which has been strongly involved on advanced reactor concepts, advanced materials and nuclear code development. In the frame of this project, to support the design and construction of CLEAR-I, KYLIN-II multi-functional LBE facility, including material, thermal-hydraulic and safety loops has been established based on the former experience on lead-lithium experimental technology.

One of the most interesting topics in thermal-hydraulic fields is the turbulent heat transfer, especially for complex geometries, e.g. fuel assemblies and high power spallation target. Owing to the lower prandtl number, the heat transfer to heavy liquid metal is very different from the heat transfer to water or gas (X. Cheng, 2006). That is, compared to the contribution from convection, the contribution from thermal conductivity in low prandtl number fluid like lead-bismuth eutectic is much higher than water even through at the conditions with high Reynolds number. Thus existing heat transfer numerical models and experimental experience are not applicable to systems cooled with liquid metal directly.

Moreover, due to the compact arrangement of rod bundle and the difficulties of heavy liquid metal measurement, the experimental data on the rod bundle cooled by heavy liquid metal is scare. Several experimental studies were conducted in the sixties using mercury (Hg) and sodium–potassium alloy (NaK), which has the similar scale prandtl number with lead or lead bismuth. Unfortunately, as mentioned by W. Pfrang (2007), in most publications about these former experiments, no detailed description was given, whether spacer or what type of spacers have been used. Recently, with the increasing interesting in heavy liquid metal technology, a series of new bundle experiments (J. Pacio, 2014; M. Tarantino, 2013) were conducted to study the thermal-hydraulic characteristics of heavy liquid metal in rod bundle. Nevertheless, the rod bundle geometries have a lot of influencing factors, e.g. pitch to diameter, wire pitch, hexagonal shell and so on, only several experiments could not summarize the complex flow and heat transfer features. For the time being, it is well agreed that there still exists big deficiency in understanding and describing special thermal-hydraulic behavior in tight rod bundles, especially related to heat transfer behavior.

2 Experimental setup

The 7-heated rod bundle test section was installed in the vertical direction with upward flow at position 4 in Fig.1. The coolant LBE enters the rod bundle at the lower end and flows upward to move the thermal generated by heated pins. A general view of this test section is depicted in Fig.2. In the flow direction, upstream of test section, the flow developing length with 1.4 meters and a flow restrictor were placed to obtain a uniform inlet flow rate around each sub-channel.

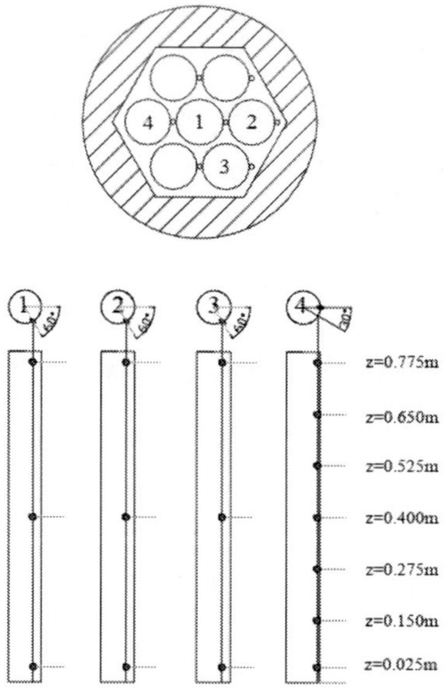

Fig.1. General view of the 7-rod bundle test section

The rod bundle were manufactured referring to the preliminary structure design of fuel assembly of CLEAR-I. Table 1 shows the main parameters of the experimental bundle. Each rod has a heated length of 800mm, and an unheated length of 1100mm. The nominal input voltage is 220V while the power of each pin is 5000W. To obtain the accurate power of experimental bundle, a direct-current power supply system consisting of a thristor rectifier and a power distribution system was employed.

To evaluate the thermal-hydraulic performance of the bundle, the wall temperature of heated rod and pressure drop of wire-wrapped bundle were measured in the test section. The bulk temperature in the bundle section was not measured. This parameter would be calculated from an energy balance throughout the test section (from inlet until axil position z). The wall thermocouples with 0.5mm diameter were embedded in the heated wall to measure the heated wall temperature which depicted in Fig.3. The number of the wall thermocouples is 16, and the detailed position is also depicted in Fig.2.

Table1 Main characteristics of the rod bundle

Parameters	Symbol	Valve（mm）
Rod outer diameter	D	15
Pitch	P	16.74
Rod length	L	1900
Rod heated length	L_h	800
Apothem	a	48
Helical wire diameter	d_w	1.64
Helical pitch	H	375

3 Results and discussion

The friction factor of wire-wrapped rod bundle was calculated according to the following formula:

$$f = \frac{2 * \Delta P * d_h}{L_B * \rho * v^2} \quad (1)$$

Where ΔP is the pressure drop of flow resistance, d_h is the hydraulic diameter of rod assembly, L_B is the length of wire-wrapped bundle, and v is the average velocity.

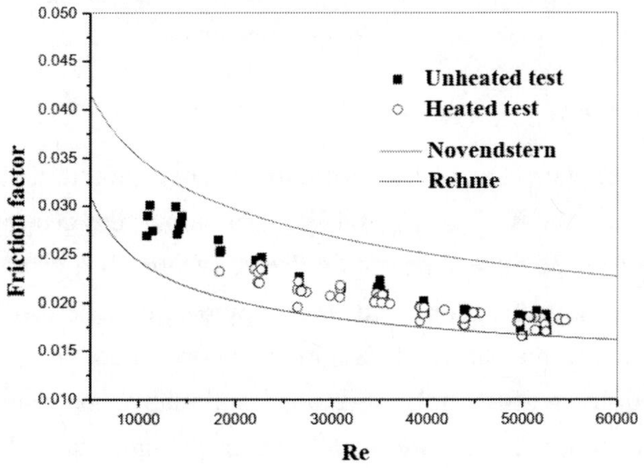

Fig.2. Evolution of the friction factor according to the Reynolds number

The experimental friction factor profiles as function of Reynolds number for the bundle are displayed in Fig.4. The unheated test was performed at ~260℃ with adiabatic running, while the heated test was conducted at ~260℃ of average inlet and outlet temperature with maximum temperature difference 40℃. From the experimental results depicted in Fig.2, there is no significant difference between the unheated and heated test due to their similar reference temperature. Furthermore, the experimental friction factor compared with empirical correlations (E.H. Novendstern, 1972; K. Rehme, 1972) which derived from the air or water experiments is also depicted in Fig.4. All the experimental values were located between the Novendstern and Rehme correlations. That could be seen that LBE has similar flow resistance characteristics of traditional fluid like water or air. In detailed the Rehme correlation presented a better agreement with the experimental results than Novendstern correlation. The probable reason is that the effect of the rod number is concluded in the Rehme correlation.

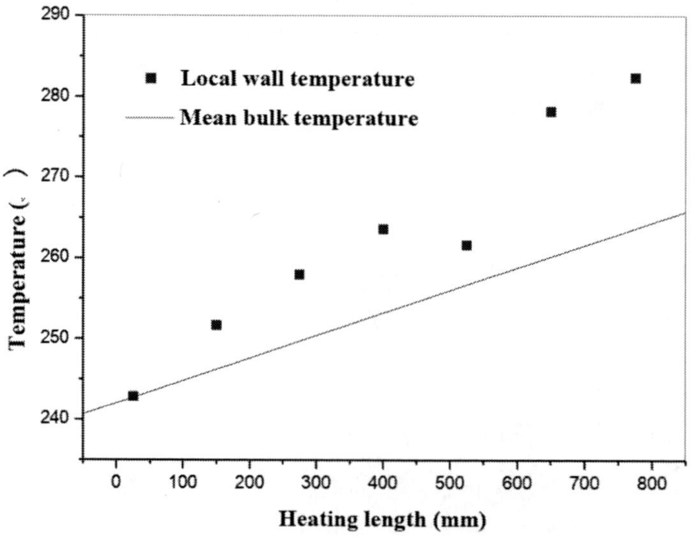

Fig.3. Thermal entrance characteristic of experimental bundle

The thermal entrance characteristic of experimental bundle is depicted in Fig.3. The blue squares are the wall temperature of PIN4 along the heating length, and the black line is the average bulk temperature derived from energy balance. It could be seen that along the flow direction the discrepancy between the wall temperature and bulk temperature became larger. This phenomenon could be concluded that the thermal field did not reach a fully developed and stable condition during the full heated bundle section, although the overall heating length reached to 140 times the hydraulic diameter. The conclusion was different with the ducted flows. This phenomenon was also predicted through CFD calculations by other publications (N. Govindha Rasu, 2013). They mentioned that the reason was the weak communication among the sub-channels in tightly packed bundle. In addition, the wall temperature at axial position z=525mm was abnormal lower. The reason was that the thermocouple probably didn't embed well in the heated wall in this experiment.

Fig.6 shows the temperature deviations in the cross section of the bundle, respectively for two different Peclet numbers. To characterize the temperature deviations, a non-dimensional parameter Θ (M. Berthoux, 2010) is defined as following:

$$\Theta = \frac{T_{w\max} - T_{w\min}}{T_{zf} - T_{in}} \quad (2)$$

Where $T_{w\ max}$-$T_{w\ min}$ is the measured maximum wall temperature deviation, T_{zf}-T_{in} is the temperature difference between the local bulk temperature (calculated on heat balance) and inlet bulk temperature.

The wall temperature deviations between center rod and periphery rod could be detected obviously in Fig.6. The temperature deviations were caused by the adiabatic hexagonal shell and the different local hydraulic diameters between the inner and periphery sub-channels. Furthermore, the comparison between the Fig.6 (a) and (b) revealed that the temperature non-uniformity was more significant for lower Peclet number. That is because in the higher Peclet regimes, the coolant mixing is stronger, which can induce the temperature distribution more uniform.

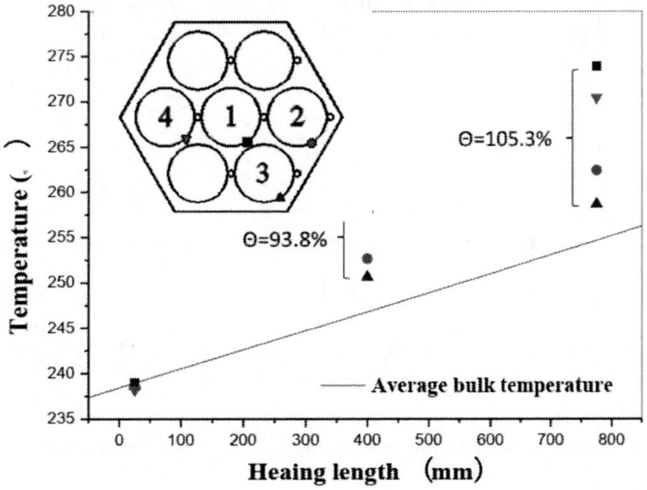

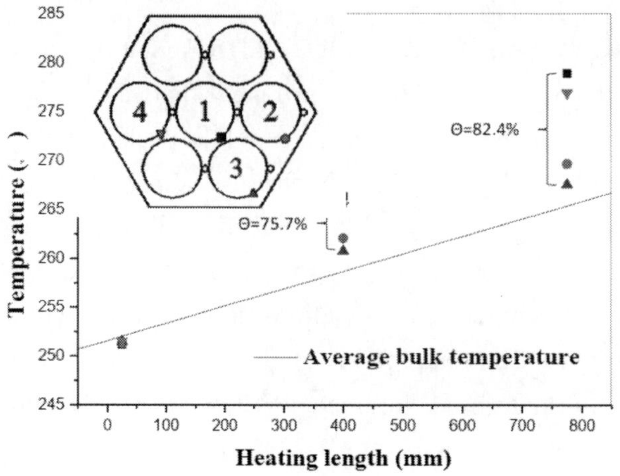

(a) Pe=570 (b) Pe=1150

Fig.4. Temperature deviations in the cross section of bundle

The wall temperature of the center rod was used to evaluate the heat transfer coefficeient. The local Nussult number (Nu) is defined as in Eq(3), where d_s is the hydraulic diameter of sub-channel, T_b is the average bulk temperture driven by energy balance, T_w is the measured wall temperature of center rod.

$$Nu = \frac{q^{"} * d_s}{(T_w - T_b) * \lambda} \tag{3}$$

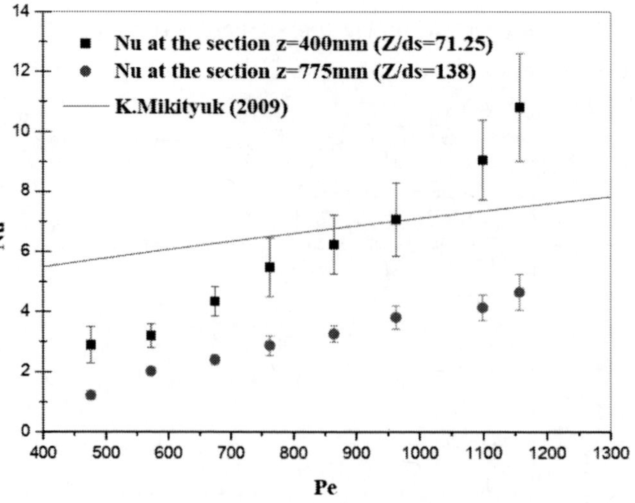

Fig.5. Evolution of the local Nussult number according to the Peclet number (The error bars were based on the standard deviation)

The Nu profiles as function of Peclet number (Pe) are depicted in Fig.7. At the given Pe, the Nu at axial position z=400mm ($z/d_s \approx 71.25$) was remarkably larger than the value at axial position z=775mm ($z/d_s \approx 138$). The phenomenon could be attributed to thermal developing condition which has discussed in Fig.3. Furthermore, the deviation became larger and larger along the flowing direction. The empirical correlation K. Mikityuk (2009) which is valid for full developed condition of infinite rod bundle is also depicted in Fig.7. It could be seen that the Nu at section z=775mm had a similar rising slope of K. Mikityuk correlation, but the Nu at section z=400mm had a relatively steep slope. Moreover, the magnitude of experimental Nu at section z=775mm is smaller than K. Mikityuk correlation. That is because of the non-uniform distribution of wall temperature which discussed in Fig.6. Thus the local Nu which based on center rod wall temperature and average bulk temperature would be lower than the average Nu like K. Mikityuk. This indicates that the hexagonal shell has a great influence on the heat transfer coefficient of rod bundle geometry. For this reason the application of empirical correlation should be kept cautious.

Acknowledgements

We would like to thank the other members of FDS Team for their help on experimental works.

References

C. Rubbia, 1996. Status of the Energy Amplifier Concept. Proceeding of the 2nd International Conference on Accerator-driven Transmutation Technologies and Applications, Kalmar, Sweden, June 3-7.

E.H. Novendstern, 1972. Turbulent flow pressure drop model for fuel rod assemblies utilizing a helical wire-wrap spacer system[J]. Nucl. Eng. Des. 22: 19-27

J. Pacio, M. Daubner, F. Fellmoser, et al, 2014. Heavy-liquid metal heat transfer experiment in a 19-rod bundle with grid spacers. Nucl. Eng. Des. 273, 33–46.

K. Rehme, 1972. Pressure drop performance of rod bundles in hexagonal arrangements. Int. J. Heat Mass Transfer 15 (12), 2499–2517.

K. Mikityuk, 2009. Heat transfer to liquid metal: review of data and correlations for tube bundles. Nucl. Eng. Des. 239 (4), 680–687.

L. Qiu, Y. Wu. B. Xiao, et al, 2000. A low Aspect Ratio Tokamak Transmutation System. Nucl Fusion. 40: 629-633.

M. Berthoux, T. Cadiou, 2010. The Thermal Hydraulics in a Rod Bundle Representative if the Start-up Core of the ALLEGRO Gas Cooled Fast Reactor-Experimental and Numerical Approaches. Nucl. Eng. Des. 240, 3372–3386.

M. Tarantino, I. Di Piazza, P. Agostini, et al, 2013. Thermal-hydraulic Assessment of HLM-cooled Pin Bundle in CIRCE Pool Facility. International Conference on Fast Reactors and Related Fuel Cycles: Safe Technologies and Sustainable Scenarios FR13. Paris, France, March 4-7, 2013.

N. Li, 2008. Lead-alloy Coolant Technology and Materials-technology Readiness Level Evaluation. Progress in Nuclear Energy. 50, 140-151.

N. Govindha Rasu, K. Velusamy, T. Sundararajan, et al, 2013. Investigations of flow and temperature field development in bare and wire-wrapped reactor fuel pin bundles cooled by sodium.

OECD/NEA, 2007. Handbook on Lead-Bismuth Eutectic Alloy and Lead Properties, Materials Compatibility, Thermal–Hydraulics and Technologies. 2007 edition: 15-21.

Q. Huang, S. Gao, Z. Zhu, et al. 2009. Progress in Compatibility Experiments on Lithium-Lead with Candidate Structural Materials for Fusion in China. [J]. Fusion Eng. Des. 84: 242-246.

Q. Huang, C. Li, Y. Li, et al, 2007. Progress in Development of China Low Activation Martensitic Steel for Fusion Applicayion. J. Nucl. Mater. 367-370:142-146.

T. Takizuka, K. Tsujimoto, T. Sasa, et al, 2002. Design Study of Lead-bismuth Cooled ADS Dedicated to Nuclear Waste Transmutation. Progress in Nuclear Energy. 40,505-512.

W. Gudowski, V. Arzhanov, C. Broeders, et al, 2001.Review of the European project — Impact of Accelerator-Based Technologies on nuclear fission safety (IABAT). Progress in Nuclear Energy. 38,135-151.

W. Pfrang, D. Struwe, 2007. Assessment of correlations for heat transfer to the coolant for heavy liquid metal cooled core designs. Technical Report FZKA 7352, Forschunszentrum Karlsruhe GmbH.

X. Cheng, N. Tak, 2006. Investigation on Turbulent Heat Transfer to Lead-bismuth Eutectic Flows in Circular Tubes for Nuclear Applications. Nucl. Eng. Des. 236, 385-393.

Structural geological mapping of the Cenozoic sediments of the Brest region using GIS technologies

Anna Maevskaya[1*], *Nikolay* Sheshko[2], *Natalia* Shpendik[2] and *Maksim* Bogdasarov[1]

[1]Brest State University named after A.S. Pushkin Cosmonauts Boulevard, 21, 224016, Brest, Belarus
[2]Brest State Technical University Moskovskaya St., 267, 224017, Brest, Belarus

Abstract. Cenozoic sediments of the territory of the Brest region is the object of research in this work. The aim of this work is to detail the structure of the Cenozoic stratigraphic deposits by creating a set of structural geological maps. The process of creating maps included several sequential stages implemented using the ArcGIS 10.5 software product. In general, a set of maps for each period of the Cenozoic era was made according to the implemented method. As a result of mapping, the features of the geological structure of the Cenozoic sediments were detailed (based on the use of the most complete materials on the drilling exploration of the territory during the construction). The use of geoinformation systems in the process of building will allow for quick updating of cartographic materials in the future.
Keywords: Brest region, cenozoic sediment, gis mapping, big data, structural-geological maps.

Introduction

Cenozoic deposits have become quite widespread within the territory of the Brest region, which due to their lithological diversity can be considered as a promising regional resource base of minerals, primarily building materials. This necessitates a serious detailing of the nature of the surface of the buried horizons of the Cenozoic as a basis for a qualitative forecast and assessment of the prospects for the development of the mineral resource potential of the territory as well as optimization of the organization of engineering and construction activities.

It is possible to obtain information about the nature of the surface of buried geological horizons by analyzing structural-geological maps. However, the present cartographic materials reflecting the nature of the surface of the horizons of Cenozoic sediments made for the territory of the Republic of Belarus and covering the territory of the Brest region as well as maps made for separate tectonic structures located within the boundaries of the region [1–4] have a number of disadvantages. First, the preparation of cartographic

* Corresponding author: maevskaya.anna@inbox.ru

materials was carried out at different time periods using different construction methods. Secondly, the creation of structural geological maps was often carried out manually with subsequent digitization which does not exclude the presence of distortions on such maps associated with the subjective vision of the compilers. Thirdly, the use of traditional methods of drawing up structural maps limited the ability to use the entire amount of available data in the construction and identify errors at intermediate stages of modeling. Fourth, updating the existing cartographic materials due to the peculiarities of their creation is a laborious process. All of the above indicates the impossibility of using existing cartographic materials to obtain high-quality and reliable information about the relief structure of the Cenozoic stratigraphic units within the territory of the Brest region.

At present, in the work of a number of Belarusian authors we can find examples of creating structural geological maps using GIS technologies. But in these works, the structural-geological maps are compiled for the territory of separate regions of Belarus and do not include the territory of the Brest region [5, 6]. In the works, where the territory of Belarus was the object of mapping, the maps were implemented for the horizons of the Quaternary system of the Cenozoic era which does not allow getting an idea of the nature of the relief structure of the Paleogene-Neogene deposits [7].

Based on the foregoing, the creation of a set of digital geological maps of the Cenozoic deposits of the Brest region is becoming urgent, implemented according to a unified technique using geoinformation systems allowing to obtain a high-quality detailed picture of the structure of the underground relief of the Cenozoic stratigraphic units.

Analysis of a number of literary sources in the field of creating digital geological models showed that at present there are many examples of the implementation of structural geological maps for various territorial levels and areas of tasks being solved [8–15]. However, there is no unified approach in the literature regarding the methodology for creating such maps. As a rule, the preliminary stage of modeling is the collection of all available thematic information and its introduction into the GIS-shell which allows you to create a universal structured data model with the preservation of topological integrity as well as a properly organized subordination of geological units [16, 17].

In the future, the construction of structural geological maps using well data is carried out with the help of various interpolation methods which are used to solve a wide range of geographic problems [18, 19]. At the same time, in most of the works describing the methodology for constructing structural geological maps the main emphasis is placed on the algorithm for constructing interpolation models, on the features of the choice of the interpolation method taking into account the specifics of the available data [6, 20]. Less common are methods that include, along with the description of approaches to surface modeling, stages associated with the preparation of data for construction including the processing of initial information [21]. When modeling underground geology, 3-D construction and various proprietary methods are actively used today which are mainly aimed at solving rather highly specialized problems (for example, solving problems in the oil geology of a particular field). At the same time, such methods are quite difficult to implement and require the collection of a significant amount of diverse geological information [22–23].

In this paper, on the basis of approaches existing in the literature using GIS technologies, a method for creating structural geological maps in the geoinformation system ArcGIS 10.5 is proposed. The presented technique has been tested on the example of creating a set of digital geological maps for separate horizons of Cenozoic sediments in the Brest region. The difference between the proposed approach and those existing in the scientific literature is as follows. This technique is a sequential series of stages of creating structural geological maps from the stage of data preparation to the implementation of the finished layout. When developing the methodology, the fact was taken into account that geological data are large amounts of information and the amount of initial information can increase over time, i.e. the developed methodology is focused on working with large amounts of information. Data processing is carried out both at the "input" in order to eliminate distortions associated with errors in the source materials and at the "output" to assess the correctness of the implemented models. While in most studies, processing is carried out only at the stage of data preparation.

In general, the algorithm presented in this work simplifies the process of drawing up structural geological maps while allowing to qualitatively visualize the features of the relief structure of buried stratigraphic horizons and can be used to create structural geological maps for other territories.

The implemented digital cartographic material has a high-precision spatial reference and can be quickly changed taking into account the latest literature data, information from engineering and geological surveys which will significantly simplify the procedure for mapping geological deposits based on drilling knowledge. Various operations can be carried out with the surfaces obtained in the course of the worked out technique including the combination of several maps in order to perform subsequent analysis.

Methods

Initial data. The initial data for the compilation of a set of structural geological maps presented in this work were the materials of drilling exploration of the territory of the Brest region provided by the State Scientific Institution "Institute of Nature Management of the National Academy of Sciences of Belarus" and the State Enterprise "Research and Production Center for Geology".

The information presented in the initial database is characterized by rather large volumes and can be classified as geospatial data obtained as a result of prospecting and exploration work [24–26].

The original database is presented in Microsoft Access format and includes several related tables: well assignment, work objectives, lithology, stratigraphy, wells. In addition, coordinates are presented for each record in the table which allows it to be embedded in software GIS shells. In this case, the desktop software ArcGIS 10.5 was used. The preliminary stage of importing the database into the GIS environment was the selection of information by building a query in the query designer on the basis of which a pivot table was created containing the information selected most necessary for the subsequent construction of maps from various tables presented in the database.

In general, the database implemented in the GIS contains 40,590 records providing information on more than 5,000 wells. Several records are associated with each well on the map for each of the Cenozoic era horizons that were penetrated by this well. The database also contains 22 fields which provide information about the number of the well, the date of drilling, the local structure to which it is confined and its territorial and administrative affiliation, coordinate reference, marks of the mouth, bottom hole and depth of the well, the purpose of its drilling, stratigraphy and lithology.

It is worth noting that the database contains indicators of the depths of the horizons but at the same time the absolute marks of the top and bottom of the layers which form the basis for building maps of the deep geological structure have not been calculated. The absolute marks of the bottom of the horizons were calculated by subtracting the marks of the depth of the well from the marks of its mouth; the level of the base of the overlying layer is taken as the mark of the top of the stratigraphic units. It should also be noted that the network of wells drilled within the territory of the Brest region is uneven. In some areas, it is quite dense, in others - sparse (Fig. 1).

Considering the significant volumes of initial data as well as their belonging to the type of geospatial information their qualitative processing is possible with the use of geographic information systems which will allow: 1) to perform a preliminary analysis of data and identify errors that are present in them; 2) carry out interpolation in automatic mode with minimal time costs despite the large amount of data presented; 3) apply various construction methods to the available data and compare the resulting models; 4) carry out mathematical operations with the constructed surfaces 5) perform high-quality design and layout of the resulting maps.

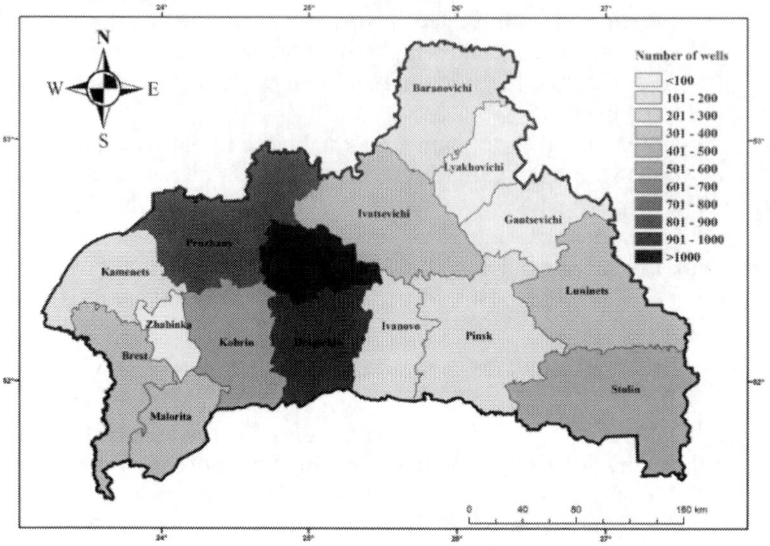

Fig. 1. Information on the number of boreholes in the context of the administrative districts of the Brest region

The methodology for constructing structural geological maps of the Cenozoic deposits of the Brest region included several successive stages (Fig. 2) they were implemented using the desktop software ArcGIS 10.5. We will dwell on them in more detail below.

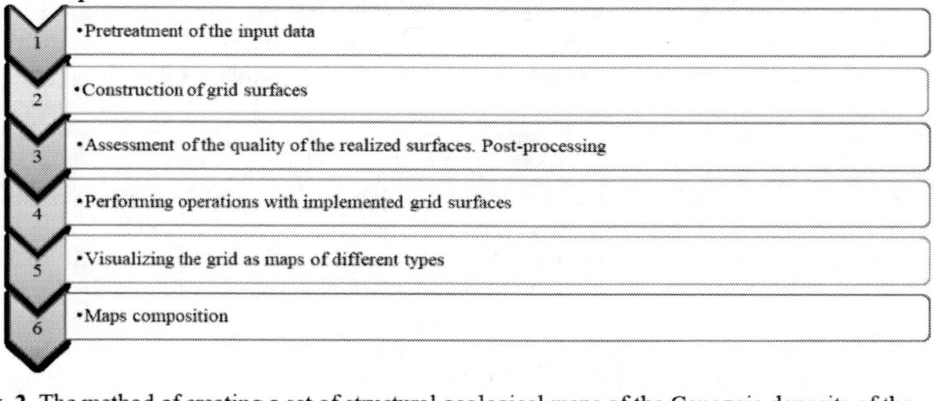

Fig. 2. The method of creating a set of structural geological maps of the Cenozoic deposits of the Brest region

Pretreatment of the input data. In order to eliminate errors that may be present in the initial data at the initial stage, preliminary processing was carried out which included the search for "outliers" in the initial data (points that are strongly out of sequence and do not fit into the model for any reason). Many methods are proposed (as a rule, these are various graphical tools, as well as statistical methods; most of the methods for finding data with outliers are based on distance measures, clustering, and spatial methods) to detect outliers in data in the scientific literature [27, 28].

Kriging method was used to detect anomalies in the existing dataset. We used values obtained as a result of interpolation errors using the ordinary kriging method to identify outliers. Subsequently, the interpolation results were added to the original attribute table using the Extract tool. After that, in the field calculator, the obtained indicators were subtracted from the original data, thus the error indicators were found between the interpolated value and the value at the starting point.

The advantages of using this method, in contrast to other common methods of finding errors in data (for example, the use of tools for geostatistical analysis of ArcGIS), allows, on the one hand, to detect data with errors (Fig. 3) and perform their quick removal and on the other hand, to preserve data with natural anomalies (for example, areas of manifestation of karst processes). Such areas are displayed as errors when using other methods.

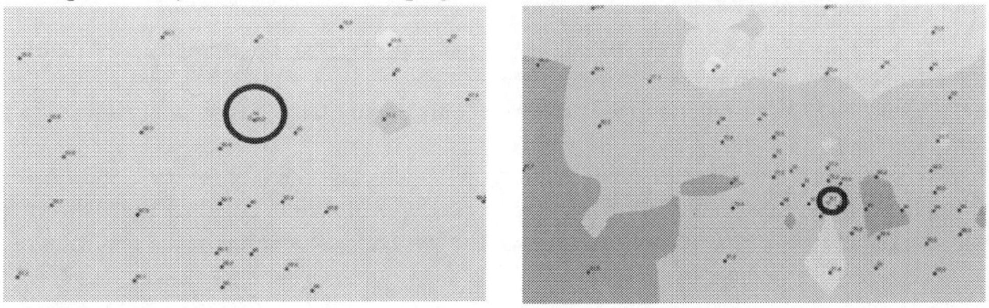

Fig. 3. Data errors identified using the kriging tool

Construction of grid surfaces. Currently, when creating interpolation models, as a rule, two main forms of surface representation are used: grid and tin [29, 30]. The grid method was chosen for the construction of structural geological maps presented in this work and it is the most common type of model of structural surfaces. In addition, grids are easy to transform and perform mathematical operations.

Creation of grid models of the top and bottom of beds of the Cenozoic sediments of the Brest region included several sub-stages [31]:

1. *The choice of the option for constructing a geological model* (the construction was carried out according to the principle "from the general to the particular", i.e. the construction from the larger stratigraphic units, in this case the "system", to the smaller ones – the "horizon".

2. *Preparation of input data.* This stage included work on the unification of stratigraphic units represented in the original database to the system level as well as sampling of layers from the unified database where stratigraphic units were used as a criterion. Three layers were formed according to the results of the sampling and they correspond to three periods of the Cenozoic era.

3. *Choice of interpolation method*. The application of several interpolation methods proposed in ArcMap was considered: IDW, Natural Neighbor, Kriging, Topo to Raster (Figure 4) to construct grid models of the top and bottom of beds presented in this paper. In general, all types of interpolation presented in this set of tools show correct results and it is due to a rather dense network of wells. But given the fact that in this case we are talking about creating terrain models, the Topo to Raster tool was chosen which allows to create hydrologically correct digital elevation models (DEM).

4. *Creation of interpolation grid-models of the top and bottom of beds*. Based on the layers obtained during the sampling (Paleogene system, Neogene system, Quaternary system) as well as the selected interpolation method (Topo to Raster), the surfaces of the top and bottom of beds were constructed.

Assessment of the quality of the realized surfaces. Post-processing. The correctness of the constructed grid models was checked within this stage. At the first stage, for the subsequent identification of modeling errors, the created grid models (having the same raster resolution) were reclassified according to the roof / bottom values (for example, the roof of the Neogene system / the bottom of the Quaternary system). The reclassified surfaces were then subtracted using the Map Algebra tool (Spatial Analyst toolbox). The "processing extent" was set (as for a larger raster layer) when subtracting raster surfaces. This is due to the fact that the deposits of some stratigraphic units do not cover the entire territory of the Brest region. After subtraction of rasters, cells in places where deposits of one of the stratigraphic subdivisions are absent automatically are assigned the value "no data", instead of which, after processing by the isNull function, the necessary indicators from another raster can be substituted.

Two variants of surfaces were formed as a result of raster processing:

1. Surfaces with no construction errors, i.e. the model is correct and no further processing is required.

2. There are build errors. The surface is not modeled correctly. An adjustment operation is required.

The adjustment (averaging) of surfaces with errors can be implemented using two options that allow to get equally correct results.

1. Using the Cell Statistics tool which calculates statistics based on cell values from multiple rasters.

2. Using the Con (Conditional) tool with specifying the necessary conditions. For example, if the values of the roof raster cells differ from the values of the bottom raster cells, the mean of the cells of both rasters will be written as the value in the resulting raster.

a) an example of using the inverse distance weighted method

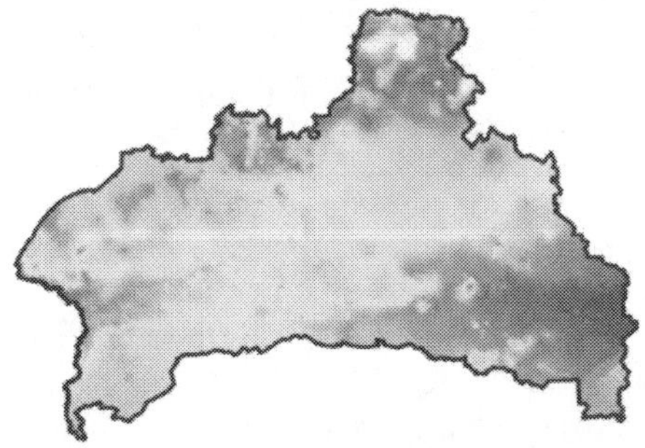

b) an example of using the kriging method

c) an example of using the natural neighborhood method

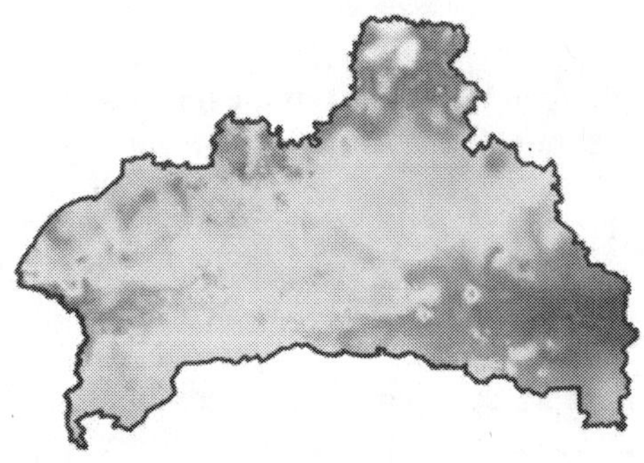

d) an example of using the topo to raster method

Fig. 4. An example of using different interpolation methods when building a map of the top of Quaternary deposits

Performing operations with implemented grid surfaces. At this stage, based on the obtained grid surfaces using the Map Algebra tool (Spatial Analyst module), the vertical and true thickness for each Cenozoic horizon was calculated. The vertical thickness represents the vertical distance between the top and the bottom of the bed and was calculated as the difference between the structural maps of the horizons, the thickness between which is calculated. True thickness is the shortest distance between the bottom and the top of the bed and was calculated by multiplying the vertical power by sine of the angle of incidence of solids.

Visualization of the grid as maps of various types. At this stage, based on the obtained grid surfaces, they were visualized in the form of maps of various types. In this study, 3 types of maps were implemented:

1. Contour maps. These maps are maps in contours with a certain step. Isolines can be smoothed and the space between the lines can be filled with a color palette.
2. Raster maps. Map data is a raster with smooth transitions from minimum to maximum values.
3. Shaded relief. They are a surface with a light source fixed at some point.

Maps composition. The final stage in compiling a set of maps was the development of a layout of composition. For its compilation, taking into account the type of maps being implemented, all the elements were placed, a general view was formed, taking into account the main recommendations for map compilation. The development of design techniques for the design and symbolization of maps was carried out using a method for displaying a cartographic image. Class intervals, colors, line types and other graphic elements were selected.

Results and Discussion

Thus, using the above technique, a set of structural-geological maps of the horizons of the Cenozoic deposits of the Brest region was compiled based on the interpretation of data on the geological structure of the Cenozoic deposits. In general, the following map compositions were formed for each of the Cenozoic units (table).

Table. Sets of cards made for Cenozoic sediments of the Brest region

Paleogene system	- hypsometric maps of the roof and bottom of the Paleogene system - hypsometry maps of the roof and bottom of the Kiev suite - hypsometry maps of the roof and bottom of the Kharkiv suite - map of sediment thickness of the Paleogene system - map of the thickness of the Kiev formation - map of the thickness of the Kharkiv Formation deposits - maps of overburden thickness - maps of sediment thickness combined with maps of overburden thickness - maps of the thickness of sediments combined with maps of the roof of the horizons
Neogene system	- hypsometric maps of the roof and bottom of the Neogene system - hypsometric maps of the roof and bottom of the Brinevsky horizon - hypsometry maps of the roof and bottom of the Antopol horizon - map of sediment thickness of the Neogene system - map of the thickness of sediments of the Brinevsky horizon - map of the thickness of sediments of the Antopol horizon - maps of overburden thickness - maps of sediment thickness combined with maps of overburden thickness - maps of the thickness of sediments combined with maps of the roof of the horizons
Quaternary system	- hypsometry maps of the roof and bottom of the Quaternary system - hypsometric maps of the top and bottom of the Pleistocene - maps of hypsometry of the roof and bottom of the Holocene - map of the thickness of sediments of the Quaternary system - map of thickness of Pleistocene sediments - map of Holocene sediment thickness - maps of overburden thickness - maps of sediment thickness combined with maps of overburden thickness - maps of the thickness of sediments combined with maps of the roof of the horizons

The analysis of the implemented cartographic materials made it possible to detail the features of the geological structure of the Cenozoic sediments in the territory of the Brest region.

Paleogene sediments occupy significant areas within the study area and lie on Cretaceous formations underlying the Neogene and in some cases directly Quaternary. In lithological terms, sediments of the upper part of the Middle Eocene, Upper Eocene and Oligocene are distinguished here.

Analysis of the hypsometry of the bottom of the described formations shows that they occupy the highest position in the southwestern parts of the region (about 140 m) and the lowest in the northern (about 50 m) (Fig. 5). Their depth on average is 60–70 m, in the south of the region – 25–30 m.

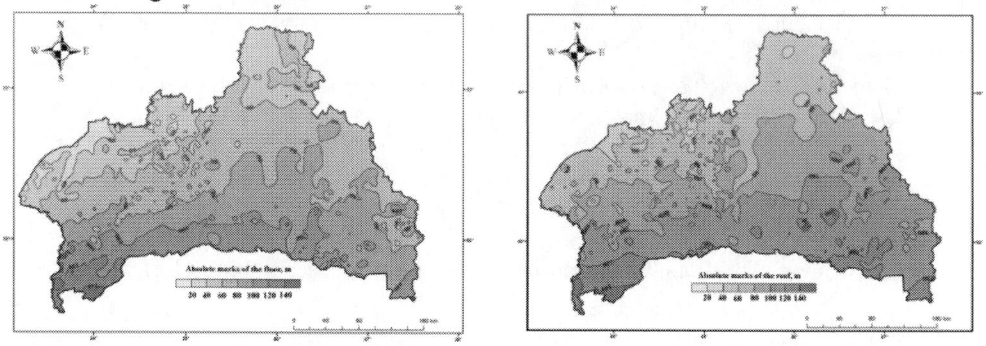

a) structural map of the bottom of the Paleogene sediments

b) structural map of the roof of the Paleogene sediments

Fig. 5. Structural maps made for the Paleogene sediments of the Brest region

Fig. 6 gives an idea of the thickness of the Paleogene sediments. The average thickness of the Paleogene sediments is 20–30 m. The absolute marks of the roof vary within 20–150 m. The overburden thickness varies from 20–30 m in the south of the region to 120–130 m in the north-west.

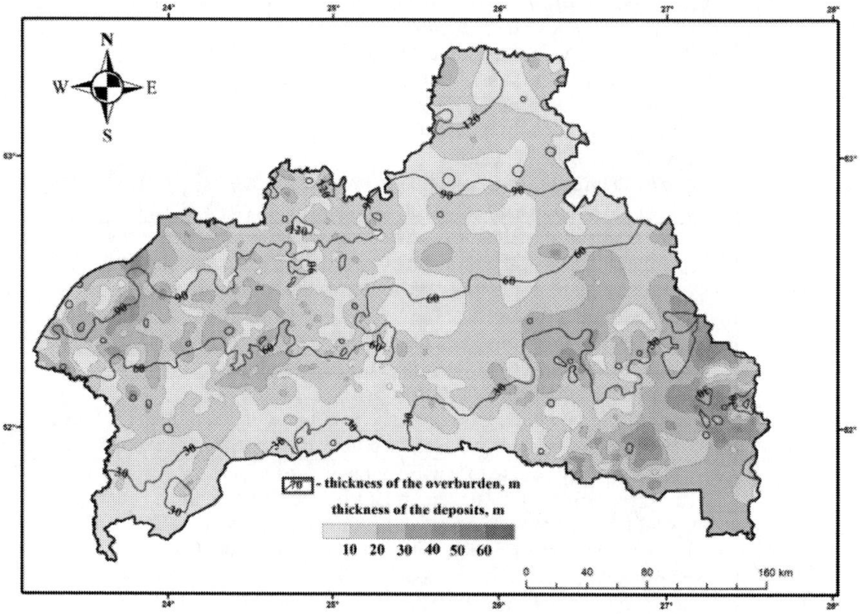

Fig. 6. Thickness of Paleogene sediments of the Brest region

Neogene sediments within the Brest region also occupy significant areas, directly underlying the Quaternary. According to the data of spore-pollen analysis, sediments of the Lower, Middle, and Upper Miocene and deposits of the Lower and Upper Pliocene were distinguished in the Neogene strata. The absolute marks of the bottom of the studied sediments vary in the interval of 10–150 m (Fig. 7). Average depth of occurrence is 90 m. The average thickness of the Neogene sediments is 15–20 m. The maximum values (60 m and more) are recorded in the north-western parts of the region (Fig. 8).

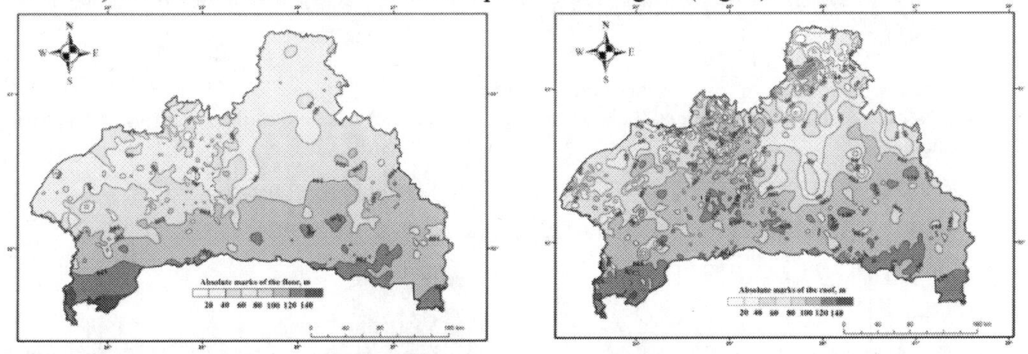

a) structural map of the bottom of the Neogene sediments

b) structural map of the roof of the Neogene sediments

Fig. 7. Structural maps made for the Neogene sediments of the Brest region

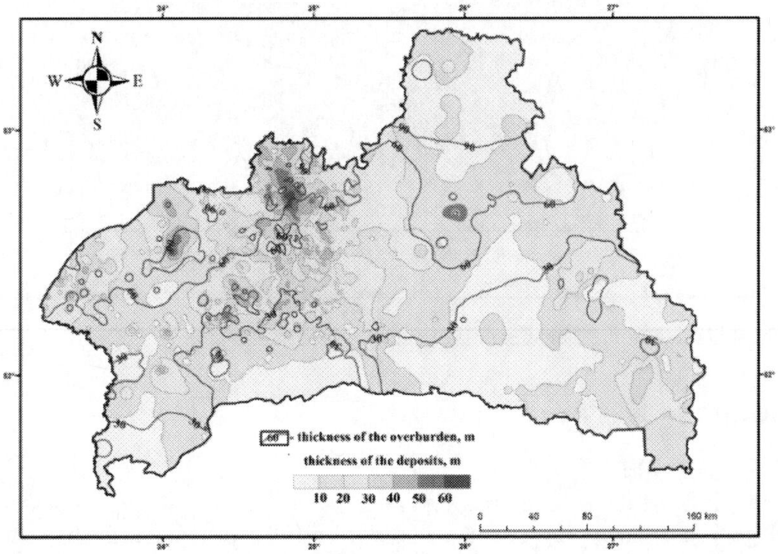

Fig. 8. Thickness of the Neogene sediments of the Brest region

Analysis of the hypsometry of the bottom of the described formations shows that the highest elevations are characteristic of the southern districts of the region (120 m and more), the lowest are in the northern parts, where they are about 40 m (Fig. 9). The average depth of their occurrence ranges from 80 to 110 m.

Quaternary sediments are characterized by uneven power distribution. The average thickness of the Quaternary strata is 80 m. The maximum values are recorded in the northern parts of the region, the minimum - in the southwest and southeast (Fig. 10).

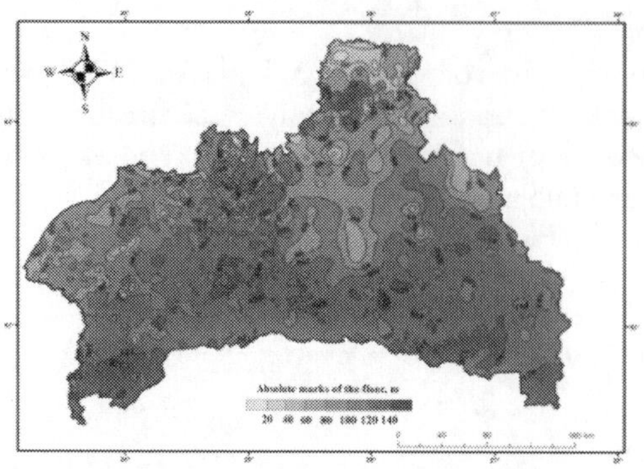

a) structural map of the bottom of the Quaternary sediments

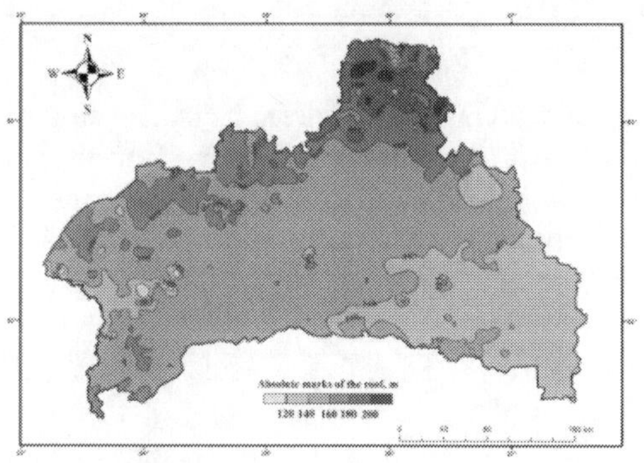

b) structural map of the roof of the Quaternary sediments

Fig. 9. Structural maps made for the Quaternary sediments of the Brest region

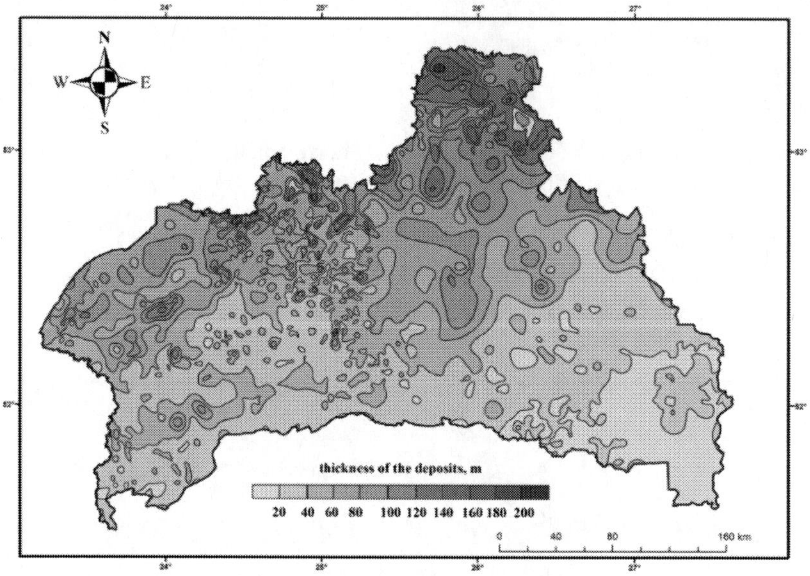

Fig. 10. The thickness of the Quaternary sediments of the Brest region

Conclusions

Thus, in the course of the study:

1. The analysis of the existing theoretical and methodological approaches in the field of structural geological mapping showed that currently in the scientific literature there are many approaches to the creation of structural-geological maps (both traditional and based on the use of GIS technologies). However, in most of the existing methods, insufficient attention is paid to the issue of preliminary processing of spatial data.
2. The selection of the information necessary for the construction of horizontal structural geological maps of the Brest region presented in the original Microsoft Access database (containing data obtained from the results of geological drilling of the territory) has been performed. The selected information based on the existing gridding is introduced into the GIS.
3. Taking into account the peculiarities of the available data set as well as the methodological approaches presented in the scientific literature, a method is proposed for

creating a set of structural geological maps based on drilling knowledge using GIS technologies (for example, mapping the horizons of Cenozoic deposits in the Brest region).

4. The approaches presented in the scientific literature in the field of preliminary processing of large geological data are considered. For the existing dataset, the best error search results were achieved using kriging which preserves areas of natural anomalies.

5. The construction of grid models of the top and bottom of the layers was implemented using various interpolation methods presented in the Spatial Analyst toolbox. In general, all the presented methods allow creating sufficiently high-quality models. However, given the fact that in this study interpolation is used to create terrain models, the Topo to Raster method was chosen.

6. To improve the quality of the created cartographic materials, the analysis of the constructed grid surfaces was carried out which makes it possible to identify errors that arise at the modeling stage.

7. The calculation of the thickness of the sediments for each stratigraphic horizon in the course of performing mathematical operations with the constructed surfaces.

8. Based on the realized grid surfaces, a set of maps was made for each period of the Cenozoic era, including the following compositions: hypsometry of the top and bottom of the horizons, the thickness of sediments, the thickness of overburden, combined maps of the thickness of sediments and overburden as well as the thickness of sediments and the top of the horizons. The completed cartographic materials made it possible to detail the features of the geological structure of the Cenozoic stratigraphic units in the Brest region.

In general, in the course of the implemented study, a number of new cartographic materials were obtained on the structural features of the stratigraphic horizons of the Cenozoic in the Brest region as well as information was updated and more accurate results were obtained on the structure of Cenozoic deposits within the territory of the Podlyassko-Brest depression. The use of geoinformation technologies in the course of creating makes it possible to quickly and timely update them, in case of receiving new information about the structure of this territory.

Bibliography

1. Bogdasarov MA [Geology and minerageny of Quaternary deposits in the territory of the Podlasie-Brest depression]. Brest, BrGU Publ., 2011, 166 p. (In Russian).

2. Makhnach A. S., Garetskii R. G., Matveev A. V. [Geology of Belarus]. Minsk, Institut of NAN Belarusi Publ., 2001, 716 p. (In Russian).

3. Grechanik N. F., Matveev A. V., Bogdasarov M. A. [Relief of the territory of the Podlasie-Brest Depression]. Brest, BrGU Publ., 2013, 154 p. (In Russian).

4. Miasnikovich M. U. et al. [National Atlas of Belarus]. Minsk, Belkartografiia Publ., 2002, 292 p. (In Russian).

5. Kurlovich D. M. [Spatial differentiation and dynamics of morphostructures of the Belarusian Poozerye]. Minsk, BGU Publ., 158 p. (In Russian).

6. Kroshinskii V. A. [Geological mapping of the northern section of the Minsk Upland based on GIS technologies]. Materials [Proccedings of the International Scientific Conference dedicated to the 100th anniversary of the birth of academician K. I. Lukashev]. Minsk, 2017, pp. 36-38. (In Russian).

7. Onoshko M. P., Kroshinskii V. A., Podruzhaia M A, Shidlovskaia A.V. [GIS-technologies in the construction of horizon-level structural-facies maps of Quaternary deposits of Belarus]. Materials [Proccedings of the International Scientific Conference Problems of regional Geology of the West of the Eastern European platform and adjacent territories]. Minsk, 2019. (In Russian).

8. Cushing E. M. Building a three dimensional model of the active Plio-Quaternary basin of Argostoli (Cephalonia Island, Greece). Engineering Geology, 2020, vol. 265, pp. 1-22. DOI: 10.1016 / j.enggeo.2019.105441.

9. Borisov D V [Application of GIS in the tasks of imaging overliying seams to determine promising petroleum territories]. INTEREKSPO GEO-SIBIR ', 2016, vol. 2, no.2, pp. 123-126. (In Russian).

10. Kazanskaia D. A., Aleksandrov V. M., Belkina V. A. [Geological modeling of Vikulovskaya suite production sediments]. , 2019, vol. 330, no 7, pp. 195-207. (In Russian).

11. Nugmanov B. H. [3D Structural-Tectonic Modeling of Geological Structure of the Deposit of "Kalamkas" Field]. SOCAR Proceedings, 2017, no 1, pp. 17-23. (In Russian).

12. Prokudin V. G., Valitov M. G., Kononets S. N. [The Cenozoic sediments structure of the Amur Bay depression]. Vestnik DVO RAN, 2018, no 1, pp. 121-127. (In Russian).

13. Sivtsev A. I., Zueva I. N., Chalaia O. N. [Structural map by thickness and depth of the top of the Kuonamskaya combustible shale formation]. Materials [Proccedings of the International Scientific and Practical Conference]. Iakutsk, 2017, pp. 233-237. (In Russian).

14. Ryzhov A. E. [Allocation of new promising objects in the saline sediments within Kovykta zone of gas accumulation and its contiguous territories]. , 2017, no 3, pp. 100-111. (In Russian).

15. Vasilenok E. A. [Use of GIS-technologies to identify oil-bearing structures within the Girovskaya area of the Pripyat trough]. Materials [Proccedings of the GIS projects of students and graduate students of higher education institution of the Republic of Belarus, held in celebration of International GIS Day]. Minsk, 2015, pp. 18-22. (In Russian).

16. Vouillamoz N. et al 3D cartographic modeling of the Alpine arc. Tectonophysics, 2012, vol. 579, pp. 131-143. DOI: 10.1016 / j.tecto.2012.06.012.

17. Gasselt S., Nass A. Planetary mapping - The datamodel's perspective and GIS framework. Planetary and Space Science, 2011, vol. 59, pp. 1231-1242. DOI: 10.1016 / j.pss.2010.09.012.

18. Mendonca M. Three-dimensional GIS cartography applied to the study of the spatial variation of soil horizons in Swiss floodplain. Geoderma, 2000, vol. 97, pp. 351-366. DOI: 10.1016 / S0016-7061 (00) 00045-8.

19. Mironov O. K. [Geoinformation technologies for drawing up large-scale geological maps of Moscow]. , 2011, no 3, pp. 198-214. (In Russian).

20. Kurlovich D. M. [The use of GIS technologies for the development of geodatabase and information projects of brown coal and oil shale deposits in the Republic of Belarus]. Materials from Proxings of the International scientific Congress on Informatics information systems and technologies]. Minsk, 2011), pp. 188-193. (In Russian).

21. Khrushchov D. P. Regional structural-lithological modeling of sedimentary cover. Geologic journal, 2015, no 2, pp. 27–38.

22. Mangkhemthong N. et al. Geological model and development of the Cenozoic Wiang Pa Pao Basin, Chiang Rai Province, Northern Thailand, based on gravity data modelling and surface structural interpretation. Tectonophysics, 2020, vol. 786. DOI: 10.1016/j.tecto.2020.228454.

23. Attwa M., Henaish A. Regional structural mapping using a combined geological and geophysical approach – A preliminary study at Cairo-Suez district, Egypt. Journal of African Earth Sciences, 2018, vol. 144, pp. 104–121. DOI: 10.1016/j.jafrearsci.2018.04.010.

24. Przulj N., Malod-Dognin N. Network analytics in the age of big data. Science, 2016, vol. 353, pp. 123–124. DOI: 10.1126/science.aah3449.

25. Jin H., Wah B., Cheng X., Wang Y. Significance and Challenges of Big Data Research. Big Data Research, 2015, vol. 2, pp. 59–64. DOI: 10.1016/j.bdr.2015.01.006.

26. Lee J., Kang. M. Geospatial Big Data: Challenges and Opportunities. Big Data Research, 2015, vol. 2, pp. 74–81. DOI: 10.1016/j.bdr.2015.01.003.

27. Lyutikova L. Logical Analysis of Data for outliers detection. Procedia Computer Science, 2020, vol. 169, pp. 330–336. DOI: 10.1016/j.procs.2020.02.192.

28. Benjelloun F. Improving outliers detection in data streams using LiCS and voting. Journal of King Saud University, 2019, vol. 2, pp. 1–9. DOI: 10.1016/j.jksuci.2019.08.003.

29. Muzik J. et al. Creation of 3D Geological Models Using Interpolation Methods for Numerical Modelling. Procedia Earth and Planetary Science, 2015, vol. 15, pp. 25–30. DOI: 10.1016/j.proeps.2015.08.007.

30. Kebloutia A., Ouerdachia L., Boutaghanea H. Spatial Interpolation of Annual Precipitation in AnnabaAlgeria – Comparison and Evaluation of Methods. Energy Procedia, 2012, vol. 18, pp. 468–475. DOI: 10.1016/j.egypro.2012.05.058.

31. Maevskaya A. N. [Algorithm for constructing models of structural geological surfaces using geoinformation technologies]. [Actual problems of Earth Sciences: proccedings of IV international. scientific and practical Conference, dedicated to the 1000th anniversary of Brest]. Brest, 2019, pp. 92-96. (In Russian).

Energy efficiency of multi-apartment residential houses with individual heat supply

Vladimir Melnikov[1] , *Uladzimir* Navaseltsau[2*] and *Dzina* Navaseltsava[2]

[1]Stoletov State University of Vladimir, Institute of Architecture, Construction and Energy, Gorky str., 87, 600000, Vladimir, Russian Federation
[2]Brest State Technical University, Faculty of Engineering systems and Ecology, Maskouskaya str. 267, 224017, Brest, Republic of Belarus

Abstract. Centralized hot water systems widely used in Russia and Belarus are characterized by a considerable length and branching which inevitably leads to increased heat losses and to an unstable hydraulic system. The operation of the domestic hot water system in the circulation mode can be characterized by several parameters; one of which is the specific ratio of the cost of thermal energy for heating a cubic meter of hot water. The parameter is often regulated by law in Russia; exceeding this parameter is considered as administrative violation. The aim of the research is to determine the design and actual costs of thermal energy for hot water supply (hot water heating) and their comparison, analysis of the data obtained. The methodology for determining the design and real costs of thermal energy for hot water supply was to study the operation of the hot water supply system of a residential 144-apartment 9-storey building. The research showed that the actual circulating flow rate is much less than the calculated circulating flow rate. The authors note that in order to optimize the standard for heating a cubic meter of water it is necessary to observe the calculated circulation modes. This will require stabilization of the hydraulic systems of both the external and internal networks which is a difficult but feasible task. The research results are supposed to be taken into account when setting up existing hot water supply systems.
Keywords: centralized heat supply, hot water supply, circulation rate, heat calculator, heat energy consumption, hot water consumption

Introduction

There is an expression that the greater part of hot water consumption in the total volume of water consumption is observed, the higher the civilization of society is. Indeed, a civilized person during the day repeatedly uses hot water for various purposes and the costs of this invention are becoming more tangible.

Different countries carry out the solution of technical and economic issues of hot water supply (hereinafter DHW) by their own methods, which is quite obvious, given the

* Corresponding author: vgnovoseltsev@yandex.ru

features, history, mentality. European countries serve as a definite model and example for Russia, taking into account the specifics of the country.

In literature you can find a lot of data on the successful implementation of energy-efficient buildings and monitoring the operation of their engineering systems [1-4]. Experience has been gained both in the design and operation of the hot water supply systems of residential buildings, but, nevertheless, many problems have not yet been resolved, one of which is considered in this article.

The modern district heating system of a residential building is based on an energy-efficient and resource-saving individual heating station, which is designed and manufactured by well-known manufacturers of heat energy equipment [5, 6]. Hot water boils at such a point and supplied to the consumer. Regulatory documents and practical wisdom provide for a domestic hot water system with circulation in modern buildings, i.e. when the consumer opens the tap, the water is almost hot. This big advantage has a significant side effect - necessity to organize the circulation process, which requires additional costs, both labor and material [7].

Dead-end DHW systems, which are typical for old buildings, however, exist at present. The advantage is simplicity and relative cheapness, the disadvantage is the useless discharge of chilled water into the sewer while waiting for warm. The transformation of the deadlock system into a circulation (reconstruction) is unlikely, because the layout of old buildings does not provide for the installation of additional pipelines and the placement of appropriate equipment. Plus, the issue of financing reconstruction, search and determination of investment conditions.

A separate niche in Russia is occupied by centralized hot water supply systems as part of the four-pipe design of a heating system for residential buildings with a heat source in the form of a boiler room or central heating station - a central heating center. The presence of the supply and return pipes in the DHW system automatically creates the basis for circulation, but the different distance of the objects from the heat source and their heat-hydraulic heterogeneity creates difficulties in balancing. Such a branched system combines the features of both a dead end and a ring network. The dead end is characterized by the branching and unambiguous flow direction, the ring one includes the interaction of the supply and return branches with an unstable flow of hot water.

Despite the complexity of such networks, they are very successfully operated in single-industry towns, small settlements, often with intuitively found system parameters and an optimal radius of heat supply, both for heating and hot water. At the same time, overflows do not occur precisely in the period of transitional temperature, approximately +8 ^{0}C – -1 ... -5 ^{0}C, when the temperature in the return line is less than 70 ^{0}C, and the temperature of hot water must be provided at least 60 ^{0}C according to the norms.

But in most situations when four-pipe heat supply systems operate from a single heat source, problems arise and can be solved both by general recommendations and by applying an individual approach. One of the problems is taking into account the amount of thermal energy for heating hot water, assessing the accuracy of such measurements, comparing the results with standard specific indicators [8].

Many residential buildings are equipped with multi-channel heat meters SPT 961 [9], which measure electrical signals corresponding to the parameters of the heat carrier transported through pipelines and calculate the thermal characteristics and amount of heat carrier. It is also feasible to process data, archive and process it using standard interface tools for the purpose of system analysis.

From the variety of technical and operational parameters that fully comply with operating conditions, we can distinguish metrological indicators characterizing the accuracy of measuring time, pressure, flow and thermal energy, shown in table 1.

Table 1. Metrological parameters of the heat calculator SPT961

Name of the parameter of the heat calculator SPT961	Parameter value
watch error (fractional)	±0,01 %
parameters' processing (fractional error)	±0,02 %
thermal energy calculation (fractional error)* *according to the results of input signals measurements	±(0,5+3/ΔT) %

Attention should be paid to the calculation of the relative error of thermal energy, which is an integrated indicator of input signals with different errors, both relative and absolute. The parameter ΔT must be understood as the temperature difference in the mains of the supply and return heat network (or hot water supply network).

Consumers of hot water in apartments install conventional meters such as SV-15G, which are universal and widespread, [10]. According to the parameters of maximum flow rates, pressures and temperatures, they correspond to the hydraulic mode of the DHW network. For the research, the minimum overhead costs, depending on the installation of the meter, are interesting, table 2. Horizontal installation halves the minimum consumption.

Table 2. Metrological parameters of the individual residential meter SV-15G

Name of the meter's parameter SV-15G	Parameter value
Water consumption Q, m^3/h: Minimal Q_{min} Class B (horizontal set) Class A (vertical set)	 0,03 0,06
Transition Q_t Class B Class A	 0,12 0,15
Nominal nomenclature Q_n Maximum Q_{max}	1,5 3,0
The limits of permissible relative error, % in the range of consumption from Q_{min} to Q_t in the range of consumption from Q_t to Q_{max} included	 ±5 ±2
Threshold of sensitivity	not more than 0,5 Q_{min}

As an object of research, we consider a residential 144-apartment 9-floor building with a hot water circulation system, equipped with a common heat meter SPT 961, manufactured by AOZT NPF Logika, St. Petersburg. This device is included in the Federal Information Fund for Ensuring the Uniformity of Measurements, verified and equipped with all the necessary interface. The aim of the research is to determine the design and real costs of thermal energy for hot water supply (hot water heating) and their comparison, analysis of the data obtained.

1 Methods

1.1 Methods of research

The methodology for determining the design and real costs of thermal energy for hot water supply was to study the operation of the facility's hot water supply system. Stages of implementation: project documentation study; inspection of the system in order to identify their technical condition, compliance with design decisions and proper functioning; processing the received data.

1.2 Description of design choice for the investigated object.

The DHW system of the facility provides the following solutions. The hot water supply system is made with individual circulating risers at each water riser, with sequential connection of heated towel rails installed on the circulation riser. The connection of the domestic hot water system of a residential building is carried out directly to the external DHW networks in a two-pipe version. External DHW networks are connected to the central heating center (DHC), which provides direct supply of hot water of the required quality to consumers.

1.3 Determination of heat energy consumption.

The methodology for determining the thermal energy consumption at the initial stage was based on the well-known approach combining a total sum of hydraulic calculation, empirical data, and elements of probability calculus. This methodology was given in SNiP 2.04.01-85 * "Internal water supply and sewerage of buildings", which is not currently used, but designers and testers still prefer this technique, because it gives good convergence and is tested by time and practice.

This calculation is necessary at the preliminary stage of determining the diameters of the DHW supply pipelines under the assumption that there is no circulation. For dead-end systems, such a calculation is sufficient; for systems with circulation, further adjustment is necessary taking into account the circulation method. The basic input data and calculation results are shown in table 3.

Table 3. Consumption of hot water and heat for its heating

Total number of hot water consumers in the building, person	U	387
The total number of water intake devices in the building, inits	N	432
The likelihood of sanitary appliances in the building	P	0,0124
NP (N times P)	NP	5,38
Coefficient α (selected from empirical tables or approximated)	α	2,693
The maximum second water flow rate at the DHW inlet to the building, l/s.	q^h	2,524
The probability of using the total number of devices served by the designed system	P_{hr}	0,045
NP (N times P)	NP_{hr}	19,35
Coefficient α (selected from empirical tables or approximated)	α_{hr}	6,374
Maximum hourly consumption of hot water, m³/h.	q_{hr}^h	6,374
Average hourly consumption for the period (day) of maximum water consumption, m³/h.	q_T^h	1,935
Average hourly water consumption for an average day, m³/h.	$q_{T,m}^h$	1,693
The average hourly heat flow for the period (day) of maximum water consumption (taking into account heat losses),kW.	Q_T^h	148,1
Maximum heat flux during an hour of maximum heat consumption, kW.	Q_{hr}^h	488,0
Average hourly heat flux for an average day, kW.	$Q_{T,m}^h$	129,6

The design scheme for one entranceway of 36 apartments with 9 floors is shown in Figure 1. Each supply riser is looped with a circulation riser, heated towel rails are located on the circulation branches.

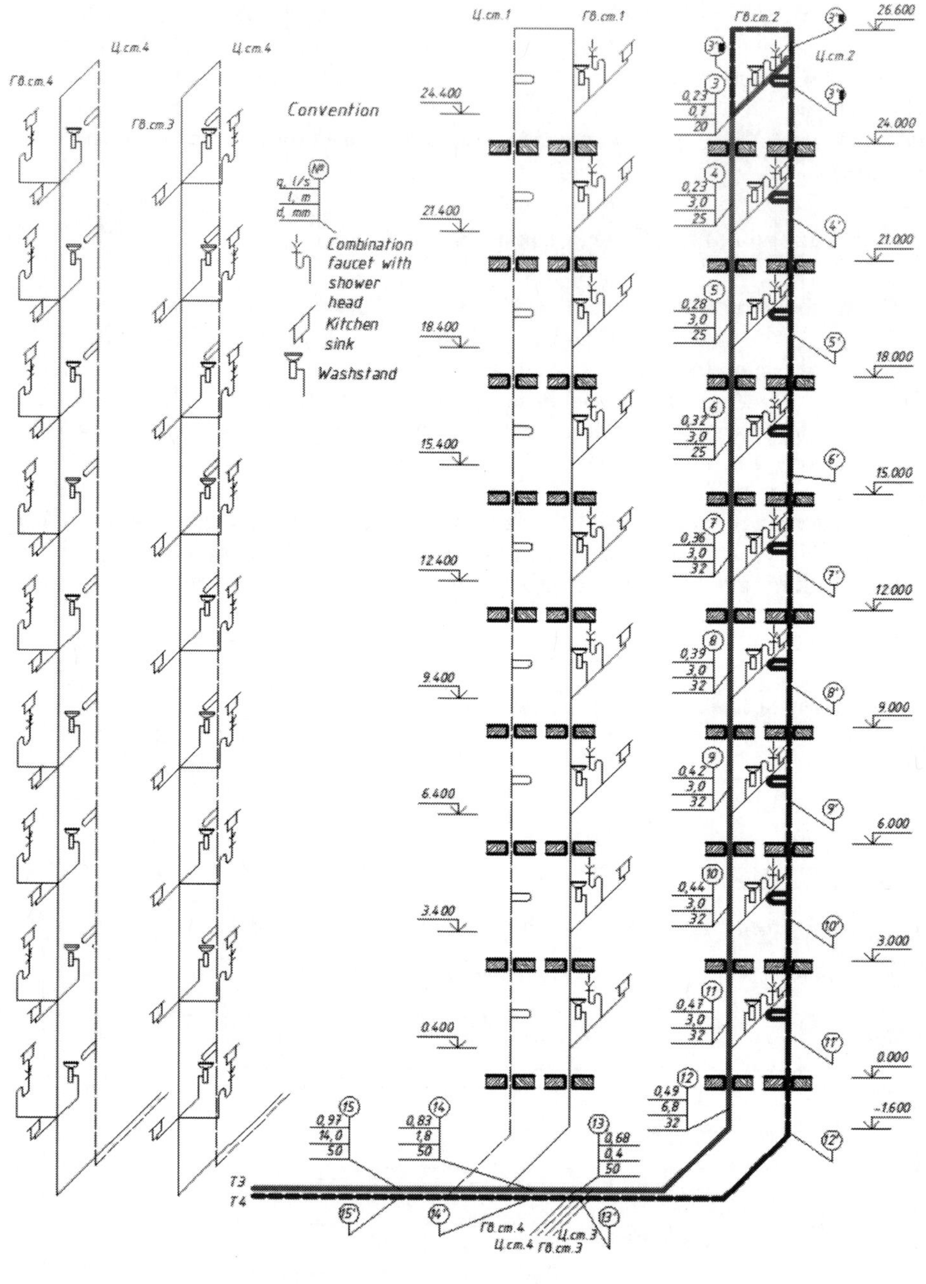

Fig. 1 Design scheme of the hot water supply system of one entranceway of a 9-floor building

After calculating the main parameters, the result of table 3, the calculation of the circulation costs of the DHW network. The circulation flow rate is calculated according to standard formulas and depends on the length, diameters of the pipelines, the circulation pattern, the type and state of thermal insulation, the temperature difference in the supply and return (circulation) lines, the results are shown in table 4.

Table 4. Calculation of the circulation costs of the 36 apartments' entranceway of a 9- floor building

№ land	L, m	d_y, mm	Loss of heat			Circulation rate q^{cir}, l/s	Circulation rate q^{cir}, l/s
			Specifically q, W/m	On a land Q_{ht}, W	Sum ΣQ^{ht}, W		
In-house wiring							
1-2-3	2,3	20	24	1405,20	1405,20	0,034	0,05
4	3	25	12,6	37,80	1443,00	0,035	0,05
5	3	25	12,6	37,80	1480,80	0,036	0,05
6	3	25	12,6	37,80	1518,60	0,037	0,05
7	3	32	13,9	41,70	1560,30	0,038	0,05
8	3	32	13,9	41,70	1602,00	0,039	0,05
9	3	32	13,9	41,70	1643,70	0,040	0,05
10	3	32	13,9	41,70	1685,40	0,041	0,05
11	3	32	13,9	41,70	1727,10	0,042	0,05
12	6,8	32	19,1	129,88	1856,98	0,045	0,05
Ст,3					1822,60	0,044	
13	0,387	50	24,1	9,33	3688,91	0,089	0,10
Ст,4					1822,60	0,044	
14	1,8	50	24,1	43,38	5554,89	0,135	0,15
Ст,1					1822,60	0,044	
15	14	50	24,1	337,40	7714,89	0,187	0,20

Consequently, the circulating flow rate for the house under study (4 entranceway) will amount to a small fraction $G_{ц} = 4 \times 0{,}2 = 0{,}8$ l/s $= 2{,}88$ m³/h $= 2{,}88 \times 0{,}9835 = 2{,}83$ tph.

2 Results and discussion

The above SPT 961 heat meter installed in the building measures the following parameters of the DHW system:

Gп, Go – mass flow rate for supply and return, tph;
Vп, Vo – volumetric flow rate for supply and return, m^3/h;
Tп, To – flow and circulation temperatures respectively, °C;
Pп, Po – supply and return pressure, kg/cm^2;
dM – network mass flow, tph;
dW – thermal power, Gcal/h;
dV – volume difference, m^3/h;
Δ – quotient from dW divided dV, $Gcal/m^3$

Below there are the results of measuring the main parameters of the DHW system during the day. The time interval is selected for the period of the normal rhythm of life, a working day in the middle of the week in order to exclude non-standard mode of water consumption. The measurement results are shown in table 5.

Table 5. The results of the main parameters of the DHW system measurements during the day

Measure-ment time	Gп	Go	Vп	Vo	Tп	To	Pп	Po	dM	dW	dV	Δ
h	tph	т/h	m^3/h	m^3/h	^0C	^0C	kg/cm^2	kg/cm^2	tph	Gcal /h	m^3/h	Gcal /m^3
00:00:00	3,57	3,16	3,63	3,20	59,31	49,44	6,28	6,18	0,41	0,05	0,43	0,1163
01:00:00	3,38	3,25	3,43	3,28	59,10	49,30	6,31	6,23	0,13	0,04	0,15	0,2667
02:00:00	3,19	3,19	3,24	3,22	58,88	49,13	6,32	6,24	0,00	0,03	0,02	1,5000
03:00:00	3,19	3,19	3,24	3,23	58,72	49,00	6,33	6,25	0,00	0,03	0,01	3,0000
04:00:00	3,21	3,19	3,26	3,22	58,63	48,88	6,33	6,25	0,02	0,03	0,04	0,7500
05:00:00	3,20	3,19	3,25	3,22	58,62	48,85	6,33	6,24	0,02	0,03	0,03	1,0000
06:00:00	3,44	3,12	3,49	3,16	58,68	48,88	6,31	6,22	0,31	0,05	0,33	0,1515
07:00:00	4,11	3,05	4,17	3,08	58,93	48,99	6,23	6,13	1,06	0,09	1,09	0,0826
08:00:00	4,07	3,08	4,13	3,11	59,29	49,32	6,23	6,13	0,99	0,08	1,02	0,0784
09:00:00	3,94	3,11	4,01	3,15	59,42	49,47	6,25	6,15	0,83	0,07	0,86	0,0814
10:00:00	3,88	3,13	3,94	3,16	59,47	49,59	6,25	6,15	0,75	0,07	0,78	0,0897
11:00:00	3,76	3,17	3,81	3,20	59,34	49,48	6,26	6,16	0,59	0,06	0,61	0,0984
12:00:00	3,62	3,23	3,68	3,27	59,22	49,48	6,27	6,17	0,39	0,05	0,41	0,1220
13:00:00	3,77	3,24	3,83	3,27	59,21	49,44	6,27	6,16	0,54	0,06	0,56	0,1071
14:00:00	3,68	3,20	3,74	3,24	59,18	49,41	6,28	6,18	0,47	0,06	0,50	0,1200
15:00:00	3,49	3,20	3,54	3,23	59,10	49,34	6,28	6,19	0,29	0,04	0,31	0,1290
16:00:00	3,53	3,17	3,58	3,20	59,02	49,18	6,28	6,18	0,36	0,05	0,38	0,1316
17:00:00	3,61	3,16	3,67	3,20	59,04	49,14	6,27	6,18	0,45	0,06	0,47	0,1277
18:00:00	3,70	3,17	3,76	3,20	59,09	49,17	6,25	6,15	0,54	0,06	0,56	0,1071
19:00:00	4,06	3,06	4,12	3,09	59,23	49,08	6,24	6,13	1,00	0,08	1,03	0,0777
20:00:00	4,10	3,10	4,16	3,14	59,36	49,19	6,22	6,11	0,98	0,08	1,02	0,0784
21:00:00	4,31	3,03	4,38	3,06	59,49	49,30	6,20	6,09	1,29	0,10	1,32	0,0758
22:00:00	4,45	2,96	4,52	2,99	59,63	49,42	6,21	6,10	1,49	0,10	1,53	0,0654
23:00:00	3,84	3,14	3,90	3,17	59,53	49,51	6,25	6,15	0,71	0,07	0,73	0,0959

Based on the results of table 5, graphical dependencies are plotted as shown in Figure 2.

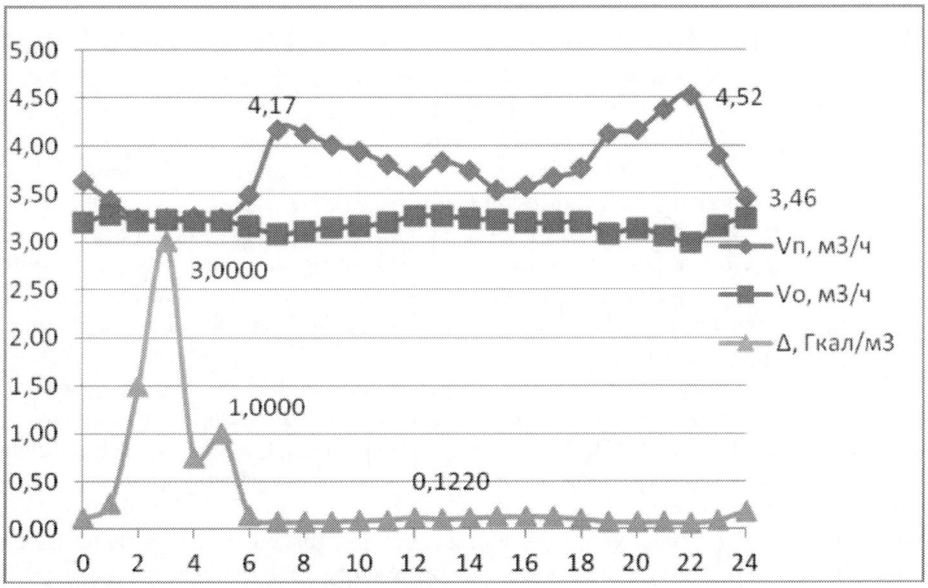

Fig 2. Change of Vп, Vo, Δ during the day

The same dependencies are also built for other houses and for different periods of time, which can characterize the mode of consumption of hot water and the features of the domestic hot water system. For an example the most characteristic house has been selected.

The hourly average flow rate of hot water is taken as the base point, as second consumption is not of practical value for operating organizations. And registering apparatus are not tuned to such delicate and accurate measurements.

The average hourly volumetric flow rate of hot water is determined as the difference Vп - Vo. The values of Vп and Vo are taken according to the readings of heaters SPT 961. On the other hand, the average hourly volumetric flow rate of hot water is the sum of the costs of water collection and circulation, and the terms constantly change over time and are determined by the nature of hot water consumption for a particular building. This is the volume, number of floors, the degree of improvement, the number of residents and their structure (age, mentality, degree of employment, etc.).

The peculiarity lies in the fact that the water draw-off is taken into account by individual water meters with a sufficient degree of accuracy, and the circulation flow is taken into account by a house-building, the degree of accuracy of which may be insufficient. Any consumption of hot water requires heating and pump circulation, i.e. material and financial costs. Not only technical, but economic and, in some way, political issue appears.

On the technical side, the analysis of the presented dependencies made it possible to distinguish the following modes during the average consumption day:

1. Circulation, time 01.00 - 06.00. There is almost no hot water draw-off, the maximum circulation is [0.01 - 0.33] m^3/h. The indicator Δ [0,1515-3,000] $Gcal/m^3$.

2. Water intake - circulation, time 14.00 - 16.00, 18.00. Water intake and circulation are carried out simultaneously within the dynamically changing range of [0.26 - 0.66] m^3/h. The indicator Δ [0.1053–0.1282] $Gcal/m^3$.

3. Water intake, time 7.00 - 13.00, 17.00, 19.00 - 00.00. Water intake and circulation are carried out simultaneously, and the circulation is minimal [0.06 - 0.1] m^3/h. The indicator Δ [0,07527-0,09589] m^3/h.

Conclusion

Centralized DHW systems are characterized by a significant length and branching, which inevitably leads to increased heat losses, and to an unstable hydraulic regime.

Domestic hot water systems of buildings have different number of storeys and sizes, variable hydraulic resistance, are directly hydraulically connected to external networks, being their continuation. Hydraulic balancing should be carried out both along the internal circuit and the external, which is a difficult task.

The modes of water analysis of the domestic hot water supply network are variable in time, circulation, circulation-water analysis, water analysis can vary by week, month and season, and may not coincide in time with other internal systems. Heat losses are almost constant throughout the day, the flow of hot water varies from the minimum circulation flow to the maximum flow rate.

The operation of the domestic hot water system in the circulation mode can be characterized by several parameters, one of which Δ is the specific ratio of the cost of thermal energy for heating a cubic meter of hot water. This parameter is often regulated by law at the levels of the city, region, i.e. at the local level. Exceeding this parameter is administratively punishable.

The full circulation time, as a rule, is night hours, constant heat loss with a minimum consumption of hot water. Heat losses can be reduced, but there is the concept of the optimal thickness of thermal insulation, normalized heat flux density. Then it is necessary to increase the consumption of hot water to reduce the standard for heating a cubic meter of water.

The analysis has showed that in present situation the circulation flow does not exceed 0.43 m^3/h, or 0.42 t/h. As noted above, the estimated circulating flow rate is at least 2 t/h. Therefore, to optimize the heating standard of a cubic meter of water, it is necessary to observe the calculated circulation modes. This will require stabilization of the hydraulic modes of both the external and internal networks, which is a difficult but feasible task.

References

1. L. Zhang, O. Gudmundsson, J.E. Thorsen, H. Li, S. Svendsen 2014 *Energy Procedia* **64** 2833-2840
2. T. Kitzberger, D. Kilian, J. Kotik, T. Pröll 2019 *Energy and Buildings* **195** 126-138
3. J. An, D. Yan, G. Deng, R. Yu 2016 *Energy and Buildings* **133** 321-334
4. S. Werner 2017 *Energy* **126** 419-429
5. V.M. Melnikov, M.V. Shenogin 2019 *Innovations and Investments* **1** 209-212 (in Russian)
6. E.A. Biriuzova, K.I. Ogurtsova 2012 *Herald of civil engineers* **4 (33)** 188-192 (in Russian)
7. U. Navaseltsau, D. Navaseltsava, I. Chernikov 2016 *Bulletin of the Brest State Technical University: Water construction, heat power engineering, geoecology* **2** 81–84. (in Russian)
8. V.S. Mokrousov, A.G. Manov, V.M. Melnikov (2012) The system of accounting and regulation of heat energy consumption by subscribers *Russian Federation Utility Model Patent No. RU 112410 U1, 10.01.2012* Moscow: The Russian Federal Service for Intellectual Property
9. Heat meter SPT961 (mod. 961.2) (2020) Retrieved from http://www.логика.рф/catalog/1/370/
10. SV-15G, SV-15X hot and cold water meters (2020) Retrieved from http://теплоприбор.рф/catalog/sv-15/

CLIMATE RESOURCE POTENTIAL TO DEVELOP SOLAR POWER IN BELARUS

Aleh Meshyk[1,*], *Maryna* Barushka[1], *Viktoryia* Marozava[1], *Erbol* Sarkynov[2] and *Anastasiya* Meshyk[1]

[1]Brest State Technical University, Moskovskaya str. 267, Brest, 224017, Belarus
[2]Kazakh National Agrarian University, 8 Abai avenue, Medeu district, Almaty, 050010, Kazakhstan

Abstract. The work analyses climate resources that can potentially be used to develop solar power in Belarus efficiently. The authors determine space-time variability of radiation regime including such parameters as solar irradiance, atmosphere transparency, sunshine duration, cloud cover patterns, etc. The efficiency of solar power generators is assessed by taking into account the number of clear days with low cloud cover per year, sunshine duration per month, and solar irradiance of a horizontal surface in the daytime.
Keywords: solar irradiance, sunshine duration, cloud cover, solar power potential.

1 Introduction

Solar power is being developed in the United States, Western Europe, China, Japan, and South Korea most intensively. Bloomberg New Energy Finance (BNEF) predicts that the share of renewable energy sources will account for almost 3/4 of the world's investment in power generation by 2040. BNEF estimates that in the next 22 years $ 10.2 trillion will be spent on electricity production in the world. $ 7.4 trillion of it will be spent on clean energy [1]. $ 2.8 trillion is expected to be invested in solar energy by 2040, resulting in a 14-fold increase in capacity. As a result, by 2040 wind and solar electricity will account for 48% of the world's installed capacity and 34% of electricity production, compared with 12 and 5% at present. Renewable energy is also expected to reach 74% in Germany by 2040, 38% in the US, 55% in China and 49% in India [1]. The EU has begun to generate and consume 50 times more solar energy over the past ten years.

The European Union supports Belarus' transition to solar energy by implementing the EU4Energy initiative. Developing solar power allows us to reduce partially our dependence on hydrocarbons and suppliers-monopolists while providing maximum environmental friendliness of energy production. Modern equipment, including the equipment of home manufacture, even now allows achieving a return on capital investments within a specified time frame. At present, the share of renewable energy sources in Belarus is 5.1% [2]. It is planned to bring this figure to 9% by 2035. Today, 108 power plants with the total capacity of about 250 MW are already in operation in Belarus converting solar energy into

* Corresponding author: omeshyk@gmail.com

electricity. It is a rather modest figure but it has significant reserves for growth. In order to provide high efficiency of solar technology in Belarus it is necessary to scientifically study such features of our climate as spatial variability of factors that affect solar irradiance of the ground surface.

2 Method

To estimate how feasible it is to develop solar energy in a particular area, we should take into consideration thermal energy resources of its climate, i.e. characteristics of radiation regime which result from solar radiation [3]. Many factors influence distribution of solar energy and its transformations in the atmosphere. The main ones [4] are: types of cloud cover; profiles of temperature, water vapour and ozone; presence of dust and haze in the atmosphere; spectral properties of the underlying surface; concentration of CO_2 in the atmosphere; air pressure at the level of the Earth's surface, etc.

There is a lot of research that estimates solar energy resources of climate in Belarus [5, 6, etc.], regions of Russia [7, 8, 9, 10, etc.] and other countries. All these works present similar statistical generalizations of radiation regime parameters. They conclude that the conditions are quite suitable to develop solar power despite the significant territorial remoteness and quantitative differences in the parameters under consideration. Space-time variability of the estimated parameters needs to be detailed. As actinometric observation data are often limited, it is necessary to apply methods of analytical calculations and forecasting more widely. At present it is also essential to estimate environmental impact of developing solar power in a region by taking into account current and predicted climatic fluctuations. Many researchers [11, 12, 13, 14, 15, 16, etc.] focus on this issue.

This research is based on the data which characterize radiation regime, cloudiness and other atmospheric phenomena on the territory of Belarus [17]. The time series are taken for the 41-year representative period (1979–2019) at 11 meteorological stations where actinometric parameters are recorded. Certain problems related to representativeness of data may occur when we need to generalize spatial information concerning large areas. In this regard, it makes sense to employ external resources and data from neighbouring countries or from such information sources and databases as NASA SEE; ESRA 1996, 2000; WRDC; Meteonorm 4.0; S@tel-Light, etc. All these databases differ in forms of presenting information, sets of characteristics, data cost, periods of averaging and the number of weather stations presented. The authors of this research used such methods of statistical processing of experimental data as regression analysis, time series analysis, spatial generalization of meteorological information, etc. Methods of analytical calculations and mapping are also applied here.

It is noted that the construction, operation and eventual decommissioning of solar energy facilities affect the environment and ecosystem biodiversity. In work [11] the researchers say that the installation of solar power plants on the ground leads to removal of vegetation and landscape fragmentation. Transmission corridors create barriers to the movement of species and their genes [18]. An analysis of water use for maintaining solar power plants in the southwest of the EU indicates that water for dust control is the main component (60-99%) of total water consumption [15]. It should also be borne in mind that water is often used in water cooling systems necessary for solar power facilities. In this regard, there may be a certain hydrochemical and hydrological impact on the environment. The installation of solar power plants may require extensive landscape modification which involves removal of vegetation, land transformation, soil compaction and infrastructural construction. All these activities increase risks for water and wind erosion [13]. Photovoltaic cells must be properly disposed of at the decommissioning stage to prevent environmental contamination with toxic materials contained in the cells. If improperly handled, industrial waste can pose hazards to air quality, health of the public and the environment, surface and ground water, in particular [15].

Strong heating of air caused by solar radiation concentrated by mirror reflectors and passing through the air leads to changes in humidity, heat balance and wind direction. This results in adverse environmental impacts on ecosystems in the areas where large power plants are located. The radiation balance at the level of the ground surface can change when the albedo of photovoltaic cells differs from the original background albedo. With their absorbing capacity, photovoltaic panels have an effective albedo (on average 0.18-0.23). Some studies [14] have shown that the installation of solar panels leads to a decrease in the surface albedo and an increase in the surface temperature by 0.4 ° C. Albedo in cities is usually 0.15-0.22. Therefore, the photovoltaic panels installed on rooftops have a potential to increase albedo resulting in the cooling effect. We already mentioned the importance of this problem in relation to thermal pollution of urbanized areas [19]. It was found out that the air temperature dropped by 0.2 ° C under the panels with higher efficiency [14].

Everything mentioned above allows us to conclude that climatic resources play a crucial role in developing solar power both for making decision to install and maintain solar energy facilities in a certain area and for assessing their impact on the environment. The development of "green" energy is the future of mankind, though [20].

3 Results and discussion

Feasibility to operate solar energy facilities in a certain area and their efficiency depend on a number of meteorological factors including the intensity of solar radiation (kW / m^2) and air temperature (° C). Solar radiation reaching the ground surface can be characterized by the following indicators: direct, diffuse and total radiation, reflectivity of the underlying surface, radiation balance, sunshine duration (SD), cloud cover patterns, the number of cloudy and clear days with different cloud gradations, etc.

Solar radiation data characterize different time periods: a year, a season, a month, a decade and a day. In paper [6] it is noted that different-scale variability of solar radiation reaching the ground surface is not taken into account in the same way in solar energy calculations. However, we believe that diurnal differentiation is the most acceptable. Daily values allow easy switching to ten-day, monthly and annual calculations.

As we are short of actinometric data, we developed a method to calculate daily values of total solar radiation [21, 22]. The calculated monthly and annual amounts are shown in Table 1.

Table 1. Total solar radiation, MJ / m^2.

Month / Meteorological station	January	February	March	April	May	June	July	August	September	October	November	December	Annual
Minsk	58	127	276	420	571	615	608	494	329	190	79	37	3804
Brest	73	143	291	431	576	618	610	500	341	206	94	51	3933
Homel	71	140	288	429	575	618	610	499	389	203	92	49	3912
Grodno	59	129	278	422	573	619	610	494	330	191	80	37	3822
Vitebsk	46	116	265	414	571	621	610	488	328	177	67	25	3729
Mogilev	57	127	277	422	574	619	609	492	327	188	78	35	3805

Our comparing the calculated and measured values of the total solar radiation at Minsk and Vasilevichi meteorological stations revealed that the annual radiation amounts differ by no more than 1–2%. There is a high convergence of the results in most months of the year. However, in January, October, and November deviations can reach 16–25 %. This can be explained by the fact that the last official data of generalizations were registered in 1980 and do not take into account the temporal variability of total solar radiation over the past forty years.

In order to analyse a natural solar energy potential of an area, it is customary to represent it in its intensity units, kW * h / m^2. In this regard, we applied the ratio of 1 kW * h / m^2 as equal to 3.6 MJ / m^2. According to B.P. Weinberg [23], the total solar radiation is considered "technically acceptable" when its intensity is 0.60 kW / m^2 or higher. The performed calculations showed that in December and January this level of radiation intensity is not reached on the territory of Belarus. For example, in Minsk it reaches 0.34 and 0.54 kW * h / m^2 per day, respectively. The highest rate of 5.69 kW * h / m^2 per day is in June (Table 2).

Table 2. Intensity of total solar radiation per day, kW * h / m^2.

Month / Meteorological station	January	February	March	April	May	June	July	August	September	October	November	December	Annual
Minsk	0.5	1.3	2.5	3.9	5.1	5.7	5.5	4.4	3.0	1.7	0.7	0.3	2.9
Brest	0.6	1.4	2.6	4.0	5.2	5.7	5.5	4.5	3.2	1.8	0.9	0.5	3.0
Homel	0.6	1.4	2.6	4.0	5.2	5.7	5.5	4.5	3.1	1.8	0.8	0.5	3.0
Grodno	0.5	1.3	2.5	3.9	5.1	5.7	5.5	4.4	3.1	1.7	0.7	0.3	2.9
Vitebsk	0.4	1.1	2.4	3.8	5.1	5.8	5.5	4.4	3.0	1.6	0.6	0.2	2.8
Mogilev	0.5	1.3	2.5	3.9	5.1	5.7	5.5	4.4	3.0	1.7	0.7	0.3	2.9

Our calculations show that "technically acceptable" radiation on the territory of Belarus occurs from 289 (Braslav) to 321 (Bragin) days a year (79–88%). Its distribution depends on the latitude. The intensity of solar radiation of 0.60 kW / m^2 is not provided in Belarus from November 14-30 to January 12-28. In general, however, it is necessary to state that the solar power potential of Belarus is quite high, despite certain seasonal fluctuations.

Most researchers estimate the total solar radiation by using empirical relationships with sunshine duration and cloudiness grade. The intra-annual variation of total and low cloud cover, averaged over the territory of Belarus, is shown in Fig. 1.

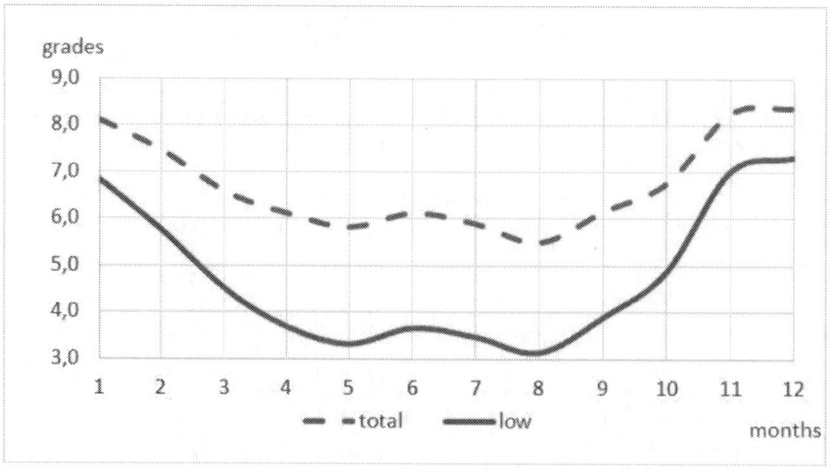

Fig. 1. The course of total (blue) and low (brown) cloud cover in Belarus within a year.

Fig. 1 shows the most typical picture for northern Europe, i.e. the lowest values of cloud cover are characteristic of the warm period with a slight increase in June and a sharp growth in winter when the Atlantic influence is great. The average number of clear and cloudy days per year follows the intra-annual course of cloudiness (Fig. 2).

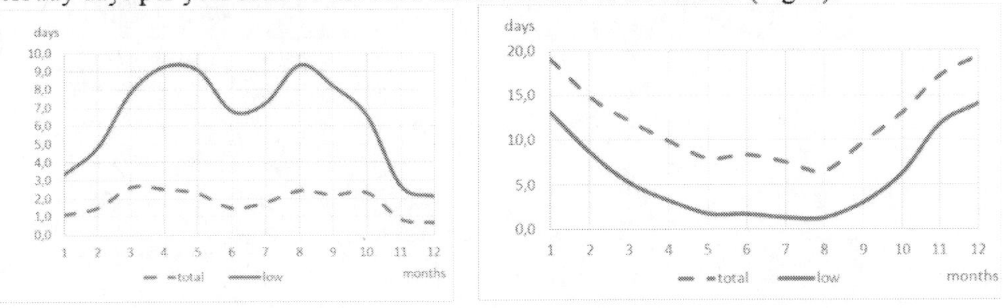

(a) (b)

Fig. 2. The number of clear (a) and cloudy (b) days (with different types of cloudiness) per year in Belarus.

The largest number of clear days falls on the warm period from April to September with a slight decrease in June due to the prevailing precipitation regime. The number of clear days per year may indicate indirectly the prospects to develop solar power. Some scientists note that the highest efficiency of solar panels is provided when the number of clear days with low clouds per year exceeds 200 [7]. However, in Belarus this indicator is a little more than 60 on average, reaching 100-105 days at certain meteorological stations only in some years. Thus, the number of clear days with low cloud cover cannot be taken as a criterion to estimate solar power potential in Belarus.

The number of clear days with total cloud cover increases from the north, north-west to the south, southeast across Belarus: from 20 (Grodno, Polotsk, Mogilev) to 30-35 days (Mozyr, Bragin) and with low cloudiness from 50 (Vysokoje) to 100 days (Mozyr, Pinsk, Zhlobin) (Fig. 3). There is also a decrease in the number of cloudy days with total cloud cover from 160 (Polotsk, Sharkovshchina) to 120 days (Bragin, Mozyr) and with low cloud cover from 120 (Borisov, Lepel, Senno) to 60 days (Vasilevichi, Pruzhany) (Fig. 4). It should be noted that, unlike the duration of sunshine, the cloudiness parameters are characterized by significant spotting on the maps, which implies a search for a relationship with the landscape patterns.

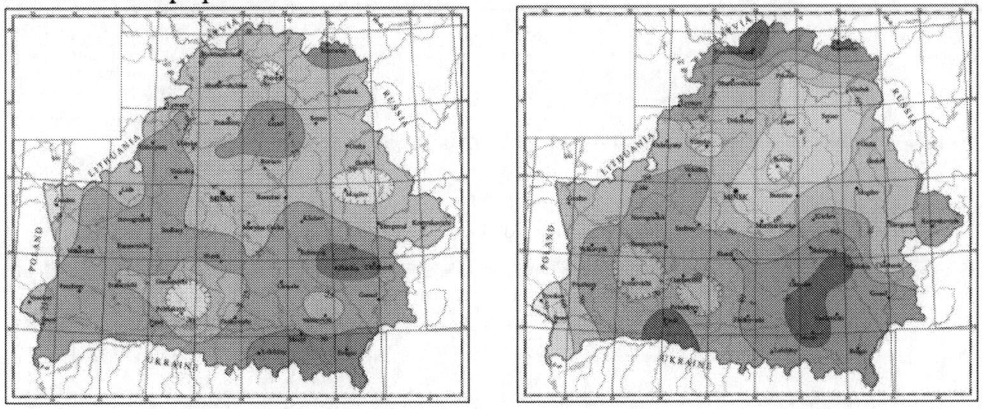

Fig. 3. The number of clear days with total cloud cover (left) and low cloud cover (right) in Belarus.

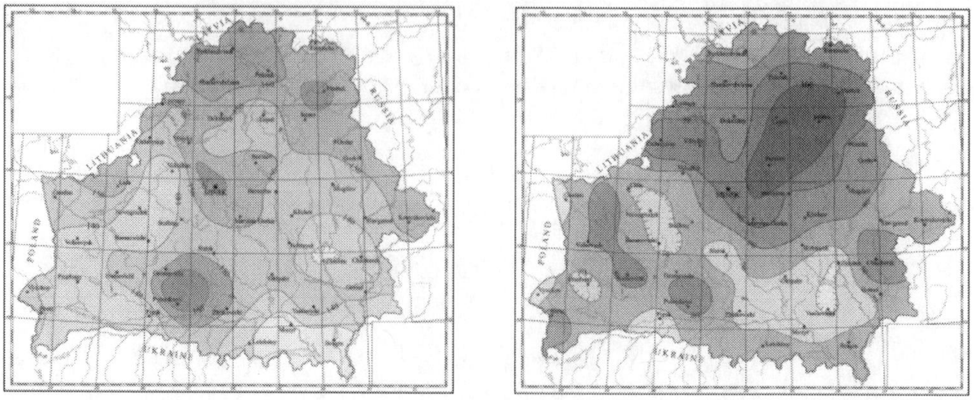

Fig. 4. The number of cloudy days with total cloud cover (left) and low cloud cover (right) in Belarus.

There is a certain correlation between the increase in the average annual duration of sunshine, the number of clear days with total and low cloudiness and decrease in the number of cloudy days with total and low cloud cover from the north, northwest to the south, southeast [4]. Cloudiness reduces the annual amount of total solar radiation by 2.5-3 times. For example, in Minsk, with no cloud cover, the annual amount of solar radiation can reach 4485 MJ / m^2. Annual sums of total radiation are reduced by about 40% compared with what they would be in a clear sky. At the same time, the amount of diffuse radiation under average cloudy conditions is about 40% higher than that in the clear sky [4].

There is a certain correlation between the annual amount of sunshine duration and the average annual cloud cover but their statistical significance is not proved (Fig. 5).

The correlations between the number of clear and cloudy days per year and low cloudiness are statistically significant (Fig. 6).

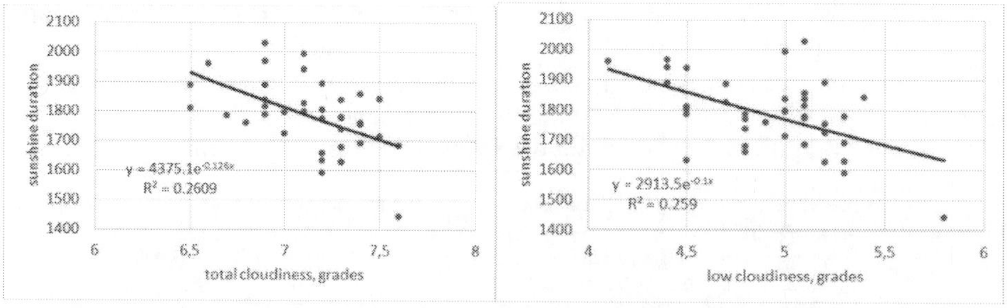

Fig. 5. Dependence of sunshine duration on different types of cloudiness in Minsk.

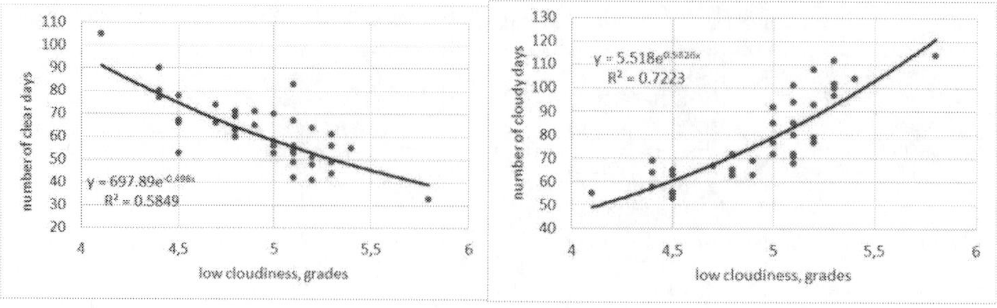

Fig. 6. Dependence of the number of clear and cloudy days on low cloudiness in Minsk.

The research has shown that characteristics of low cloud cover can be a fairly good tool to assess solar energy resources if actinometric data are limited or insufficient.

The intra-annual course of the possible and actual sunshine duration on the territory of Belarus is shown in Fig.7.

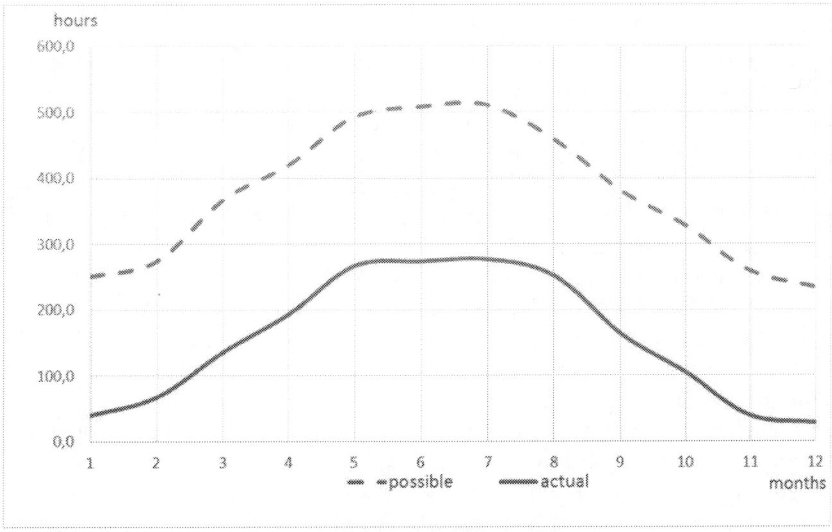

Fig. 7. Possible and actual duration of sunshine in Belarus.

The curves (Fig. 7) follow almost synchronously and allow us to assess the relationship between the actual and possible values of sunshine duration (Table 3).

Thus, Figure 7 shows that the duration of sunshine is minimal in December - February and maximal in May - August. Some researchers [8] state that the optimal efficiency of almost all solar power plants is achieved when the sunshine duration is more than 250 hours per month.

Moreover, it is not the distribution of radiation regime characteristics within a day that matters but their monthly and annual amounts are important to judge about the real solar energy potential. 220-225 hours of sunshine duration per month also allow us to consider the efficiency of solar power facilities satisfactory. This criterion is provided in Belarus from April to September. Sunshine duration of over 250 hours a month is observed from May to August. In particular years, the actual duration of sunshine can exceed 420 hours in July, June but sometimes it decreases to less than one hour in December.

Table 3. The ratio of actual duration of sunshine to the possible one, %.

Month / Meteorological station	January	February	March	April	May	June	July	August	September	October	November	December	Annual
Verchnedvinsk	13.5	23.4	36.8	45.7	55.0	53.1	55.5	53.1	40.9	28.5	12.7	10.6	40.5
Minsk	14.7	23.3	36.5	46.0	52.3	52.3	52.3	53.6	42.3	30.0	15.1	10.7	39.9
Brest	16.9	24.8	36.7	45.9	53.9	53.4	54.1	56.7	44.7	36.9	18.6	14.5	41.8
Homel	18.1	27.8	39.2	46.8	57.5	56.3	56.3	57.5	46.4	34.9	16.2	13.1	43.3
Vasilevichi	17.5	26.2	37.5	47.3	56.8	56.2	56.8	57.8	45.7	34.6	15.2	12.5	47.3

The longest actual duration of sunshine corresponds to the summer months. It reaches its maximum (277 hours) in July and decreases to 28 hours in December. In May-August period, the actual sunshine duration exceeds 50% of the possible one and it is less than 15%

in December-January. This distribution is predetermined by the cloudiness pattern in these months (Fig. 1). On average, the ratio of the actual and possible sunshine duration is about 40% within a year and it differs insignificantly around Belarus. In some months of the cold season territorial differences increase. They are connected with the latitude.

The amount of solar radiation is predetermined by the geographic location of Belarus and depends on the sunshine duration and cloud cover pattern as well as on the height of the sun above the horizon at different time of the year. The longest day is 2.5 times longer than the shortest one in the north of Belarus and it is 2.1 times longer in the south. The difference in the day length is about one hour both in the north and south, both in summer and in winter. The day in the north of Belarus in summer is longer than in the south but the sun is lower; this somewhat reduces the differences in climatic conditions between the southern and northern regions. In winter, when both the length of the day and the height of the sun's standing above the horizon in the south are greater than in the north, the south is in more favourable conditions than the north [24].

Possible sunshine duration in Belarus is 4495 ± 10 hours per year. It is longer in the north due to refraction. Therefore, differences in the actual sunshine duration are determined by the cloud cover pattern. The average annual duration of sunshine increases from the north, northwest to the south, southeast by about 7%: from 1740 hours (Lida, Oshmyany) to 1870 hours (Bragin) (Fig. 8) [22].

Figure 8 coincides with the state cadastre of renewable energy sources where the territory is zoned according to the possibilities for practical implementation of solar energy potential.

There is an increase in the annual sunshine duration at all meteorological stations around Belarus by 46 hours over 10 years on average (from 20 hours in Verkhnedvinsk to 77 hours in Oshmyany) (Fig. 9). This suggests that climatic conditions to develop solar energy are becoming more favourable. A significant increase in the annual values of sunshine duration in Moscow is also noted in the work [25] with a trend of about 35 hours over 10 years.

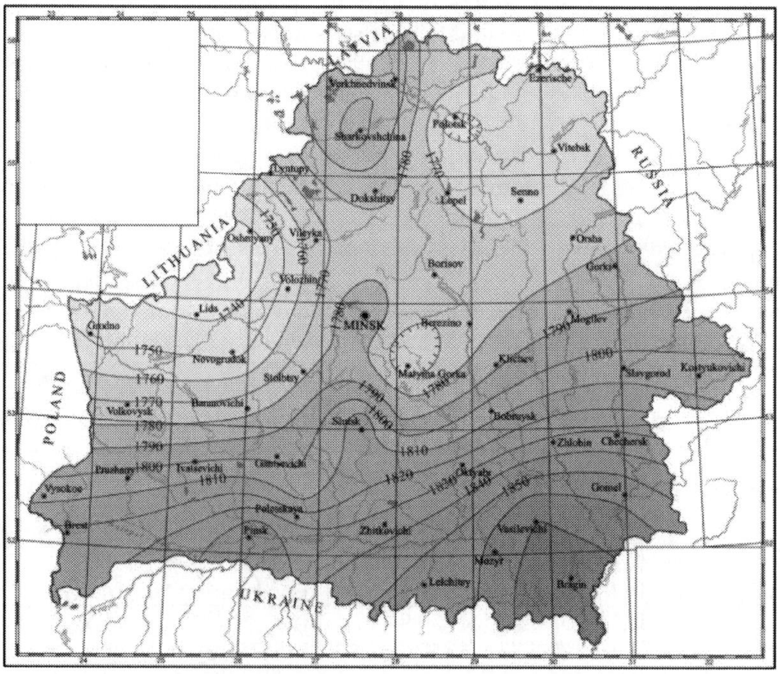

Fig. 8. Sunshine duration in Belarus, hours per year.

 https://doi.org/10.1051/e3sconf/202021201012

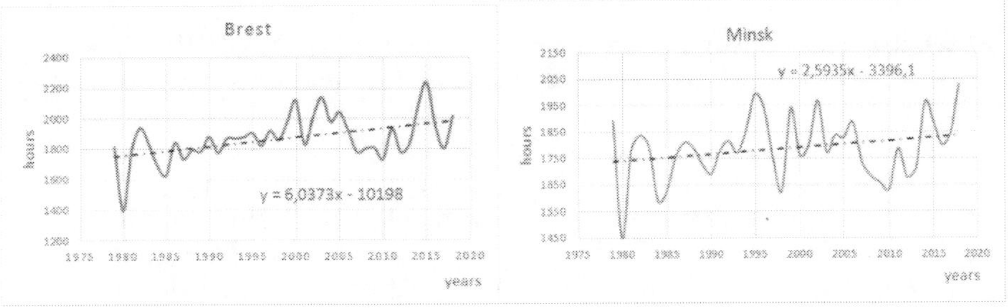

Fig. 9. Time variability of sunshine duration in Belarus.

Figure 10 shows time variability of sunshine duration in Minsk in particular months. We observe a positive trend in most months of the year with the largest one in September and June. However, there is a decrease in monthly sunshine duration in February, October, November, and December. Such trends are observed at all meteorological stations in Belarus. They result from a change in the pattern of atmospheric circulation.

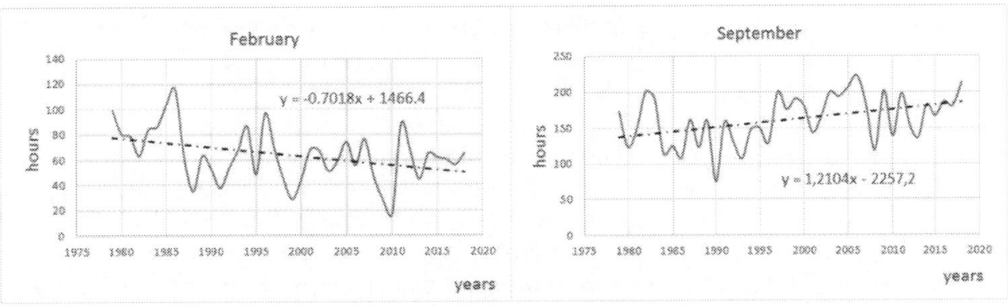

Fig. 10. Time variability of sunshine duration in Minsk (per months and per year), in hours

4 Conclusion

At present, a surplus of electricity is predicted in Belarus due to developing nuclear power and commissioning the first Belarusian nuclear power plant. Nevertheless, it is necessary to diversify our energy sources. This will strengthen the country's national security and reduce dependence on external factors, minimize market risks and energy failures. Currently, the dependence on imported hydrocarbons is great, that is why we need to develop alternative "green" energy which will also contribute to environmental safety.

The results obtained in this research allow us to conclude that there are enough solar energy resources in Belarus to develop solar power industry but they are distributed unevenly throughout the year. Our analysis of time series of observations over cloudiness patterns, sunshine duration, total solar radiation and its intensity over a 40-year period from 1979 to 2019 revealed that the most productive period for operating solar power plants is the period from February to November. In December and January, total solar radiation does not reach 0.60 kW / m^2, so generating electricity by solar power facilities is impractical. Since the number of meteorological stations that register actinometric data is limited in Belarus, the researchers have to apply analytical calculation methods to determine relationships between total solar radiation, its intensity and other widely observed parameters such as cloud cover, sunshine duration, etc.

References

1. *EEnergy Informer. The International Energy Newsletter*. – August 2017. – № 8. – Vol. 27.
2. Врублевский, Б. И. Направления использования возобновляемых и нетрадиционных источников энергии в Республике Беларусь (Directions of use of renewable and non-traditional energy sources in the Republic of Belarus) / B. I. Vrublevsky, I. V. Senko // *Consumer cooperation*. - 2015. - No. 2. - P. 27–32. (In Russian).
3. Meshyk, A. Thermal Resources of the Climate of West Polesie, Belarus / A. Meshyk, M. Sheshka, M. Barushka // *7th International Congress on Energy and Environment Engineering and Management (CIIEM7)* : Abstracs Book, Canary Islands, Spain, 17-19 July 2017 ; edited by Science Know conferences. – Las Palmas (Spain), 2017. – P. 94–95.
4. Мешик, О. П. Перспективы развития солнечной энергетики в Республике Беларусь (Prospects for the development of solar energy in the Republic of Belarus) / A. P. Meshyk, M. V. Barushka // *Actual problems of earth sciences: studies of transboundary regions: collection of articles and materials IV Int. scientific. - practical. conf., to the 1000th anniversary of Brest*, 12-14 Sept. 2019 / Brest; ed. A. K. Karabanov [and others]. - Brest: BrSU, 2019. - Part 2 - P. 250–253. (In Russian).
5. Камлюк, Г. Г. Гелиоэнергетические ресурсы и перспективы развития гелиоэнергетики в Республике Беларусь (Solar energy resources and prospects for the development of solar energy in the Republic of Belarus) / G. G. Kamlyuk // *Energy strategy*. - 2012. - No. 6 (30). - P. 35–37. (In Russian).
6. Пашинский, В. А. Оценка падающей солнечной радиации на горизонтальную поверхность территории в условиях Республики Беларусь (Assessment of incident solar radiation on the horizontal surface of the territory in the conditions of the Republic of Belarus) / V. A. Pashinsky, A. A. Butko, A. A. Cherkasova // *Ecological Bulletin*. - 2015. - No. 2 (32). - P. 77–82. (In Russian).
7. Невидимова, О. Г. Климатические условия развития гелиоэнергетики на территории Западной Сибири (Climatic conditions for the development of solar energy in the territory of Western Siberia) / O. G. Nevidimova, E. P. Yankovich // *Modern problems of science and education*. - 2015. - No. 1. - Part 2. (In Russian).
8. Севастьянова, Л. М. Потенциальные ветро- и гелиоэнергетические ресурсы в Алтайском крае (Potential wind and solar energy resources in the Altai Territory) / L. M. Sevastyanova, Yu. N. Nikolchenko // *Earth Sciences*. - 2012. - No. 4. - P. 187–193. (In Russian).
9. Носкова, Е. В. Природный гелиоэнергетический потенциал Забайкальского края (Natural solar energy potential of the Trans-Baikal Territory) / E. V. Noskova // *Geographical Bulletin*. - 2017. - No. 4 (43). - P. 105–112. doi: 10.17072 / 2079-7877-2017-4-105-112. (In Russian).
10. Бутолин, А. П. Возобновляемые источники энергии Южного Урала (Renewable energy sources of the South Urals) / A. P. Butolin, V. A. Shcherba, E. A. Abramova // *Ecology of urbanized territories*. - 2018. - No. 4. - P. 20–25. doi: 10.24411 / 1816-1863-2018-14020. (In Russian).
11. Lovich, Jeff & Ennen, Joshua. (2011). Wildlife Conservation and Solar Energy Development in the Desert Southwest, Unites States. *BioScience*. 61. 982-992. doi: 10.1525/bio.2011.61.12.8.
12. Hu, Aixue & Levis, Samuel & Meehl, Gerald & Han, Weiqing & Washington, Warren & Oleson, Keith & van Ruijven, Bas & He, Mingqiong & Strand Jr, Warren. (2015). Impact of solar panels on global climate. *Nature Climate Change*. 6. doi: 10.1038/nclimate2843.

13. Li, Junran & Okin, Gregory & Alvarez, Lorelei & Epstein, H. (2007). Quantitative effects of vegetation cover on wind erosion and soil nutrient loss in a desert grassland of southern New Mexico, USA. *Biogeochemistry*. 85. 317-332. doi: 10.1007/s10533-007-9142-y.
14. Hernandez R. R., Easter S. B., Murphy-Mariscal M. L. et al. (2014). Environmental impacts of utility-scale solar energy. *Renewable and Sustainable Energy Reviews*, 29, 766-779. doi: 10.1016/j.rser.2013.08.041.
15. Mani, Monto & Pillai, Rohit. (2010). Impact of dust on solar photovoltaic (PV) performance: Research status, challenges and recommendations. *Renewable and Sustainable Energy Reviews*. 14. 3124-3131. doi: 10.1016/j.rser.2010.07.065.
16. Crook, Julia & Jones, Laura & Forster, Piers & Crook, Rolf. (2011). Climate change impacts on future photovoltaic and concentrated solar power energy output. *Energy Environ. Sci*. 4. 3101-3109. doi: 10.1039/C1EE01495A.
17. Климатический кадастр Республики Беларусь. Метеорологический ежемесячник (Climate cadastre of the Republic of Belarus. Meteorological Monthly). - Minsk: Republican Center for Hydrometeorology, Radioactive Contamination Control and Environmental Monitoring, 1979–2019. (In Russian).
18. Saunders, Denis & HOBBS, RICHARD & Margules, Chris. (1991). Saunders D, Hobbs R, Margules C. Biological Consequences of Ecosystem Fragmentation: A Review. Conservation Biology. *Conservation Biology*. 5. 18-32. doi: 10.1111/j.1523-1739.1991.tb00384.x.
19. Волчек, А. А. Выявление микроклиматических аномалий урбанизированных территорий методами дистанционного зондирования (на примере г. Бреста) (Revealing of microclimatic anomalies of urbanized territories by remote sensing methods (for example, Brest)) / A. A. Volchak, A. P. Meshyk, A. O. Meshyk // *Bulletin of Brest State Technical University*. - 2018. - No. 2 (110): *Water management construction, heat power engineering and geoecology*. - P. 36-40. (In Russian).
20. Barushka, M. Thermal resources of climate of Belarus and their application in "Green" energy industry / M. Barushka // Actual environmental problems : proceedings of the IX International Scientific Conference of young scientists, graduates, master and PhD students, Minsk, 21–22 November 2019 / editor : S. A. Maskevich, N. A. Lysukha. – P. 142.
21. Мешик, О. П. Исследование и моделирование составляющих теплоэнергетических ресурсов климата Беларуси (Research and modeling of the components of heat and power resources of the climate of Belarus) / O. P. Meshik // *Rational use of natural resources: proceedings of the Intern. conf. "Scientific aspects of the rational use of natural resources"*, Brest, 20-22 October. 1998 / Brest Polytechnic. in-t; editorial board : V. E. Valuev [et al.]. - Brest, 1998. - P. 40-50. (In Russian).
22. Мешик, О. П. Оценка гелиоэнергетических ресурсов климата Беларуси (Assessment of solar energy resources of the climate of Belarus) / A. P. Meshyk, M. V. Barushka, V. A. Marozava // *Bulletin of BrSTU*. - 2020. - No. 2: *Water management construction, heat power engineering and geoecology*. - P. 93–99. https://doi.org/10.36773/1818-1212-2020-120-2.1-93-99. (In Russian).
23. Вейнберг, Б. П. Гелиоэнергетические ресурсы СССР (Solar energy resources of the USSR) // *Atlas of energy resources of the USSR*. - M.; L.: United Scientific and Technical Publishing House, 1935. - V. 1. - Part III. - Hydropower, wind energy, solar energy resources. - 127 p. (In Russian).
24. Климат Беларуси (Climate of Belarus) / Academy of Sciences of Belarus, Committee on Hydrometeorology of the Ministry of Emergencies of the Republic of Belarus; ed. V. F. Loginov. - Minsk: Institute of Geological Sciences of the Academy of Sciences of Belarus, 1996. - 234 p. (In Russian).
25. Горбаренко, Е. В. Изменчивость солнечного сияния в Москве за период 1955–2017 гг. (Variability of sunshine in Moscow for the period of 1955–2017) / E. V. Gorbarenko // *Meteorology and Hydrology*. - 2019. - No. 6. - P. 24–36. (In Russian).

MAPPING THE CHARACTERISTICS OF SNOW COVER IN BELARUS

Aleh Meshyk[1*], *Viktoryia* Marozava[1] and *Maryna* Barushka[1]

[1]Brest State Technical University, Moskovskaya str. 267, Brest, 224017, Belarus

Abstract. The paper substantiates a necessity to create maps of snow cover characteristics in Belarus. The designed maps can be used by engineers to forecast spring floods on the rivers of Belarus and to assign snow load limits imposed on buildings and structures.
Keywords: creating maps, snow cover characteristics, snow height, snow density, snow water equivalent.

1 Introduction

Snow cover is one of the most important climate forming factors. Snow has a particularly great influence on the climate in the middle latitudes of the northern hemisphere where Belarus is located. It predetermines the pattern of calendar seasons, the annual course of air temperature as well as weather changes during the day. Much of current research is devoted to both the study how climate warming influences the characteristics of snow cover and the study of snow cover as a factor of climate change [1–8, etc.].

Snow cover is a layer of snow on the ground surface that forms as a result of precipitation. Snow cover also includes ice layers that form on the surface of snow and soil and melt water accumulating under the snow.

Different countries have their own history of observations over snow cover, for example [9]. Systematic observations over snow cover in Belarus began in 1891. In the 1930s snow surveying was introduced in addition to observations conducted with permanent snow stakes. Observations done with permanent snow stakes registered every day give us an idea of changes in snow height in winter but they do not show a pattern of its distribution on the ground [10, 11].

In order to characterize a snow cover over all territory, additional snow surveying is carried out in two kinds of areas: 1) open (field, meadow, etc.), 2) protected (in the forest under tree crowns). As a result, we obtain the following data: 1) average values of snow height, its density and water equivalent in the snow cover; 2) patterns of snow cover distribution on various kinds of relief and land surface (within a meteorological station); 3) patterns of changes in snow accumulation and snow melting in time.

The main characteristics of snow cover are its height, density, snow water equivalent and the degree of snow coverage of the corresponding area. Height and density allow us to determine water equivalent in the snow cover. They serve as a basis for hydrological calculations and forecasts, play an important role agricultural decision-making and are also

* Corresponding author: omeshyk@gmail.com

widely applied in solving a number of scientific and practical problems. Many researchers point out certain difficulties in determining snow water equivalent directly at meteorological stations. They, therefore, propose to apply methods of remote sensing of the earth [12–14, etc.]. Today, however, in Belarus we can calculate snow water equivalent with sufficient accuracy only with the help of snow surveys carried out on routes near the weather stations.

2 Method

The aim of this research is to study and map the characteristics of snow cover recorded at meteorological stations in Belarus within the representative period of 1944-45–2019-20 and to assess space-time variability of these characteristics and to forecast their dynamics. The study object of this work is such characteristics of snow cover as snow height and snow water equivalent observed at 48 meteorological stations of Belarus within the representative period mentioned above. The study subject of this work is quantitative assessment of the snow cover characteristics and patterns of their space-time distribution over the territory of Belarus.

The methods applied in the research include a space-time analysis of observation data, analytical calculations, mapping.

Mapping snow cover characteristics is of practical significance to us because the maps created in this research can supplement the ones already published, like in [15]. They also can become a basis for creating maps of snow areas [16] which can be applied by engineers to assign snow load limits to buildings and structures. They are also relevant for environmental engineers to predict spring floods in rivers, which is an extremely important issue in Belarus [10, 17].

3 Results and discussion

The initial stage in mapping the characteristics of snow cover is preliminary processing of observation data with the optimization of the location of control points, their quantity and quality. The main task is to process and visualize two-dimensional data sets described by the function z = f (x, y). The logic of work can be represented in the form of three main functional blocks: a) creation of a digital surface model; b) auxiliary operations with digital surface models; c) visualization of the surface.

One of the common tasks in mapping is to study how points are distributed on a 2D surface or map. These points can correspond to the places of taking snow samples, obtaining observations, etc. The task may be to study the uniformity of the distribution of observation points, the density of the distribution or to study the relationship of points with each other. All these questions arise among researchers. Moreover, field observations related to the analysis of the position of points always lead to these or similar problems. The developed methodology is applicable directly to the study of natural phenomena.

It is convenient to divide existing schemes of location of points on maps into three categories: uniform, random and group. Of course, most maps are characterized by patterns of distribution of points that occupy an intermediate position between the listed extreme types, and usually the task is to classify the observed pattern to one of these types.

The optimal number of control points should be specifically justified for each mapped characteristic. If there is a lack of points in the meteorological network, it is necessary to take into account factors of climate formation and indirect physical and geographical signs, by introducing a distribution function of the characteristic under study [18]

$$M_{ij}=f(\varphi_j, \lambda_j, H_j), \qquad (1)$$

where M_{ij} is the value of the physical and geographical characteristics at (j)-point, for (i)-period; φ_j, λ_j, are geographic latitude and longitude, respectively; H_j is absolute elevation of the earth's surface.

Optimization of control points is based on minimizing errors resulting from the construction of certain maps. A map constructed with insufficient data provides only a generalized representation of information.

The assessment of representativeness of the spatial location of observation points can be performed using the criterion (χ^2). In this case, the study area is divided into a certain number of areas containing control points. The sizes of the areas are determined based on the premise of combining the studied characteristics into space-time fields, approximated by spatial correlation functions. Within the selected areas (regions), estimates of representativeness are usually carried out under the assumption that the optimal distance (step) between meteorological stations is 20 km. With the existing density of the meteorological network in local areas (20x20 km), there may be no weather stations at all. On this basis, the boundaries of the isocorrelate fields of the studied characteristics can be used as the boundaries of the regions [10]. The criterion (χ^2) is theoretically independent of the shape and orientation of the regions in space. If the existing meteorological stations are located evenly over the territory, then each selected area will contain an equal number of points. The obtained values (χ^2) are compared with the critical ones, and appropriate conclusions about the representativeness of the spatial location of the observation points are made. This conclusion concerns only the uniformity of the distribution of points over areas of a certain size. It is quite possible that there is a variant of the size of the square (especially if it is smaller than the chosen one), in which the hypothesis of uniformity will be rejected.

The choice of a rational method for mapping the characteristics of the snow cover is based on the following factors: individual characteristics of the object (phenomenon, information), its location, structure; tasks of mapping information; content and type of initial data, methods of their processing, etc.

The most widely known methods of cartographic display of information: icons, line signs, areas, high-quality background, isolines, carto diagrams, cartograms.

Taking into account the above and the fact that maps of snow cover characteristics are intended not only for a general qualitative assessment of the current situation in specific territories but also as a source of quantitative information for researchers and designers, the main method of creating maps is that of isolines and high-quality background. Map created in isolines allow for quick and quite accurate assessment of the meteorological situation in the area where projects of various purposes are designed. They also help to prepare initial data for engineering calculations which are declared by the regulatory and technical documents of the Republic of Belarus but which often lack a basic database. Using the method of high-quality background allows visualizing information even more [15, 18].

The adopted method of mapping should provide a uniquely accurate representation of data at a control point, be continuous within the mapping area. Autocorrelation is assumed at a distance bigger than the average distance between control points. Autocorrelation indicates that the values at adjacent points are closely related. This is confirmed by the performed studies of snowfall synchronicity. In this study we determined some regions that include a certain number of control points [10].

To construct a map it is necessary to prepare a mathematical surface which is divided into square cells that cover the mapped area completely. The smaller the cell, the higher the resolution of the card will be. The task of surface preparation is to determine the values of the studied indicators at the nodes of the adopted grid according to the data of nearby control points. Contours are drawn not from the data of the control points of observation but from the calculated values in the nodes of the network. If the mapped area is divided into large cells then with a smooth slope of the surface, there may be a displacement of individual isolines from the control points with which it should be consistent (the passage of the isoline on the wrong side of the control point).

The majority of known methods for constructing maps in contours evaluate values from nearby control points. In this case, when building a map on a prepared mathematical surface, most of the interpolation nodes will lie in the interval between the values of the control points. The values at the nodes that are outside the control points are obtained by extrapolation and will be close in magnitude to the values of the extreme observation points. The control points in relation to the network node are weighted, and the weights are set depending on their distance from each other. The sum of the weights of the control points of one node is equal to one. In this regard, it is possible to establish the most distant control point from the node of the regular network participating in the assessment. The weighting function of the inverse values of the squares of the distances is scaled, the values taken by it are within $0 \leq w \leq 1$. This process is written in the form [18]

$$w = \left(1 - \frac{D}{D_{max}}\right)^2 \Big/ \left(\frac{D}{D_{max}}\right) \quad (2)$$

where D is the distance to the estimated point; D_{max} is the distance to the most distant point.

A reliable criterion for the correctness of the chosen method is the form of isolines, which shows how accurately the adopted mathematical model describes the control points. It is statistically impossible to establish which method of contouring is more reliable. The decision is made in each specific case, depending on the type of mapped characteristics, the quality of the initial data, their density and the uniformity of the spatial distribution. The main purpose of building a map in contours is to ensure the maximum correspondence of the initial data to the values set by the map.

The most promising, in our opinion, are the following methods for constructing contour maps: that of inverse distances, Kriging, minimal curvature, radial basis functions, Shepard, triangulation. We use all of them but we justify the most acceptable method on the results of map comparison.

When mapping the characteristics of the snow cover on the territory of Belarus, we used a sample of experimental meteorological data at 48 empirical points with distances between them from 48 to 585 km. The degree of spatial continuity of a regionalized variable is expressed by a variogram. If there is data in a scattered set of points and a known variogram form, an independent surface value is estimated at any point not belonging to the sample (Z).

We substantiated Kriging method as the most appropriate for mapping the characteristics of snow cover in isolines because it has optimal statistical properties (measuring the error or uncertainty of the surface depicted by isolines, using a variogram to find the optimal set of weights, estimating the surface at points other than empirical, as a function of distance and weight, varying according to the geographical location of the observation points for the components of the snow cover).

Maps constructed with the use of Kriging method are characteristic of a statistically stationary variable. In fact, in natural processes, it is customary to single out the trend component, in which the calculated values will be systematically underestimated or overestimated which depends on the actual location of observation points for the characteristics of the snow cover and the direction of the trend plane. In this case, the non-stationary regionalized variable is considered as having two components. The trend is the average or expected value of a regionalized variable within an area and it changes slowly, characterizing a non-stationary part of the surface. The remainder is the difference between the actual change and the trend. If the trend is removed from the regionalized non-stationary variable, then the residuals become stationary and Kriging method can be applied to them. In this case, Kriging consists of the following procedures:

- identification of the trend component and its removal;
- carrying out Kriging for the obtained residues at points outside the limits of instrumental observations;
- grouping of the obtained residuals with trend components and obtaining the true surface.

We have widely used mapping methods adapted to the research tasks in the spatial generalization of the main characteristics of the snow cover [19]. Fig. 1a shows a map of the distribution of the maximum water equivalent in snow (mm) on the territory of Belarus, constructed by using Kriging method. Map (Fig. 1a) characterizes the absolute maximums in the formation of snow water equivalent over a long-term period (1944-45 –2019-20).

As can be seen from the map (Fig. 1a), the largest snow water equivalent is confined to the northeastern (Vitebsk, Orsha region), northwestern (Lyntupy) and central (Novogrudok, Berezino) parts of the territory of Belarus. Here, a significant contribution to the snow water equivalent is the height of the terrain as a regional factor (Minsk, Novogrudok, Oshmyany uplands). The smallest snow water equivalent is inherent in the southwest (Brest, Lelchitsy) and northwest (Sharkovshchina, Senno) of the territory of Belarus. The averaged values of the identified maximum values characterize the most typical pattern of forming snow water equivalent in the study area (Fig. 1b). In fig. 1b, there is a similar confinement of the corresponding average maximum values of snow water equivalent (north-western, north-eastern and central parts of Belarus). However, Fig. 1b shows smoother isolines and an increase in snow reserves in the southwest - northeast direction.

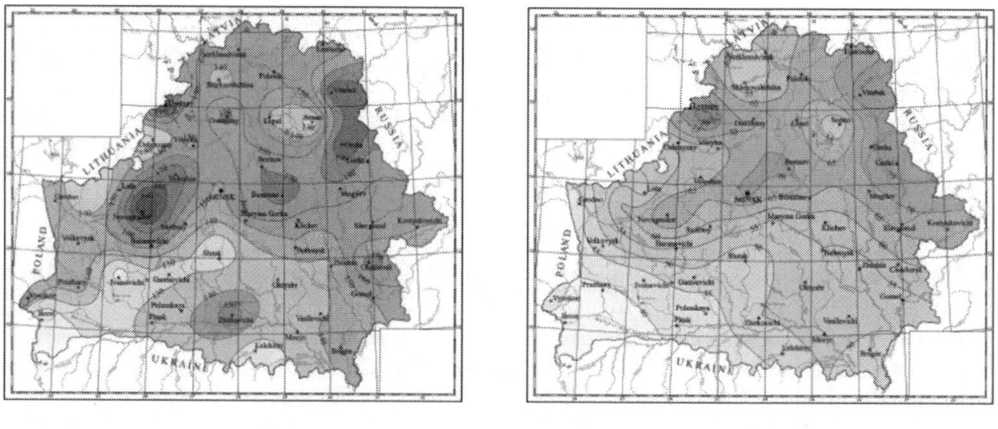

a) b)

Fig. 1. Distribution of maximum snow water equivalent in Belarus, mm: a - maximum, b - average maximum.

It is known that the water equivalent in snow as the main factor of spring floods and a determiner for snow loads imposed on buildings and structures is predetermined by the height of snow cover and its density. Moreover, these parameters are used in conjunction. There are different ratios of density and height of snow in nature, most often there is a shift of the peaks in time by 1-2 decades. The period with the maximum depth of snow cover begins earlier, and then, when the snow melts (in spring and during thaws in the cold period), its thickness decreases but its density increases. The highest snow water equivalent is observed when the values of snow cover thickness and its density are maximum. In this regard, it is necessary to analyse the maximum heights of snow cover and maximum densities.

Fig. 2a shows the maximum snow depth (cm) obtained from snow survey data in the field for the representative period. Fig. 2b characterizes the averaged maximum snow depth for the observation period of 1944-45 - 2019-20. The information presented in fig. 2a and 2b, corresponds to the previously made conclusions about the mechanisms of forming maximum water equivalent in snow. The maximum heights of the snow cover are characteristic of the highlands, their leeward slopes, in particular. They are determined by the global moisture transfer. However, there is also a random component here.

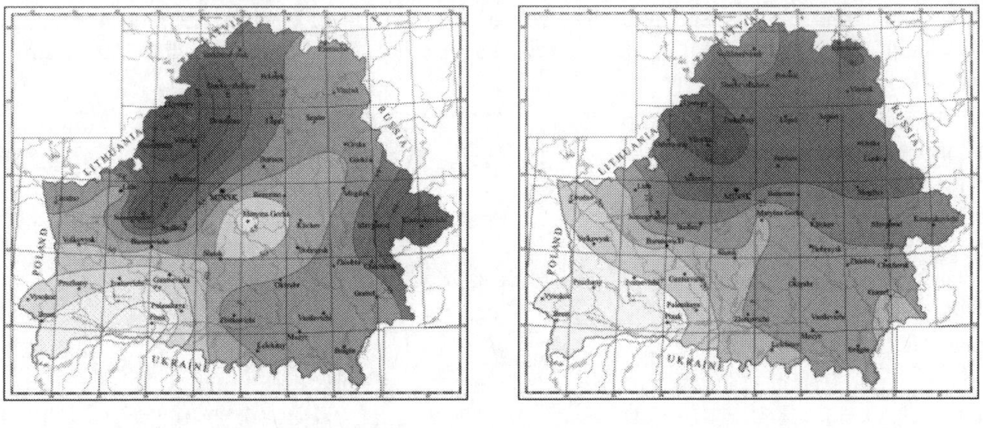

a) b)

Fig. 2. Distribution of snow cover depth in Belarus, cm: a - maximum, b - average maximum.

At the same time, the averaged maximum values of snow cover heights have a vivid latitudinal orientation on the territory of Belarus. They correlate with the temperature and wind regime of this area which depends on the radiation characteristics of Belarus' climate and turbulent heat transfer of the surface atmosphere, which affect the processes of snow melting and evaporation in the southern regions of Belarus. Southern winds prevail on the territory of Belarus in winter.

Forecasting the characteristics of the snow cover can be carried out by identifying trend components, assessing their space-time variability [19–22, etc.]. Snow water equivalent is the main contributor to spring flooding on Belarus' rivers and forming snow loads on structures and buildings. Therefore, its prediction plays an important role. To assess changes in the characteristics of snow cover, we used linear trends for 48 meteorological stations for the entire observation period, starting from 1944-45. The results for the regional centers of Belarus are shown in Table 1, where t is the ordinal number of the year.

As the result of analytical assessment of the curves of moving five-year averages of the maximum and average values of snow water equivalent, snow height, and snow density at 48 meteorological stations in Belarus, we determined regions with positive and negative trends.

Over most of Belarus, there is a tendency for snow water equivalent to decrease by 4-8 mm in 10 years. An increase in snow water equivalent is typical for the catchments of the Western Bug, Pripyat, Berezina, and Dnieper rivers (Fig. 3).

Table 1. Linear trends in snow cover characteristics.

Meteorological station	Maximum snow water equivalent, mm	Snow height, cm
Minsk	$Q = -0{,}4t + 87{,}4$	$h = -0{,}026t + 29{,}88$

Grodno	Q = –0,103t + 47,06	h = –0,020t + 19,06
Mogilev	Q = –0,261t + 77,97	h = 0,005t + 26,89
Brest	Q = 0,072t + 33,5	h = 0,047t + 13,54
Gomel	Q = 0,113t + 48,01	h = 0,110t + 17,57
Vitebsk	Q = 0,136t + 74,32	h = 0,082t + 25,89

The map in Fig. 3 shows that the area in the southwest with a positive trend includes Brest, Vysokoe, Pinsk, Volkovysk. The exception is Pruzhany, where the trend is practically unchanged throughout the study period. In the south-east of Belarus, the borders of the area where water equivalent in snow is also increasing include Bragin, Mozyr, Vasilevichi, Gomel, Zhlobin, Bobruisk, Klichev, Berezino. In the northeast, Gorki and Vitebsk also have a positive trend. The rest of the territory of Belarus has a negative trend, i.e. snow water equivalent is decreasing. The exception is Stolbtsy where the trend is positive.

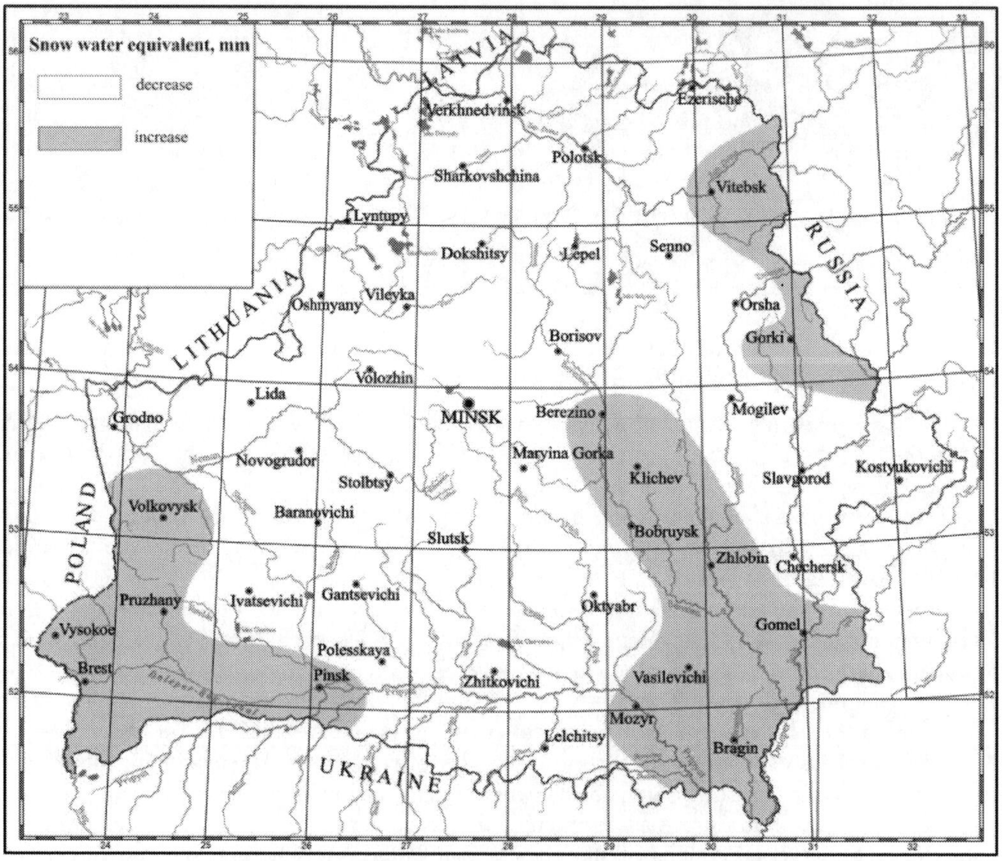

Fig.3. Transformation of maximum values of snow water equivalent in Belarus.

Fig. 4 shows the map of zoning the territory of the Republic of Belarus according to the trends in maximum values of snow cover height.

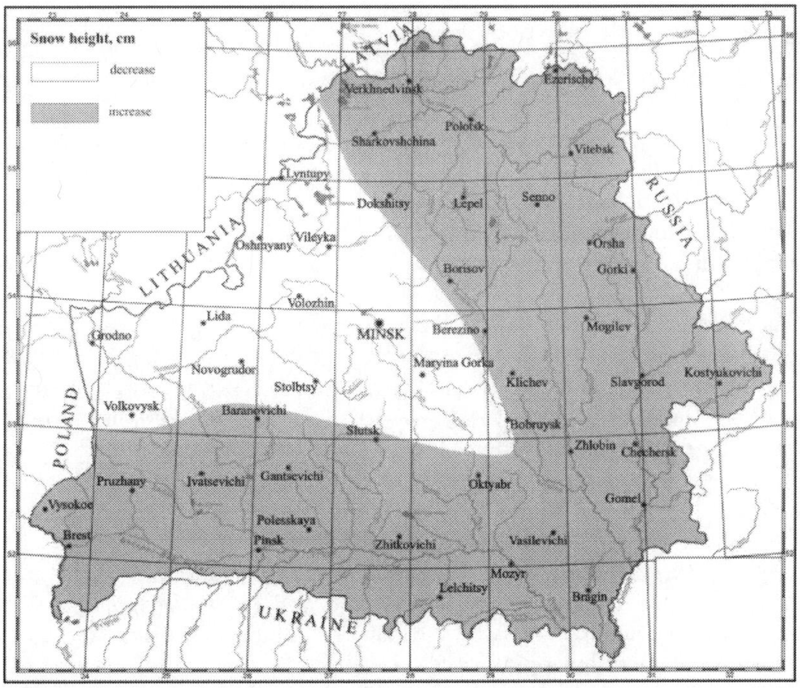

Fig.4. Transformation of maximum values of snow cover height in Belarus.

The map in Fig. 4 shows that snow cover height decreases (negative trend) in the central (Minsk, Maryina Gorka) and north-western part of Belarus (Grodno, Lida, Novogrudok, Lyntupy). However, over most of Belarus (60%), there is a slight increase in the depth of the snow cover. The density of snow cover decreases slightly throughout the study area, which is associated with current climatic changes. Thus, an important task is to determine causes of the current changes, where the leading role belongs to the planetary ones associated with the global climate warming. In this regard, it makes sense to do a joint analysis of snow accumulation with the temperature regime of the area.

4 Conclusion

Mapping the characteristics of snow cover allows us to forecast changes in snow water equivalent, as the main contributor to spring floods in Belarus. To create the maps we used data from meteorological stations together with the results of remote sensing of the earth. The constructed maps with the data about snow water equivalent and snow height at the beginning of flooding make it possible to determine areas in Belarus which are prone to spring flooding. The set of maps presented above is based on the use of representative, appropriately prepared, meteorological observation data analysed and studied in terms of the identified independent structures of space-time fields. However, the main function was assigned to the factor of synchronicity of solid atmospheric precipitation in the cold period. The revealed patterns of spatial distribution of maximum and average maximum density and height of snow cover, the outline of the isolines and the peculiarities of their spatial confinement were used to map the maximum and average maximum snow water equivalent. The proposed set of interconnected contour maps should serve as a basis for zoning the territory of Belarus according to the changes in snow reserves. The maps can also be used to assign snow load limits in the design of various construction projects.

References

1. Diro, G. T., & Sushama, L. (2020). Contribution of snow cover decline to projected warming over North America. *Geophysical Research Letters*. 47. https://doi.org/10.1029/-2019GL084414.
2. Callaghan, Terry & Johansson, et al. (2011). Multiple Effects of Changes in Arctic Snow Cover. *AMBIO A Journal of the Human Environment*. 40. 32-45. https://doi.org/-10.1007/s13280-011-0213-x.
3. Judah L Cohen et al. (2012). Arctic warming, increasing snow cover and widespread boreal winter cooling *Environ. Res. Lett*. 7. https://doi.org/10.1088/1748-9326/7/1/-014007.
4. Jonas Bhend, Janice Bathols and Kevin Hennessy (2012). Climate change impacts on snow in Victoria *The Centre for Australian Weather and Climate Research*. 42.
5. Wang, A., Xu, L., and Kong, X. (2018) Assessments of the Northern Hemisphere snow cover response to 1.5 and 2.0 °C warming. *Earth Syst. Dynam*. 9. 865–877. https://doi.org/10.5194/esd-9-865-2018.
6. Cordero, R.R., Asencio, V., Feron, S. et al. (2019). Dry-Season Snow Cover Losses in the Andes (18°–40°S) driven by Changes in Large-Scale Climate Modes. *Sci Rep*. 9. 16945. https://doi.org/10.1038/s41598-019-53486-7.
7. Callaghan, Terry & Johansson, Margareta & R.D., et al. (2011). Changing snow cover and its impacts. In book: *Snow, Water, Ice and Permafrost in the Arctic (SWIPA)*: Climate Change and the Cryosphere, Chapter: 4, Publisher: Oslo: Arctic Monitoring and Assessment Programme, pp.4 1–4 58.
8. Goodison, B. E. and Walker, A. E. (1993) Use of snow cover derived from satellite passive microwave data as an indicator of climate change. *Annals of Glaciology*. 17. 137–142.
9. Colbeck, S.C. (1987) History of snow-cover research. *Journal of Glaciology*. Special Issue. 60–65.
10. Meshyk, A., Barushka, M, Marozava, V. (2020) Snow as a Contributor to Spring Flooding in Belarus. *Environmental Science and Pollution Research*. 1–11. https://doi.org/10.1007/s11356-020-09638-8.
11. Валуев, В. Е. Изученность и статистические оценки снегозапасов (State of knowledge and statistical assessment of snow water equivalent) / V. E. Valuyev, A. P. Meshyk // *Vestnik Brestskogo Gosudarstvennoro Tehnicheskoro Universiteta: Vodohozyaistvennoye Stroitelstvo, Teploenergetika i Geoekologiya*. – 2013. – № 2. – P. 8–11. (In Russian).
12. Henkel, P., Koch, F., Appel, F., Bach, H., Prasch, M., Schmid, L., et al. (2018). Snow water equivalent of dry snow derived from GNSS carrier phases. *IEEE Trans. Geosci. Remote Sens*. 56, 3561–3572. doi: 10.1109/TGRS.2018.2802494.
13. Appel, F., Koch, F., Rösel, A., Klug, P., Henkel, P., Lamm, M., et al. (2019). Advances in snow hydrology using a combined approach of GNSS in situ stations, hydrological modelling and earth observation—a case study in Canada. *Geosciences* 9:44. doi: 10.3390/geosciences9010044.
14. Dai, L., Che, T., Ding, Y., and Hao, X. (2017) Evaluation of snow cover and snow depth on the Qinghai–Tibetan Plateau derived from passive microwave remote sensing. *The Cryosphere*. 11. 1933–1948. https://doi.org/10.5194/tc-11-1933-2017.
15. Volchak, A., et al. (2017). Atlas: Weather Hazards in Belarus. Moscow. All-Russian *Research Institute for Hydraulic Engineering and Land Reclamation*. 70.
16. Tur, V. et al. (2008) Normalization snow loads for the territory of the Republic of Belarus. *Construction Science and Technology*. 2, 27–45.
17. Volchak, A. A., Meshyk, A. P., Sheshka, M. M. et al. (2016). Floods on the territory of Polesie. *Procedia Engineering*. 162, 91–97. https://doi.org/10.1016/j.proeng.2016.11.020

18. Тур, В. В. Картографирование основных характеристик снегового покрова по результатам комплексной статистической обработки данных метеорологических наблюдений (Mapping of main snow-cover characteristics obtained in the result of an integral statistical analysis of the data of meteorological observation) / V. V. Tur, V. E. Valuyev, S. S. Derechennik, A. P. Meshyk // *Vestnik Brestskogo Gosudarstvennoro Tehnicheskoro Universiteta: Vodohozyaistvennoye Stroitelstvo, Teploenergetika i Geoekologiya*. – 2008. – № 2. – P. 2–10. (In Russian).
19. Мешик, О. П. Особенности оценки запасов воды в снеге и их пространственно-временной изменчивости на территории Беларуси (Peculiarities of assessing water reserves in snow and their spatio-temporal variability in the territory of Belarus) / A. P. Meshyk, V. A. Marozava // *Actual problems of earth sciences: studies of transboundary regions: collection of articles and materials IV Int. scientific*. - practical. conf., to the 1000th anniversary of Brest, 12-14 Sept. 2019 / Brest; ed. A. K. Karabanova [and others]. - Brest: BrSU, 2019. - Part 2 - P. 34–37. (In Russian).
20. Robert A. Metcalfe, James English and James J. Luce (2018) Variability and trends in seasonal snow cover in Ontario from 1980 to 2010 detected using remote sensing. *Ontario Ministry of Natural Resources and Forestry. Science and Research Branch*. Peterborough, ON. Climate Change Research Report CCRR-50. 27 p.
21. Don M. Gray et al. (1978) Snow accumulation and distribution. *Modeling Snow Cover Runof.* 31.
22. Stine Højlund Pedersen, Mikkel P. Tamstorf, et al. (2016) Spatiotemporal Characteristics of Seasonal Snow Cover in Northeast Greenland from in Situ Observations, *Arctic, Antarctic, and Alpine Research*. 48:4. 653-671. https://doi.org/-10.1657/AAAR0016-028.

MATHEMATICAL EXPRESSION OF THE FORM OF A HELIO CONCENTRATOR

Vitali Severyanin[1,*], *Pavel* Yanchilin[1]

[1]Brest State Technical University, Faculty of Engineering systems and Ecology, Maskouskaya str. 267, 224017, Brest, Republic of Belarus

Abstract. It is necessary to increase the efficiency of existing solar power plants both through the study of energy processes and the creation of new schematic diagrams of solar systems by reducing the cost of their design, principle of operation and operation. Authors have developed the LUCH solar plant in solving these problems. The article presents the device of the installation, a mathematical description of the shape of the solar concentrator, calculations of the parameters of the cones and their layout diagram. When developing the concentrator, the condition was accepted — the adopted form of the paraboloid as the basis of the concentrator needs to be changed so as to simplify the design and manufacturing process. The design change was made by splitting the paraboloid of revolution into separate cones. The calculated geometrical parameters of the LUCH solar concentrator are compared with the obtained parameters for an ideal paraboloid concentrator.
Keywords: solar power plants, solar radiation concentrators, solar radiation flux intensity, paraboloid, mathematical description, splitting of a paraboloid of rotation.

Introduction

The growing demand for energy — electricity and heat — forces, despite many years of large-scale research, development and application, to look for further ways to develop solar plants. To increase the efficiency of existing solar power plants complex designs and expensive materials are used which increases the payback period of solar power plants. A serious obstacle to the effective implementation of the high energy potential of solar radiation is its low density because of the great distance between the Earth and the Sun. This contradiction can only be overcome by concentrating radiation.

The long history of the development of energy-saving equipment in Europe has allowed accumulating significant experience in this area. It is possible find in the literature a lot of data on the successful implementation of projects of energy efficient systems using renewable energy sources in particular solar using concentrators [1-5].

Authors of the article have developed a solar plant "LUCH" in the research laboratory "PULSAR" of the Brest State Technical University. It refers to municipal industrial heat power engineering and can be used for heating and hot water supply systems for small con-

*Corresponding author: tgv_bstu@tut.by

https://doi.org/10.1051/e3sconf/202021201014

sumers (cottages, greenhouses, workshop premises, warehouses, various technological needs in agriculture) as a backup for fuel-using systems. The purpose of the solar plant is to capture and concentrate the sun's rays in focus on a spherical heat receiver, transfer the heat generated in the focus to the coolant and collect the heated coolant in the storage tank for subsequent consumption.

1 METHOD

1.1 Device Description

The solar plant design consists of five main parts (Figure 1):

1. Optical system (a complex of cones-mirrors on a special frame — a solar concentrator);
2. Sun tracking system (copiers, optical system turning mechanism, electric drive, control system);
3. Mechanism for lifting the optical system (levers and rods);
4. Heat receiver (spherical solar water heater, pipe-wire system, storage tank, regulators and valves);
5. The solar plant is mounted on a metal base (plate, channels, bearings).

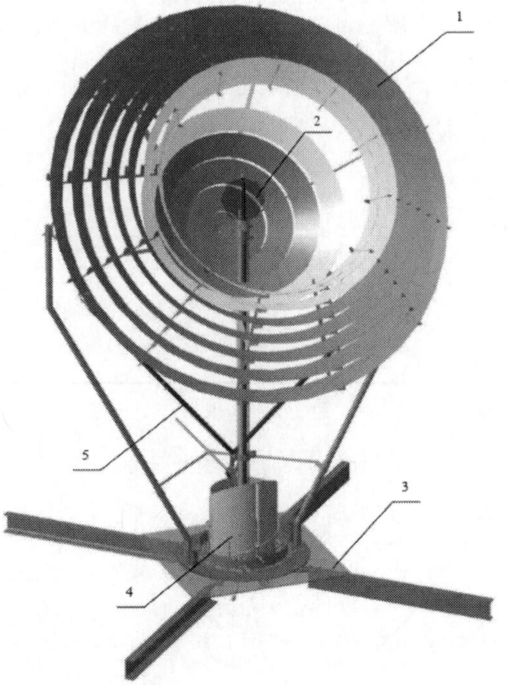

1 – optical system; 2 – heat receiver; 3 – base; 4 – solar tracking system;
5 – lifting mechanism of the optical system.
Figure 1 – Model of a solar plant

The main features (differences from known installations) of the LUCH solar plant are:

– "Splitting" of the continuous rotation of the paraboloid into separate concentric cones and construction from them of a new, more technological, solar concentrator;

- the original design of the cone solar energy concentrator allows its use in the spring-autumn and winter periods;
- the use of this concentrator of solar energy can significantly increase the utilization rate of solar energy, as a result of which the temperature of the coolant can be increased to 300°C;
- mirror part of the solar concentrator is a group of narrow, separated from each other concentric cones which reduces the wind load because the air flow passes through the distance between the cones and simplifies the manufacture and assembly of mirrors since the surfaces of the cones have I degree of curvature;
- the original design of the heat receiver allows you to effectively use concentrated solar energy and the presence of an air gap between the surface of the heat receiver and the transparent shell eliminates heat losses in a convective way;
- the heat receiver is stationary, the optical system rotates around it, this simplifies the design of the coolant communications;
- orientation of the solar concentrator axis to the Sun is carried out by a special tracking mechanism;
- movement of the optical system takes into account not only daily but also seasonal changes in the position of the Sun.

1.2 Research progress

When developing the concentrator, the condition was accepted that we take the form of a paraboloid as the basis of the concentrator and it is required to change it so as to simplify the design, manufacturing and operation process. In the course of the geometric transformations shown in Figure 2, we get a new type of concentrator, shown in Figure 3.

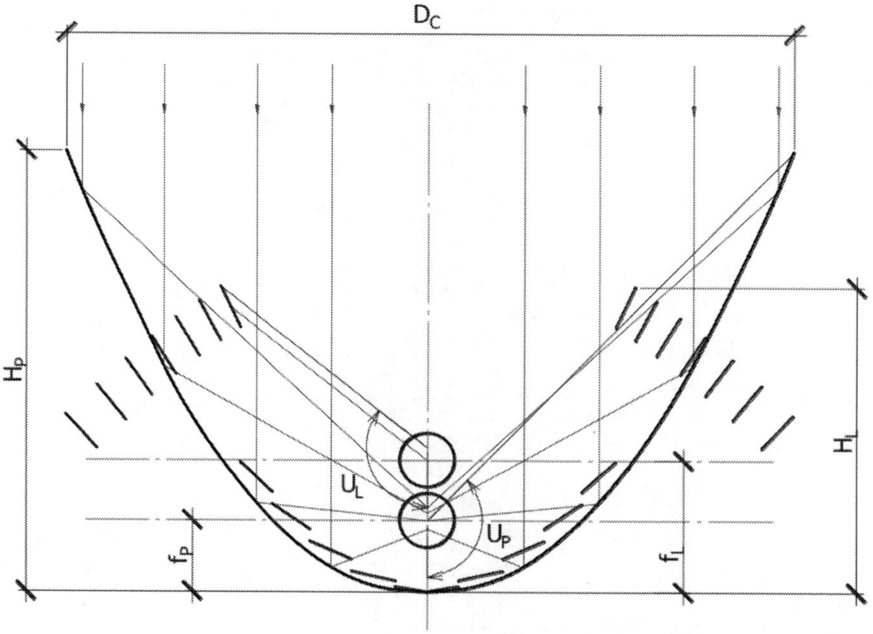

Figure 2 – Combination of two concentrators and their main dimensions

Using a graphical-analytical method based on geometric constructions we will display rays falling and reflected by a mirror. In this case, we use the law of specular reflection and we will consider a parallel beam of rays, the distribution of radiation in the beam along the directions is assumed to be uniform. We will simplify the construction by splitting the paraboloid of revolution into separate cones. We get that the concentrator is a collection of concentric cones. Each cone is a tape of sheet material bent in the form of a truncated cone, the inner surface of which is mirrored. The angle of the generatrix of the cone to its axis is chosen so that the sunbeam after reflection is directed into focus. All cones have the same focus.

The optical system or concentrator of the LUCH solar plant consists of three main elements: radii, cones, and fastening rings. Their design dimensions are determined by geometric constructions. The cones are spaced relative to each other by radii, thus forming a blown-through robust spatial structure with less windage. Radii are flat elements that have specially cut shapes for attaching the cones at the required predetermined angle, which is a prerequisite for the best reception of the flow of sunlight. The radii are fixed on support rings that are attached to the posts.

Concentrators in which the generatrix of the reflecting surface is a curve of the second order — a circle, a parabola, a hyperbola, etc. belong to the group of highly concentrated systems. An ideal paraboloid concentrator focuses a parallel beam of rays into a point, which corresponds to an infinitely high degree of concentration ($K_c \to \infty$). The geometry of a paraboloid reflector is characterized by two independent parameters: the concentrator diameter D_c (or focal length f_c) and the half-opening angle U_c (or the D_c / f_c ratio) [6]. The third parameter is always dependent and is determined from the equation:

$$D_C = {4 \cdot f_C \cdot \sin U_C}/{1 + \cos U_C} \quad (1)$$

These parameters are shown in Figure 2 and differ only in indices: ***P*** – for the paraboloid concentrator, ***L*** – for the "LUCH" concentrator.

These geometric transformations of the paraboloid concentrator were carried out in order to simplify the design for the manufacturing process and operation. Therefore, it is necessary to compare the geometric characteristics of these two concentrators. An important parameter will be the ratio of the area of the reflecting surface of the concentrator (mirror) S_{sur} to the area of the captured solar flux S_{sol} :

$$K_x = {S_{sur}}/{S_{sol}} \quad (2)$$

We will use the well-known parabola formulas to calculate the main parameters of the paraboloid concentrator and the area of the reflecting surface of the concentrator (mirror) S_{sur}. The canonical equation of a parabola in a rectangular coordinate system is:

$$y^2 = 2px, p > 0 \quad (3)$$

where p is the focal parameter of the parabola, it is equal to the distance from the focus to the directrix.

Since each point of the parabola is equidistant from the focus and the directrix, so is the vertex, so it lies between the focus and the directrix at a distance $p / 2$ from both. Hence the focal length f_c, m:

$$f_K = {p}/{2} \quad (4)$$

The height of the parabola h is found from the ratio:

$$h = {R^2}/{2p} \quad (5)$$

where R is the radius of the circle described by the branches of the parabola when rotating around the Y-axis (paraboloid), in the plane perpendicular to this axis at a distance h from the apex of the parabola, m.

The surface area of the paraboloid or the area of the reflecting surface of the concentrator (mirror) S_{sur}. can be found from the equation:

$$S_{sur} = {2\pi \cdot \left(\sqrt{(p^2 + R^2)^3} - p^3\right)}/{3p} \quad (6)$$

Let us pose the problem of finding such a parameter of a parabola (and, accordingly, a paraboloid) p at which its surface area S_{sur} tends to a minimum. To do this, you need to take the derivative of equation (6) and determine the unconditional maximum of the function, equating it to zero.

$$S'_{sur}(p) = \frac{2\pi \cdot \left[3p\sqrt{(p^2+R^2)^3} - 3p^2 - \frac{\left(\sqrt{(p^2+R^2)^3} - p^3\right)}{p}\right]}{3p} =$$

$$= \cdots = \frac{2\pi \cdot \left[\sqrt{p^2+R^2}\left(2p - \frac{R^2}{p}\right) - 2p^2\right]}{3p} = 0 \qquad (7)$$

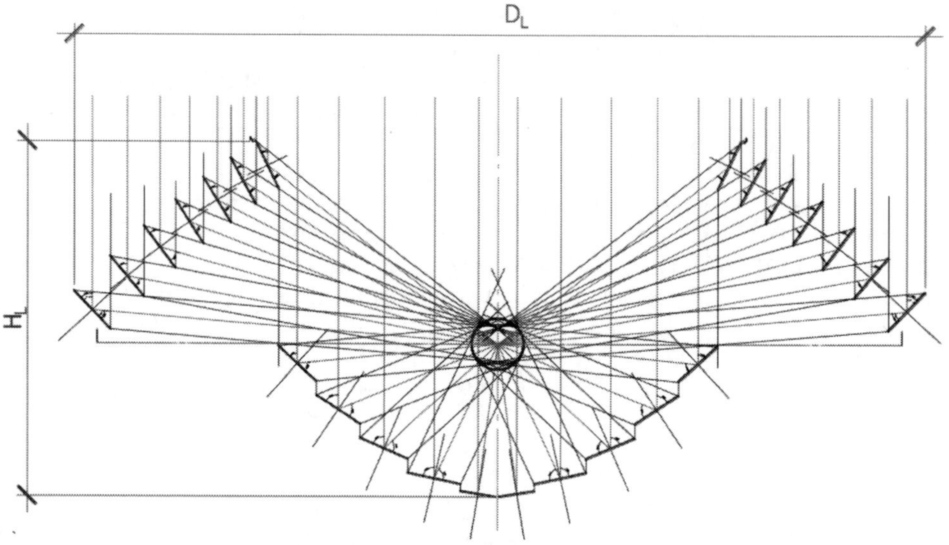

Figure 3 – Concentrator in the form of a split paraboloid

The root of this equation turns out to be $R = 0$. Therefore, it is necessary to look for a conditional (local) extremum, for this we accept the constraints established experimentally. As the radius R and the parameter p are related by the relation (5), then we will set the range of heights from 0.1 to 2 meters with a step of 0.01 m. This gap is due to the fact that the calculated ideal paraboloid should correspond as much as possible to the geometric dimensions of our composite concentrator and the height h_L of the calculated "LUCH" solar concentrator is 1.016 m. We set the value of the paraboloid radius $R_P = 1.2$ m which corresponds to the radius of the solar concentrator "LUCH" and, accordingly, the area of the captured solar flux $S_{sol} = 4.52$ m².

2 RESULTS AND DISCUSSION

Calculation sequence:

- find the area S_{sur} from formula (7),
- parameter p from (5),
- focal length f from (4).

Table 1 shows some of the data obtained.

Table 1. Results of calculating the paraboloid

S_{sur}, м²	h_P, м	p	f, м	K_x

4,56	0,1	7,2	3,6	1,01
4,59	0,15	4,8	2,4	1,02
4,63	0,2	3,6	1,8	1,03
4,71	0,25	2,88	1,44	1,04
4,79	0,3	2,4	1,2	1,06
4,89	0,35	2,06	1,03	1,08
4,99	0,4	1,8	0,9	1,10
5,11	0,45	1,6	0,8	1,13
5,24	0,5	1,44	0,72	1,16
5,37	0,55	1,31	0,65	1,19
5,51	0,6	1,2	0,6	1,22
5,66	0,65	1,11	0,55	1,25
5,82	0,7	1,03	0,51	1,29
5,99	0,75	0,96	0,48	1,32
6,16	0,8	0,9	0,45	1,36
6,19	0,81	0,89	0,44	1,37
6,23	0,82	0,88	0,44 = f_L	1,38
6,26	0,83	0,87	0,43	1,38
6,29	0,84	0,86	0,43	1,39
6,33	0,85	0,85	0,42	1,40
6,51	0,9	0,8	0,4	1,44
6,69	0,95	0,76	0,37	1,48
6,89	1	0,72	0,36	1,52
6,92	1,01	0,71	0,35	1,53
6,96	1,02 = h_L	0,71	0,35	1,54
7,00	1,03	0,70	0,34	1,55
7,04	1,04	0,69	0,34	1,56
7,08	1,05	0,68	0,34	1,57
7,27	1,1	0,66	0,33	1,61
7,31	1,11	0,65	0,32	1,62
7,35 = S_L	1,12	0,64	0,32	1,63 = K_L
7,39	1,13	0,64	0,32	1,63
7,43	1,14	0,634	0,31	1,64
7,47	1,15	0,624	0,31	1,65
7,68	1,2	0,6	0,3	1,70
7,88	1,25	0,58	0,29	1,74
8,09	1,3	0,55	0,28	1,79
8,51	1,4	0,51	0,26	1,88
8,94	1,5	0,48	0,24	1,98
9,37	1,6	0,45	0,23	2,07
9,81	1,7	0,42	0,21	2,17
10,26	1,8	0,4	0,2	2,27
10,71	1,9	0,38	0,19	2,37
11,17	2	0,36	0,18	2,45

Analyzing the data obtained, it can be seen that the minimum surface area is achieved at the minimum height of the paraboloid, which can be seen from the graph in Figure 4. With an increase in the height h, the surface area increases and the focal distance decreases which can be seen from the graph in Figure 5. It is obvious that with an increase the surface area increases and the coefficient K_x (graph in Figure 6).

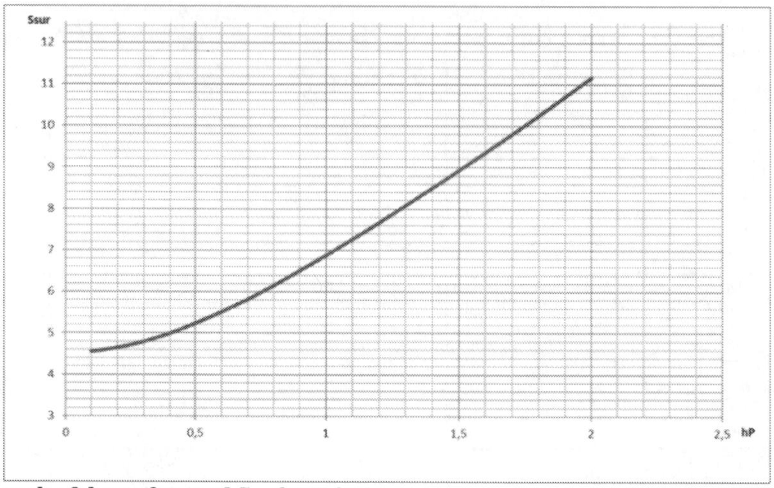

Figure 4 – Graph of dependence of S_{sur} from h_p

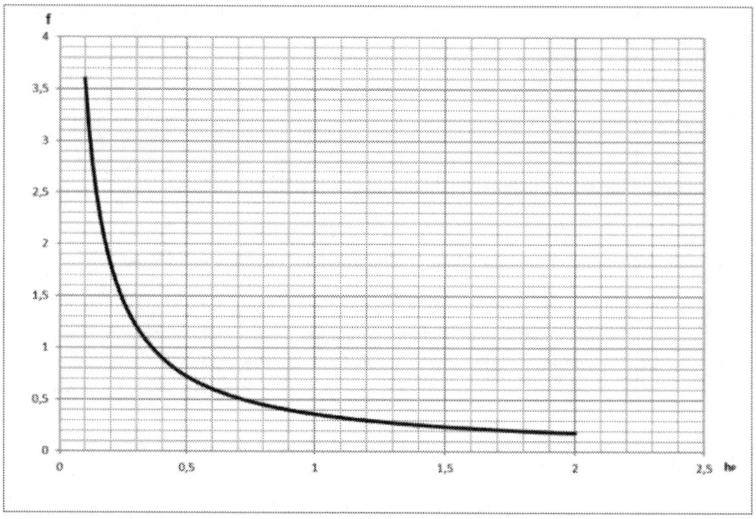

Figure 5 – Graph of f from h_p

As the radius of the paraboloid and the "LUCH" solar concentrator are equal, then the area of the captured solar flux S_{sol} is the same for these two concentrators. This means that the heat power Q received, reflected, concentrated and "utilized" (transferred to the heat carrier in the heat receiver) is also the same for them. From an economic point of view, it is unprofitable to produce a concentrator with larger geometrical dimensions, if its thermal power corresponds to a concentrator with smaller dimensions. Therefore, it is necessary to take into account the relationship between the geometric parameters of the concentrator (R, h, f) for a given thermal power of the solar plant.

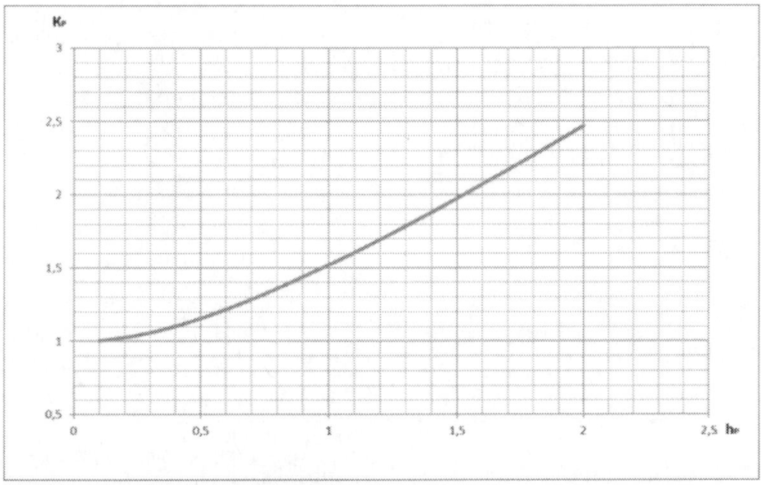

Figure 6 – Graph of dependence of *Kp* on *hp*

Let us compare the geometric parameters of the LUCH solar concentrator calculated earlier by the graphical analytical method with the parameters obtained for the paraboloid concentrator (Table 1, highlighted lines).

Calculated geometrical parameters of the LUCH solar concentrator (from table 2):

- Total height $h = 1.02$ m;
- Maximum diameter $D = 2.4$ m, radius $R = 1.2$ m;
- Focal length $f = 0.44$ m;
- Surface area of mirrors $S_{sur} = 7.36$ m^2;
- Coefficient $K_x = 1.63$.

Calculated geometrical parameters of the paraboloid concentrator at the same height h:

- Total height $h = 1.02$ m;
- Maximum diameter $D = 2.4$ m, radius $R = 1.2$ m;
- Focal length f = 0.35 m;
- Surface area of the mirror $S_{sur} = 6.96$ m^2;
- Coefficient $K_x = 1.54$.

Calculated geometrical parameters of the paraboloid concentrator at the same focus f:

- Total height $h = 0.82$ m;
- Maximum diameter $D = 2.4$ m, radius $R = 1.2$ m;
- Focal length $f = 0.44$ m;
- Surface area of the mirror $S_{sur} = 6.23$ m^2;
- Coefficient $K_x = 1.38$.

Calculated geometric parameters of the paraboloid concentrator for the same mirror surface area S_{sur} and the same K_x coefficient:

- Total height $h = 1.12$ m;
- Maximum diameter $D = 2.4$ m, radius $R = 1.2$ m;
- Focal length $f = 0.32$ m;
- Surface area of the mirror $S_{sur} = 7.35$ m^2;
- Coefficient $K_x = 1.63$.

The cone of the optical system is shown externally in Figure 7. Its main parameters are the diameter of the larger arc D_{max} and the diameter of the smaller arc D_{min}. We need a reamer to make a cone. The main parameters of the sweep of the cone: the outer radius or the radius of the sweep R_R, the length of the outer arc L, the sweep angle α and the width of the plate a.

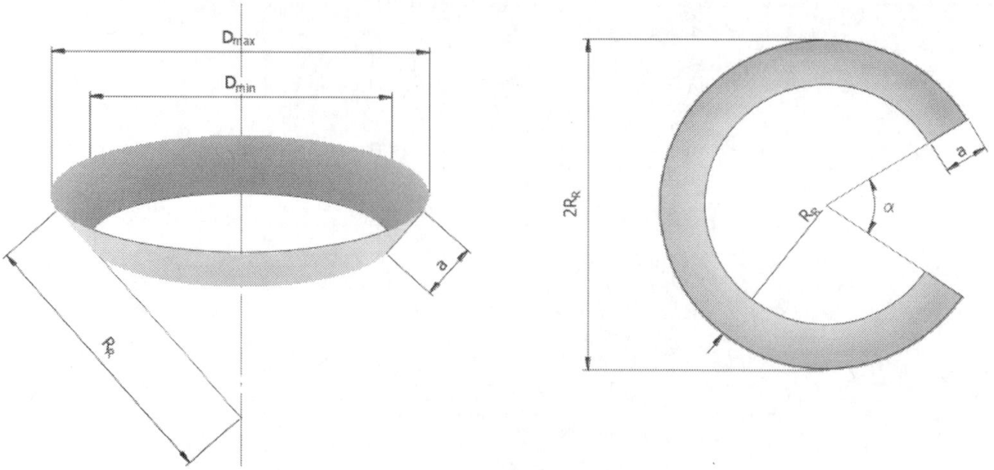

Figure 7 – Dimensions of the reflector cone of the optical system

Table 2. Calculation of the cones of the LUCH solar concentrator

N cones	Larger dia-meter, mm	Smal-ler diame-ter, mm	Width, mm	Sweep radius, mm	Sweep arc, mm	Sweep angle,	$\beta=360-\alpha$	Surface area, mm²	Surface area, m2
	D_{max}	D_{min}	a	R_R	L	α		S_{sur}	S_{sur}
«Верхние» конуса									
1	2408	2204	150	1771	7565	115,2	244,8	1086676,9	1,09
2	2204	2012	150	1722	6924	129,6	230,4	993371,6	0,99
3	2012	1836	150	1715	6321	148,8	211,2	906663,6	0,91
4	1836	1672	150	1679	5768	163,2	196,8	826553,0	0,83
5	1672	1528	150	1742	5253	187,2	172,8	753982,2	0,75
6	1528	1384	150	1592	4800	187,2	172,8	686123,8	0,69
7	1384	1252	150	1573	4348	201,6	158,4	621092,9	0,62
«Нижние» конуса									
8	1260	1028	150	815	3958	81,6	278,4	539097,3	0,54
9	1028	780	150	622	3230	62,4	297,6	426000,0	0,43
10	792	516	150	430	2488	28,8	331,2	308190,2	0,31
11	516	220	150	261	1621	4,8	355,2	173415,9	0,17
12	220	0	110	110	691	0,0	360,0	38013,3	0,04
							∑	7359180,8	7,36

Having a certain number of cones and their main dimensions D_{max} and D_{min}, one can calculate the dimensions of their sweeps. Sweep radius R_R is determined by:

$$R_R = a\,{}^{D_{max}}/_{D_{max} - D_{min}} \tag{8}$$

The length of the outer arc L is calculated by:

$$L = \pi \cdot D_{max} \tag{9}$$

To find the sweep angle α, consider the ratio: the length of the outer circle $2\pi R$ refers to the angle $360°$ since the length of the outer arc L refers to the angle $360°–\alpha$.

$$360°/_{360° - \alpha} = {2\pi R_R}/_{L} \tag{10}$$

After transformation we get:

$$360°/_{360° - \alpha} = {2R_R}/_{D_{max}} \tag{11}$$

The sweep angle α is:

$$\alpha = 360° \cdot (1 - {D_{max}}/_{2R_R}) \tag{12}$$

Using the above expressions and dependences, the optical system of a solar concentrator with a diameter of 2.4 meters was calculated [6]. The results are shown in Table 2.

Possibly the best relationship between the geometric parameters of the concentrator for a given thermal power of the solar plant takes into account the coefficient K_x (the ratio of the area of the reflective surface of the concentrator S_{sur} to the area of the captured solar flux S_{sol}). It has been experimentally established that the coefficient K_x should be in the range from 1.3 to 1.7.

The aperture area for the LUCH solar plant is 4.55 m^2 and the total surface area of all the cones or the reflection area is 7.36 m^2. The ratio of the mirror area to the aperture area is 1.62. A high concentration accuracy on a relatively small surface of the receiver is not required as the considered concentrator in the form of cones is intended for operation in "household" solar systems (low power) used for hot water supply and heating systems. The higher the concentration factor, the higher the temperature on the surface of the receiver and for these areas of application high temperatures (above 100°C) are not used. Therefore, the size of the generatrix of the solar concentrator cones can be assumed to be "relatively large", within 10-20 cm. High accuracy of the system for tracking the trajectory of the Sun is also not needed and it allows simplifying the design.

CONCLUSION

The following conclusions can be drawn analyzing the given geometric parameters of two concentrators:

1. The set goal of creating a new design of the concentrator has been achieved — to transform the paraboloid concentrator geometrically so as to simplify its design for manufacturing and operation.
2. The developed design of the LUCH solar concentrator is close to an ideal paraboloidal one, for example, the maximum deviation of the concentrator height is less than 0.1 m with a diameter of 2.4 m, with equal geometrical and optical-energy parameters.
3. It is possible to estimate the degree of "illumination" of the surface of the heat receiver, its uniformity, placed at this point by the magnitude of the focal length. The larger the focus, the smaller the surface of the heat receiver perceives concentrated radiation. Conversely, the shorter the focal length, the more part of the heat sink surface is illuminated.
4. The closest coincidence of the geometrical and optical-energy parameters of the two concentrators is obtained with the same surface area of the mirror S_{sur} and the coefficient K_x which allows us to write the mathematical expression of the "LUCH" concentrator in the form of the canonical parabola equation:
$$y^2 = 2px = 2 \cdot 0{,}6428x = 1{,}2857x \tag{13}$$
5. A clear advantage of the developed design is the improvement of the aerodynamic properties, namely, the reduction of the wind load on the building structures (high airflow of the concentrator structure). This makes it possible to produce a concentrator with large cross-sectional dimensions and to increase the heat output of one solar plant.

Bibliography

1. L. Li, B. Wang, J. Pottas, W. Lipiński 2019 *Solar Energy* **183:** 805-811.
2. S.S. Rao, H.G. Lee, Y. Hu 2014 *Journal of Mechanical Design, Transactions of the ASME* **136(9)** [091402].
3. M. Tian, Y. Su, H. Zheng, G. Pei, G. Li, S. Riffat 2018 *Renewable and Sustainable Energy Reviews* **82:** 1272-1296.
4. S.S.S.Baljit, H.-Y.Chan, V.A. Audwinto, S.A. Hamid, A. Fudholi, S.H. Zaidi, M.Y. Othman, K. Sopiana 2017 *Renewable Energy* **114:** Part B, 1258-1271.
5. M.R. Rodríguez-Sánchez, C. Marugán-Cruz, A. Acosta-Iborra, D. Santana 2018 *Solar Energy* **165:** 43-54.
6. В.М. Андреев, В.А. Грилпхес, В.Д. Румянцев. Фотоэлектрическое преобразование концентрированного солнечного излучения. – Л.: Наука, 1989. – 310 с.
7. *V.* Severyanin, P. Yanchilin. Bulletin of the Brest State Technical University: Water construction, heat power engineering, geoecology, **2**: 74–77. (2010) (in Russian).

ENVIRONMENTAL RISKS OF THE PRIPYAT RIVER IN A CHANGING CLIMATE

Aliaksandr Volchak [1], *Aleh* Meshyk [1,*], *Yury* Mazhayskiy [2], *Anatoliy* Rokochynskiy [3] and *Jerzy* Jeznach [4]

[1]Brest State Technical University, Moskovskaya str. 267, Brest, 224017, Belarus
[2]VNIIGiM of A.N. Kostiakov, Bolshaya Akademicheskaya str. 44, b. 2, Moscow, 127550, Russia
[3]National University of Water and Environmental Engineering, Soborna str. 11, Rivne, 33028, Ukraine
[4]Warsaw University of Life Sciences – SGGW, Nowoursynowska str. 166, Warsaw, 02-787, Poland

Abstract. The paper considers the hydrological characteristics of the Pripyat River. At the same time, extreme hydrological situations are described - floods and low water levels. The influence of land reclamation carried out in the catchment area on the river runoff was established. Forecast estimates of changes in the flow of the Pripyat River and its tributaries are given. For this, a method of hydrological and climatic calculations based on the joint solution of water and heat and power balances was adopted.
Keywords: river, flood, low water, runoff, climatic changes.

1 Introduction

Currently, the problem of water among the problems facing humanity is more and more often put forward in the first place as the state and development of the biosphere and human society are closely dependent on the state of water resources [1, 2]. Water problems arise when: insufficient water; its unsatisfactory quality; inconsistency of the water regime with the optimal functioning of ecosystems and economic objects; excessive moisture and floods [3]. The first three problems in the global aspect were generated by the past XX century and the fourth has been with humanity since ancient times. All these problems are to some extent inherent in the Pripyat river basin.

The Pripyat is the main river of Polesie. It is an average European river in the Black Sea basin and belongs to the plain type with a predominance of elements of snow supply. The length of the Pripyat River is 761 km, the catchment area is 173.7 thousand km^2. The general direction of the river flow is latitudinal from west to east which is not typical for the rivers of Eastern Europe. The channel at the source is canalized, the rest of the length is meandering, slightly meandering, ramified, replete with bays and adjoining old rivers. Most of the tributaries are fully or partially canalized. The largest tributaries of the Pripyat are the Yaselda, Lan, Sluch, Ptich, Pina, Bobrik, Tsna, Ippa, Stokhod, Goryn, Stviga, Ubort rivers [4, 5, 6, 7, 8].

The aim of this work is to assess the current environmental risks of the Pripyat River now and in the future.

* Corresponding author: omeshyk@gmail.com

2 Method

The methodological basis of research is the scientific provisions on the stochastic nature of the variability of the water regime of rivers which made it possible to use statistical methods for analyzing time series [9]. In addition, methods of water and heat energy balance of the underlying surface, mathematical modeling were used. The systematic analysis of the accumulated information, the comparative-geographical method made it possible to synthesize the most important key positions of the spatial-temporal fluctuations in runoff [10, 11].

We have adapted the method of hydrological and climatic calculations based on the joint solution of the equations of water and heat and power balances [12] for predictive estimates of river runoff changes in the basins. A multivariate model based on the hydrological and climatic hypothesis of V.S. Mezentsev [13] has been developed it includes a standard water balance equation for a land area with an independent assessment of the main balance elements (precipitation, evapotranspiration and climatic runoff) in an annual section. The developed model is used to assess possible changes in river water resources depending on certain hypotheses of climatic fluctuations and anthropogenic impacts on the elements of catchments.

The system of equations for the water balance of the river catchment for a certain period of time has the form

$$\left.\begin{array}{l} Y_K(I) = H(I) - E(I) \pm \Delta W(I); \\ E(I) = E_m(I)\left[1 + \left(\dfrac{\frac{E_m(I)}{W_{HB}} + V(I)^{1-r(I)}}{\frac{KX(I)+g(I)}{W_{HB}} + V(I)}\right)^{n(I)}\right]^{-\frac{1}{n(I)}}; \\ V(I+1) = V(I)\cdot\left(\dfrac{V_{cp}(I)}{V(I)}\right)^{r(I)}; \\ V_{cp}(I) = \left(\dfrac{\frac{KX(I)+g(I)}{W_{HB}} + V(I)}{\frac{E_m(I)}{W_{HB}} + V(I)^{1-r(I)}}\right)^{\frac{1}{r(I)}}; \\ H(I) = KX(I) + W_{HB}(V(I) - V(I+1)), \end{array}\right\} \quad (1)$$

where $Y_K(I)$ - total climatic runoff, mm; $H(I)$ - total resources of moisture, mm; $E(I)$ - total evaporation, mm; $\Delta W(I)$ - change in moisture reserves of the active layer of soil, mm; $E_m(I)$ - the maximum possible total evaporation, mm; W_{HB} - the smallest soil moisture capacity, mm; $V(I)$ / $W(I)/W_{HB}$ - relative humidity of soils at the beginning of the calculating period; $KX(I)$ - the sum of the measured precipitation, mm; $g(I)$ - ground component of the water balance, mm; $r(I)$ - a parameter that depends on the water-physical properties and the mechanical composition of the soil; $n(I)$ - a parameter that takes into account the physical and geographical conditions of the runoff; I - averaging interval.

The obtained values $V_{cp}(I)$ are compared with the relative value of the total moisture capacity V_{PB}. If $V_{cp}(I) \leq V_{PB}$, then the calculated value of the relative average humidity is taken, otherwise, when $V_{cp}(I) \geq V_{PB}$ is taken to the calculation $V_{cp}(I) = V_{PB}$, the difference $(V_{cp}(I) - V_{PB})\, W_{HB}$ refers to the surface runoff.

The amount of atmospheric precipitation in the months of the cold period minus the amount of total evaporation is transferred to the flood period, i.e. for the month of March. The maximum possible total evaporation is found according to the method described in [14].

The solution of the system of equations (1) is carried out by the iterative method until the value of the relative humidity of the soil at the beginning of the calculated interval is equal to the value of the relative humidity at the end of the last interval. When calculating, the initial value of moisture is taken equal to the value of the smallest moisture capacity, i.e.

$W(1) = W_{HB}$ from where $V(1) = 1$. The convergence of the solution of the system is achieved already at the fourth calculation step.

The climatic runoff is adjusted using coefficients that take into account the influence of various factors on the formation of streamflow, i.e.

$$Y_P(I) = k(I) \cdot Y_K(I), \qquad (2)$$

where Y_P *(I)* is the total channel runoff, mm; *k (I)* is a coefficient that takes into account the hydrographic characteristics of the catchment area.

Modeling the water balance of the investigated river is implemented in the form of a computer program and is carried out in two stages. At the first stage, the model is adjusted according to the known components of the water and heat balances of the investigated river. When setting up the model, the goal is to achieve the best match between the calculated climatic and channel flows. The first stage ends with the construction of graphs of the climatic and channel runoff and the output of the modeling error.

The coincidence of the measured and calculated runoff indicates the correctness of the model. The obtained parameters of the model were used in a numerical experiment.

The second stage is a direct calculation of the water balance of the investigated river using the parameters obtained during model calibration. The calculation of the elements of the water balance of the investigated river is carried out taking into account the specific features of the considered catchment [15].

The solution of the water balance equation for the catchment area is associated with the determination of the average value of those elements that are monitored at separate points of the catchment area. Therefore, one of the main components of modeling the water regime is the correct assessment of climatic characteristics and their averaging over the catchment. This problem is considered in detail in [16].

There were problems with determining the parameters for the winter months when setting up the models according to the proposed method. The fact is that the model did not accurately take into account the recent thaws that have become more frequent. Therefore, the model was adjusted to take into account thaws. The difference between the channel and climatic runoff obtained when setting up the model related to the runoff formed during thaws which was fixed in the model settings. When predicting the future runoff, this component was added directly to the runoff and its value was subtracted from the atmospheric precipitation of the corresponding month and precipitation was transferred to the flood period minus the total evaporation and runoff during the thaw period. Moreover, the amount of runoff during the thaw period was adjusted taking into account the predicted temperature of the corresponding month. In the first approximation, the value of this runoff can be taken from the ratio of monthly air temperatures and the runoff value during the thaw period obtained when setting the model.

Forecast estimates of river runoff changes were carried out according to the following scheme. The model was adjusted according to long-term average data on river runoff, atmospheric precipitation, air temperature and air humidity deficit. Then the predicted values were entered for the corresponding perspective for those weather stations that were used when setting up the model. Model settings were read and predictive estimates were made. The obtained values of the climatic runoff were compared with each other according to the ratio $\Delta_{kl} = Y_{kl}^{pr} / Y_{kl.}^{sov.} \cdot 100\%$. A direct predictive estimate of the channel runoff was found from the ratio $Q^{pr} = Q^{sov} \cdot \Delta_{kl}/100\, , m^3\ /\ s.$

As the main source materials, we used standard observational data on the hydrometeorological network, stock materials of various organizations and institutions for the period of instrumental observations as well as cartographic and published sources.

3 Results and discussion

The average annual discharge of the Pripyat River at the mouth is 450 m³ / s. The intra-annual distribution of runoff is characterized by unevenness. The runoff of the spring period averages about 61%, summer-autumn - 23%, winter - 16% of the annual runoff. The water discharge of the Pripyat River in the alignment of the city of Mozyr of various supplies is presented in table. 1. Calculations were performed based on the series of observations for more than 130 years using a three-parameter gamma distribution.

Table 1. Water flow rate of the Pripyat river with different availability, m³ / s.

Water flow view	Coefficients of variability	Q_{cp}	$Q_{p=1\%}$	$Q_{p=5\%}$	$Q_{p=95\%}$	$Q_{p=99\%}$
Annual	0.31	392	713	380	221	171
Maximum spring flood	0.87	1771	7154	4321	494	331
Maximum rainfall floods	0.78	376	1350	861	117	80.4
Minimum summer-autumn	0.49	155	359	272	80.5	64.6
Minimum winter	0.72	156	569	346	53.0	38.9

Floods

Floods rank first among natural disasters in terms of the number of casualties and damage caused. At the same time, paradoxical as it may seem, until now there are no reliable long-term forecasts of their occurrence, reliable and generally accepted methods for calculating the damage caused by them and a generally accepted concept of protection [17].

Flood usually begins on the Pripyat River in the first half of March but in some years it can shift to February or April. The average long-term duration of flooding of the floodplain of the Pripyat River is 80 - 110 days and sometimes up to 150 - 180 days. The width of the spring flood varies from 5 to 15 km, the largest in the area of Pinsk reaches 30 km. The depth of flooding is mainly 0.3 - 0.8 m, in some places 2 - 2.5 m [4]. Flooding in the floodplain of the Pripyat River from floods of varying availability is shown in Fig.1 [6].

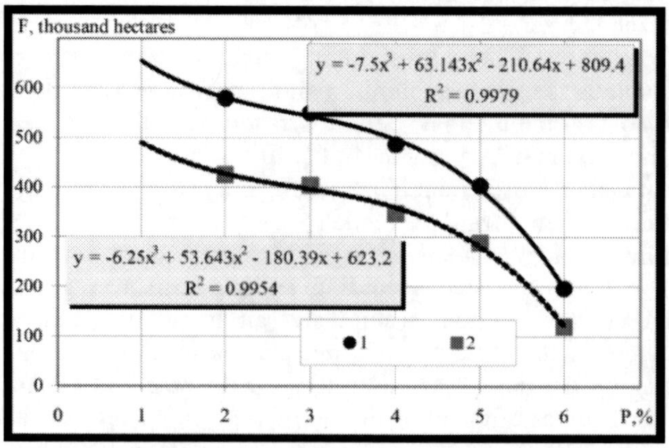

Fig. 1. Areas of flooding of the floodplain of the Pripyat River, depending on the availability of the water level: 1 - total along the floodplain; 2 - within Belarus.

The maximum flood on the Pripyat River was noted in 1845 and was so catastrophic that it probably can be attributed to the group of maximum possible ones in our climatic epoch. It is a unique hydrological phenomenon with very rare recurrence. The maximum level exceeded the zero of the graph of the modern gauging station near the city of the city of Mozyr by 675 cm and the water discharge is estimated at 11000 m^3 / s and can be approximately considered repeating no more than once every 800 years [18]. The last significant flood was in 1999. Table 2 shows the water discharge of the 10 most significant floods in Pripyat and their availability [19].

Table 2. Maximum water discharge (Q) of spring flood and supply (P) of the Pripyat River - the city of Mozyr.

Years	1845	1877	1895	1888	1889	1940	1979	1932	1970	1958
Q,m^3/s	11000	7500	5670	5100	4700	4520	4310	4220	4140	4010
P, %	0.8	1.6	2.3	3.1	3.9	4.7	5.4	6.2	7.0	7.6

Freshet level rises in contrast to floods occur irregularly and, in terms of the maximum discharge and the runoff layer, freshets are as a rule significantly less than the high water maximums. However, rain freshets in 1952, 1960, 1974, 1993, 1998 on many watercourses and sections of the Pripyat itself exceeded floods and caused significant damage to the national economy. Even local freshets on the tributaries can cause significant level rises in the lower reaches of the Pripyat River because of the downward movement of the freshet wave. The height of freshets in the middle and lower reaches of the Pripyat reaches 2 - 3.5 m above the pre-rise level [20, 21].

Low water

The high water is replaced by a summer-autumn low-water period characterized by significant variability. Summer low water is usually lower than winter and is interrupted almost every year by rain freshets. The winter low-water period is often interrupted by thaws which result in winter freshets in some years exceeding floods [22].

The conditions for the formation of low-water runoff of rivers in general can be considered favorable since the territory of Polesye is located in the zone of excessive moisture and the outflow of groundwater into the river network is more or less long-term and constant. The minimum levels and runoff of water in summer are observed at high average daily air temperatures and with prolonged periods of absence of precipitation; in winter - at low temperatures. There was a drying up of watercourses in Polesie even with catchment areas over 1000 km^2 in dry years (1939, 1951, 1952, etc.). Freezing is observed only on small rivers and for a short time.

Most low-water period of summer-autumn low water is mainly observed in July - August, at least - in September. Its duration for the Pripyat River is 85–90 days.

The winter low-water period is usually stared at the end of December. The earliest dates for the onset of low water fall on the end of October - early November and the latest - in January, the end - with the beginning of the spring flood.

Zero runoff within the catchment area of Pripyat was noted in 17 watercourses with catchment areas of 11 - 1280 km^2. The average duration of one case of zero runoff can reach 195 days in summer and 75–100 days in winter.

The values of the smallest average monthly summer discharges naturally decrease over the territory of Polesie from the northwest and north to the south and southeast, subject to geographic zoning on large and medium rivers. However, an intrazonal nature of changes is found on small rivers, depending on local hydrogeological features - the presence and thickness of groundwater horizons, the nature of their opening by river valleys and the conditions for their discharge.

The most water-abundant are aquifers in fractured and karst carbonate-sulfate rocks of the Upper Cretaceous and Neogene. Outlets of chalk waters are observed within the Polesie lowland in the form of ascending springs with a flow rate of up to 200 m^3 / h. These waters feed a number of lakes, numerous bog massifs and partially right-bank tributaries of the Pripyat - the

 https://doi.org/10.1051/e3sconf/202021201015

rivers Turya, Stokhod, Goryn, Styr, etc. The modulus of the minimum average daily runoff of these rivers 97% of availability varies from 0.07 - 0.18 l / (s·km²). Those rivers which are fed from the aquifers of alluvial and fluvioglacial sediments have low modules of minimum runoff and in dry years their runoff completely stops for a period from 15 to 120 days. Cessation of runoff on these rivers is also possible during cold, non-thaw winters [23].

Anthropogenic impacts on river runoff

There has been a discussion about the impact of land reclamation on river flow since the 50s of the last century. The main impact on the water regime of the Pripyat River was exerted during the period of large-scale hydrotechnical reclamation of the Polesie lowland. At the same time, the water resources of Polesie were more strongly affected by anthropogenic impacts than other regions. 23% of the territory was drained, the total length of the open reclamation network exceeded 65.000 km, the hydrographic network was significantly transformed, especially if we take into account the straightening and deepening of the Pripyat River itself and its large tributaries. In addition, the embankment of certain sections of the Pripyat River and the construction of polder reclamation systems which exclude flooding of the embanked floodplain sections led to the fact that the groundwater dropped by 1.0-1.5 m it, followed by a decrease in water levels in the rivers and in some - up to dryness. All this resulted in a change in the hydrological regime of the rivers. Analysis of the change in the Pripyat River runoff showed an increase in the average annual runoff of the Pripyat River during the period of active reclamation in all months of the year, except for April and May [24].

Recently, maximum losses from irrecoverable water consumption are small and are within the measurement error.

Analysis of the dynamics of the water flow of the Polesie rivers showed that starting from the mid-60s of the last century the average annual, minimum summer and winter discharges have a steady tendency to increase, at the same time, the spring flood flow decreases (Fig. 2).

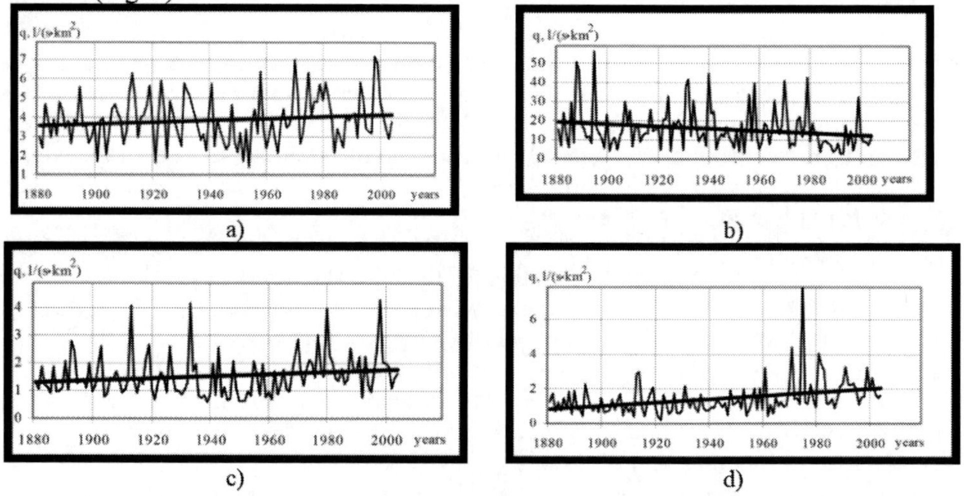

a) average annual; b) maximum spring floods; c) minimum summer-autumn; d) minimum winter

Fig. 2. Dynamics of the modules of the water flow of the Pripyat river in the alignment of the city of Mozyr.

Assessment of changes in the water regime under various future climate scenarios

Comparative analysis of river runoff changes in the Pripyat basin for hydrological stations in the monthly average and in the average annual sections for the period from 1987 to 2015 in relation to the period from 1961 - 1986 [16] showed that:

- the average annual runoff is insignificantly downward (maximum - by 9%);

• there was a significant decrease in the runoff of the spring flood by 42% with an earlier onset of its peak;

• there was an increase in runoff in winter by 20%;

• runoff in summer did not change very significantly over the entire period from 1961 to 2015, however, in recent years there has been a significant decrease in runoff which was even less than the minimum for the entire indicated period.

Predictive estimates of changes in the river runoff of the rivers of the Pripyat basin for the period up to 2035 were made according to the method of hydrological and climatic calculations proposed above. In this case, the results of assessing the actual climate change and river runoff for the period from 1961 to 2015 were used and an updated forecast of climate change for the period up to 2035 in these river basins taking into account the multi-model ensemble of four scenarios recommended by the IPCC [25] as well as regional climate variability. Taking into account the use of the most conservative climate change scenarios, the temperature rise in the basin on average per year can reach 1.9 ° C with a maximum increase in winter by 2.3 ° C, in summer - by 1.9 ° C, in spring and autumn - by about 1.7 ° C. At the same time, the annual amount of precipitation will change insignificantly (in total for the year it will decrease by 2%) with an increase in it in winter (by an average of 7%), a maximum decrease in summer (by an average of 10%), to a lesser extent in a decrease in spring (by 4%) and a slight decrease in the fall (on average by 1.6%). A generalization of the results of calculations of the predicted flow change in the Pripyat River basin for the period up to 2035 is given in Table. 3 [16].

Table 3. Forecast of changes in surface runoff until 2035 for the Pripyat River and its tributaries, in % of the current state.

River - settlement	Winter	Spring	Summer	Autumn	Average annual
Pripyat - Chernichi (Turov)	4.9	5.5	-19.2	0.6	-2.1
Pripyat - Mozyr	0.2	1.6	-20.6	-2.4	-5.3
Yaselda - Bereza	-0.3	-27.0	-41.7	-23.3	-23.1
Yaselda - Senin	-3.9	-10.6	-37.7	-11.8	-16.0
Tsna - Dyatlovichi	-3.7	-8.9	-26.9	-19.9	-14.9
Goryn - Malye Vikorovichi	-4.0	-11.8	-20.1	-16.7	-13.2
Sluch - Lenin	10.1	5.7	-15.8	1.6	0.4
Ubort – Krasnoberezhe	-13.4	-5.6	-25.2	-38.8	-20.8
Ptich – Luchitsy	10.3	-0.2	-24.0	16.7	0.70
Shat - Shatsk	-0.2	-9.2	-10.7	-4.4	-6.1
Oressa - Andreevka	-14.7	-10.7	-28.4	5.4	-12.1
average for the basin:	-1.3	-6.5	-24.6	-8.5	-10.2

With climate change according to scenario A1B, based on the results of calculations, the following conclusions can be drawn about the predicted change in runoff in the Pripyat River basin until 2035 [16]:

- a slight decrease in the average annual runoff;
- a slight increase in runoff in winter in most rivers - on average for the basins up to 2.1%, up to a maximum of 25% and for some rivers the change in runoff will be insignificant or even likely to decrease;
- in the spring period, with rare exceptions, the runoff is likely to decrease by an average of 5.5%, maximum - by 25%;
- in the summer period, a significant and maximum of all periods of the year is predicted to decrease the runoff - on average by 20%, maximum - by 40%;
- in the autumn period (especially at the beginning of autumn - until mid-October), a predominant decrease in runoff is likely - by an average of 8.5%, maximum by 35%, in the remaining autumn months the runoff will change insignificantly.

Climate change scenario B1 is more severe and leads to large changes in river flow and sharper differences between the northern and southern parts of the republic, between medium and small rivers. Unevenness and multidirectionality are noted in certain months with an increase in runoff as a whole for the year.

Detailed results of the forecast runoff for the period up to 2035 (m^3/ s), changes in runoff in absolute values (m^3 / s) and in relative values (%) are given in [16].

4 Conclusion

The water risks of the Pripyat River are far from being completely neutralized. Therefore, on the basis of the scientific results obtained, it is necessary to develop measures to minimize possible negative consequences in the event of a change in the regime of water.

Further research should be focused on the following main areas:

- prevention and reduction of negative consequences from floods;
- improving the quality of natural waters;
- protection of water sources in the design, construction and operation of public facilities;
- management of the regime of natural waters, ensuring the biosphere functioning of natural ecosystems;
- creation of basin schemes for water resources management in Polesie.

References

1. Farhan, Hussain & Wagan, & Khoso, Salim. (2013). Water shortage; its causes, impacts and remedial measures.
2. Impacts of climate change on water resources in Spain (2012) T. Estrela, M.A. Pérez-Martin, E. Vargas. *Hydrological Sciences Journal*. Vol. 57. № 6., 1154-1167.
3. Impact of climatic change on river discharge in the driest region of Poland (2014) Piotr Ilnicki [and others]. *Hydrological Sciences Journal*. Vol. 59. № 6, 1117-1134.
4. Блакітная кніга Беларусі: Энцыкл. (The Blue Book of Belarus : Encyclical.) /Belarus. Encyclical; Editorial Board: N.A. Dzisko et al. - Minsk, 1994. - 415 p. (In Russian).
5. Природообустройство Полесья (Environmental Engineering in Polesye) : monograph : in 4 books / edited by Yu. Mazhayskiy, A. Rokochynskiy, A. Volchak, A. Meshyk, J. Jeznach. – Ryazan: Meshchersk office of VNIIGiM of A. N. Kostiakov, 2017. – Book 2: Ukrainian Polesye – V. 1. – 902 p. (In Russian).
6. Природообустройство Полесья (Environmental Engineering in Polesye) : monograph : in 4 books / edited by Yu. Mazhayskiy, A. Rokochynskiy, A. Volchak, A. Meshyk, J. Jeznach. – Ryazan : Meshchersk office of VNIIGiM of A. N. Kostiakov, 2018. – Book 1 : Belarusian Polesye. – V. 1 : Natural Resources Potential. – 408 p. (In Russian).
7. Природообустройство Полесья (Environmental Engineering in Polesye) : monograph : in 4 books / edited by Yu. Mazhayskiy, A. Rokochynskiy, A. Volchak, A. Meshyk, J. Jeznach. – Ryazan: Meshchersk office of VNIIGiM of A. N. Kostiakov, 2018. – Book 2: Ukrainian Polesye – V. 2 . – 804 p. (In Russian).
8. Природообустройство Полесья (Environmental Engineering in Polesye) : monograph : in 4 books / edited by Yu. Mazhayskiy, A. Rokochynskiy, A. Volchak, A. Meshyk, J. Jeznach. – Ryazan : Meshchersk office of VNIIGiM of A.N. Kostiakov, 2019. – Book 1 : Belarusian Polesye – V. 2 : Transformation and Use of Natural Resources. – 503 p. (In Russian).
9. Liebscher, H.J. (1993) Hydrology for the water management of large river basins / H.J. Liebscher. *Hydrological Sciences Journal*. Vol. 38. №1, 1-13.

10. Wrzesiński, D. (2018) Detection of changes in flow regime of rivers in Poland / D. Wrzesiński, L. Sobkowiak. *J. Hydrol. Hydromech*. Vol. 66. № 1, 55.
11. Hannaford, J. (2012) Trends in seasonal river flow regimes in the UK / J. Hannaford, G. Buys. *Journal of Hydrology*. 475. 158-174.
12. Мезенцев, В.С. Гидрологические расчеты в мелиоративных целях (Hydrological calculations for reclamation purposes) / V.S. Mezentsev. - Omsk, 1982. - 84 p. (In Russian).
13. Мезенцев, В.С. Гидролого-климатическая гипотеза и примеры ее использования (Hydrological and climatic hypothesis and examples of its use) / V.S. Mezentsev // Water resources, 1995. - Volume 22, № 3. - P. 299-301. (In Russian).
14. Волчек, А.А. Закономерности формирования водного баланса речных водосборов Беларуси: Формирование водного баланса Беларуси (Regularities of formation of water balance of river watersheds of Belarus: Formation of water balance of Belarus) / A.A. Volchek // Saarbrucken: LAP LAMBERT Academic Publishing GmbH & Co. KG, 2011. - 387 p. (In Russian).
15. Волчек, А.А. Оценка трансформации водного режима малых рек Белорусского Полесья под воздействием природных и антропогенных факторов (на примере р. Ясельда) (Estimation of the transformation of the water regime of small rivers of the Belarusian Polesie under the influence of natural and anthropogenic factors (on the example of the Yaselda River)) / A.A. Volchek, S.I. Parfomuk // Water management of Russia: problems, technologies, management. - Ekaterinburg, 2007. - № 1. - P. 50–62. (In Russian).
16. Водные ресурсы Беларуси и их прогноз с учетом изменения климата (Water resources of Belarus and their forecast taking into account climate change) / A.A. Volchek, V.N. Korneev, S.I. Parfomuk, I.A. Bulak; under common ed. A.A. Volchek, V.N. Korneev. - Brest: Alternativa, 2017. - 228 p. (In Russian).
17. Meshyk, A., Barushka, M, Marozava, V. (2020) Snow as a Contributor to Spring Flooding in Belarus. *Environmental Science and Pollution Research*. 1–11. https://doi.org/-10.1007/s11356-020-09638-8.
18. Логинов, В.Ф. Весенние половодья на реках Беларуси: пространственно-временные колебания и прогноз (Spring floods on the rivers of Belarus: spatial-temporal fluctuations and forecast) / V.F. Loginov, A.A. Volchek, An.A. Volchek – Minsk, 2014. – 244 p. (In Russian).
19. Volchak, A. A., Meshyk, A. P., Sheshka, M. M. et al. (2016). Floods on the territory of Polesie. *Procedia Engineering*. 162, 91–97. https://doi.org/10.1016/j.proeng.2016.11.020
20. Волчек, А.А. Опасные гидрологические явления на р. Припяти (Hazardous hydrological phenomena on the Pripyat river) / A.A. Volchek, An.A. Volchek // Extreme hydrological situations / under common ed. N.I. Koronkevich, E.A. Barabanova, I.S. Zajceva. – Moscow, 2010. – P. 295-322. (In Russian).
21. Волчек, А.А. Мониторинг, оценка и прогноз чрезвычайных ситуаций и их последствий (Monitoring, assessment and forecast of emergency situations and their consequences) / A.A. Volchek, P.S. Pojta, P.V. Shvedovskij // Brest, 2012. – 423 p. (In Russian).
22. Black, A.R. (1997) Seasonality of flooding: A case study of North Britain / A.R. Black, A. Werritty. *J. Hydrol.* 195. 1-25.
23. Мониторинг, использование и управление водными ресурсами бассейна р. Припять (Monitoring, use and management of water resources in the Pripyat river basin) / under common ed. M. Yu. Kalinin and A. G. Obodovskij. – Minsk, 2003. – 269 p. (In Russian).
24. Логинов, В.Ф. Антропогенное воздействие на водные ресурсы Беларуси (Anthropogenic impact on water resources in Belarus) / V.F. Loginov, M.Yu. Kalinin, V.F. Ikonnikov. – Minsk, 2000. – 284 p. (In Russian).

25. IPCC, 2014: Climate Change 2014: Impacts, Adaptation, and Vulnerability. Part A: Global and Sectoral Aspects. Contribution of Working Group II to the Fifth Assessment Report of the Intergovernmental Panel on Climate Change [Field, C.B., V.R. Barros, D.J. Dokken, K.J. Mach, M.D. Mastrandrea, T.E. Bilir, M. Chatterjee, K.L. Ebi, Y.O. Estrada, R.C. Genova, B. Girma, E.S. Kissel, A.N. Levy, S. MacCracken, P.R. Mastrandrea, and L.L.White (eds.)]. Cambridge University Press, Cambridge, United Kingdom and New York, NY, USA, 1132 pp.

Intra-annual Runoff Distribution in the Pripyat River Basin

Alexander Volchak[1], *Sergey* Parfomuk[1,*] and *Svetlana* Sidak[1]

[1]Brest State Technical University, 224017 Brest, Republic of Belarus

Abstract. The time variability features of the intra-annual runoff distribution in the Pripyat River basin at the present stage are considered. The study used data from 10 active gauging-stations at the basin area for the period of enabled observations. Changes in the nature of natural runoff regulation of rivers are shown. An increase in the coefficient of natural runoff regulation after the period 1970-1983 for the rivers of the Pripyat River basin was revealed. The relationship between the coefficient of uneven runoff and the share of spring flood was established.
Keywords: intra-annual runoff distribution, daily runoff duration curve, coefficient of natural runoff regulation, spring flood share.

1 Introduction

The periods of cooling and warming, dry and wet periods alternate in time and the general state of water resources and their quality do not statistically change. So, studying of water resources and impact of climate change and anthropogenic activities on runoff is an urgent worldwide task [1-5].

The task of analysis and subsequent modelling of water resources is one of the most important for the state [6]. Water resources analysis was conducted in terms of studying the water regime of rivers, lakes and reservoirs. The Soil and Water Assessment Tool (SWAT) is using for simulate water, sediment, and nutrient yield in a watershed by using input data from GIS and applying different agricultural practices, climate change, and land [7-9]. The Soil and Water Integrated Model (SWIM) has been used to study various hydrological phenomena, such as runoff change causing by climate influence, extreme flood analysis, agricultural production, etc. [10-13]. Using artificial intelligence methods as an approach in solving complex nonlinear problems and lake level forecasting has been increased [14-17]. Previously, we performed the assessment of water resources of the Republic of Belarus in the beginning of 20th century [18]. Modelling of the water regime in Belarus was carried out on the basis of mathematical models [19].

The study of intra-annual runoff distribution (IARD) of rivers is an important scientific and practical aspect of organizing cost-effective and environmentally safe water use [20-23].

[*] Corresponding author: parfom@mail.ru

Water resources use is based on the IARD parameters and planned for various water management purposes, as well as for determining the main parameters of reservoirs and hydraulic structures.

Establishing patterns of runoff distribution in a year is a difficult task since the IARD is affected by a number of interrelated factors, the quantitative accounting of which is almost impossible to fully account for. First of all, the distribution of runoff depends on changes in the intra-annual course of air temperature and precipitation distribution. In addition to climatic factors, the runoff distribution is influenced by other physical and geographical parameters that express the natural runoff regulation in the basin. This group of factors includes the size and topography of the basin, hydrogeological conditions, lakes, forest cover, waterlogging, etc. [24].

In General, water users and consumers are interested in a constant stable runoff without interruptions and sharp rises. This means that the river runoff with a high degree of natural regulation is most favourable for them. A low degree of natural regulation is a danger in the form of floods that lead to economic damage and even loss of life.

This problem is particularly relevant for the Belarusian Polesie which main river is the Pripyat River and it is characterized by frequent flooding [25]. As a result of changes in factors and conditions for the formation of runoff within the Pripyat River basin due to hydraulic reclamation, as well as the impact of modern climate changes the water regime of rivers has been significantly transformed. Therefore, an urgent task is to assess changes in the intra-annual distribution of river runoff in modern conditions, which must be taken into account when solving a number of scientific and practical problems.

The purpose of the research is to analyse the current features of the time variability of the intra-annual runoff distribution in the Pripyat River basin and take into account the intra-annual unevenness of the runoff to assess the hydro-ecological safety of the territory.

2 Data and methods

The initial data were the observation materials of the Ministry of natural resources and environmental protection of the Republic of Belarus for the period of instrumental observations, published in the materials of the state cadastres. The study used data on average monthly runoff for 10 gauging-stations with the longest observation period (at least 65 years) located in different parts of the Pripyat River basin and differing in morphometric parameters. In addition, data on daily runoff was also used for a number of rivers. Statistical data processing and calculation were performed using the MS Excel package.

3 Results and discussion

An analysis of the non-calendar runoff distribution and generally accepted chronological runoff distribution graphs were used to determine the unevenness of the intra-annual runoff distribution (Fig. 1). This feature was implemented in the form of the daily runoff duration curve which is a curve that can be used to set the length of time (in days or as a percentage of the year) during which the runoff is equal to or exceeds the specified [26].

Fig. 2 shows the daily runoff duration curve of the Tsna River which is a tributary of the Pripyat River. The curve based on typical runoff does not match the actual data. Therefore, this method can lead to significant errors when calculating the parameters of the intra-annual runoff distribution.

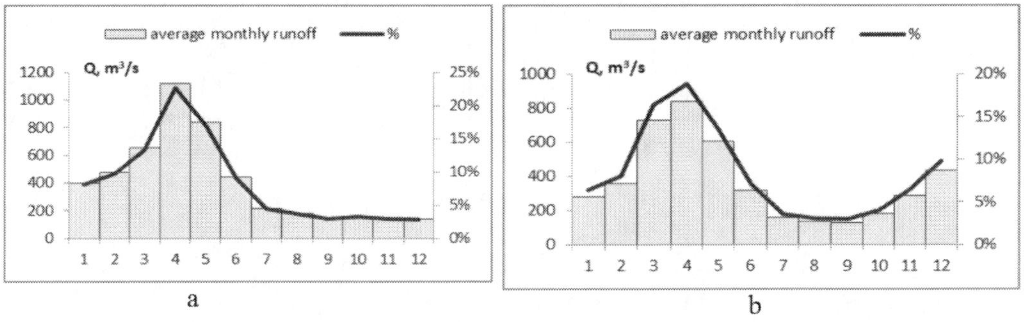

Fig. 1. The intra-annual runoff distribution of the Pripyat River at the Mozyr station (a – 1983, b – 2017)

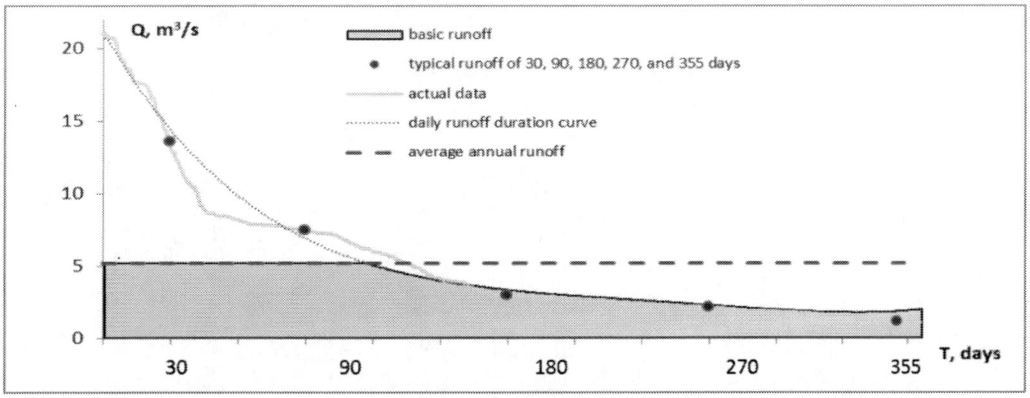

Fig. 2. The daily runoff duration curve of the Tsna River at the Dyatlovichi station, 2004

The coefficient of natural runoff regulation (φ-coefficient) which is equal to the ratio of the basic runoff Q_b to the total annual runoff is used to study the unevenness of the IARD [27]:

$$\varphi = \frac{\sum_{i=1}^{n} Q_{b_i}}{n \cdot Q_0}, \qquad (1)$$

where $Q_{b_i} = \begin{cases} Q_0, if\ Q_i \geq Q_0 \\ Q_i, if\ Q_i < Q_0 \end{cases}$, Q_0 – average annual runoff, n – number of days per year.

The data of daily and monthly runoff of the rivers in the Pripyat River basin were used to calculate the coefficient of natural runoff regulation. The monthly values of the φ-coefficient were compared with the coefficients for the daily runoff of the 10 rivers in the Pripyat River basin for the period 1971–2004 to assess the representativeness of calculated results (Fig. 3). The values of the φ-coefficient that were obtained in two ways were very close to each other. The correlation coefficient was 0.98–0.99. If you take the values of the φ-coefficient which is calculated from daily data as reference values, you can calculate the error in determining the φ-coefficient calculated from monthly data. The average absolute error was 2.96 %. The largest errors (4.37% and 3.88%) were obtained for the Sluch River at the Klepchany station and the Tsna River at the Dyatlovichi station, respectively. The results have shown that the use of average monthly runoff overestimates the value of the coefficient of natural runoff regulation. Taking into account the obtained relationship between the coefficients the φ-coefficient can be specified. Of course, using daily runoff data allows getting more accurate results, but this accuracy is unnecessary for multiple calculations.

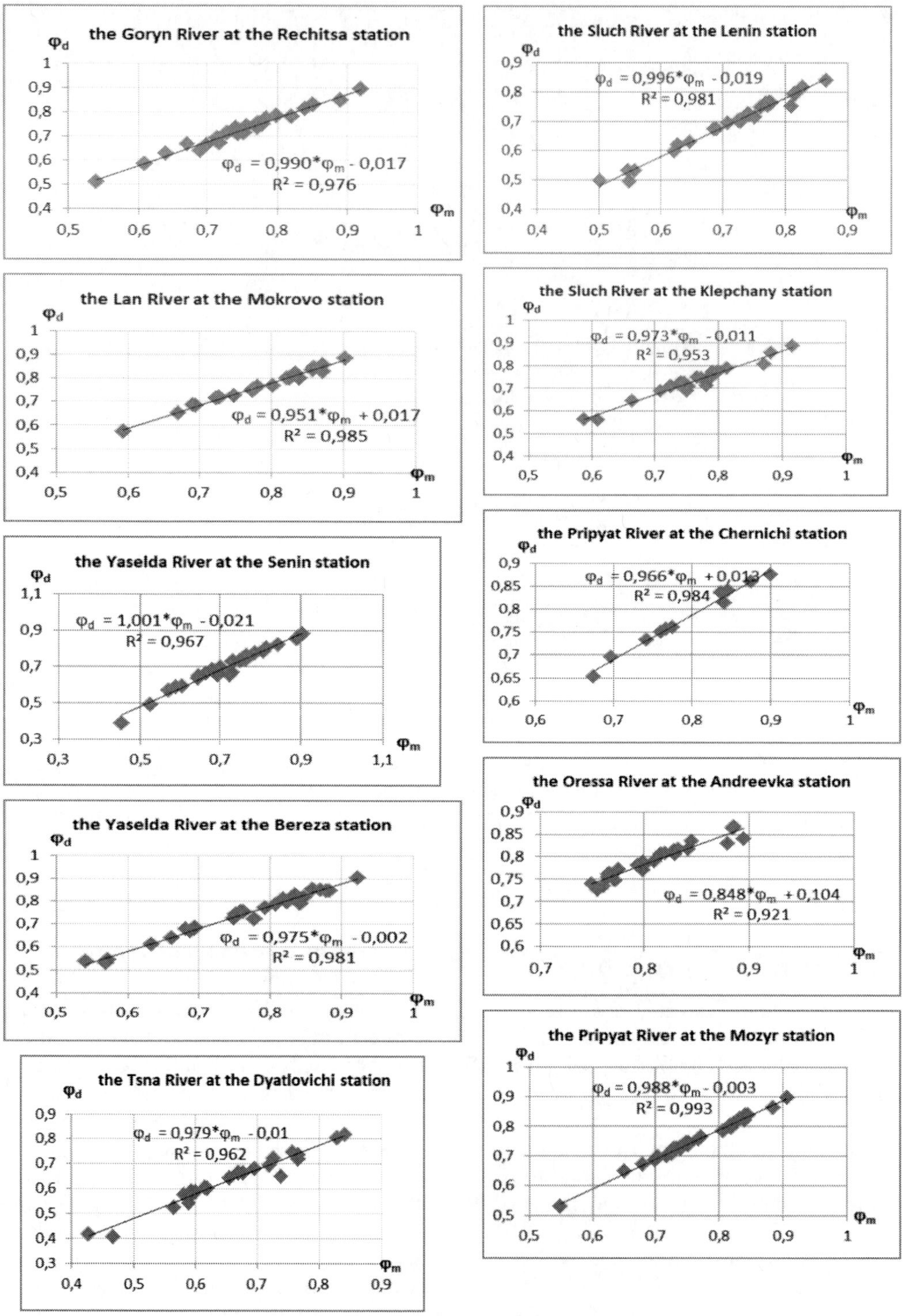

Fig. 3. The coefficient of natural runoff regulation for the daily (φ_d) and monthly (φ_m) data

The time variability analysis of the coefficient of natural runoff regulation of the Pripyat River at the Mozyr station was carried out on the basis of combined graphs of normalized difference-integral curves and chronological graphs of changes in the φ-coefficient (Fig. 4). The the integral curves ordinates are calculated as an accumulated sum

$$\sum(K_i - 1)/c_v,$$

where $K_i = \varphi_i/\varphi_0$ is the modular coefficient of the coefficient of natural runoff regulation in the i-th year (φ_i),), φ_0 is its average value, c_v is coefficient of variation, *i* is a year.

The analysis of the difference-integral curves presented in Fig. 4 has shown that the series of φ-coefficient have two sets with critical period in 1970–1983. For all 10 gauging-stations an increase in the coefficient of natural runoff regulation by 15–56% after 1970–1983 was detected.

The coefficient of intra-annual runoff irregularity d=1-φ is also used to analyze the runoff irregularity. This coefficient is equal to the control capacity required for intra-annual runoff equalization. Since the spring flood is the main parameter of the water regime in the Pripyat River basin, it is advisable to include the parameters of the main high-water phase in the generalized indicators of intra-annual runoff irregularity. The most important indicator is the proportion of spring flooding (*dvp*) in annual runoff. However, calculating the parameters of spring flooding has its own difficulties. It is especially difficult to distinguish the period of high water since the methods for determining the dates of the beginning and ending of the flood are mainly descriptive and do not have a clear algorithm for calculating. When snowmelt is very long in time or heavy rains fall on rivers then floodwaters are formed and their hydrographs have complex outlines and consist of several independent waves. In years with friendly snowmelt the slender single-peak floods are formed and characterized by a high intensity of rise and fall. The method of determining the main parameters of the spring flood including start and end dates of flood volume floods described in [28] and was used to determine the proportion of spring flooding.

In addition, it is also necessary to note the following feature: the calculation error can reach 30-40% when calculating the share of spring flood on the average monthly runoff if the average daily runoff is unavailable. The coefficient d is completely determined by the *dvp* value [29]. Thus, the results to periods for which the information about average daily runoff is not available can be extended by establishing a link between the d and *dvp* indicators for the period with the available data. To do this a graph that reflects the relationship between the coefficient of natural runoff regulation and the share of spring flood runoff for the period with the available data on average daily runoff (1968–1997) for the Pripyat River at the Mozyr station was constructed (Fig. 5).

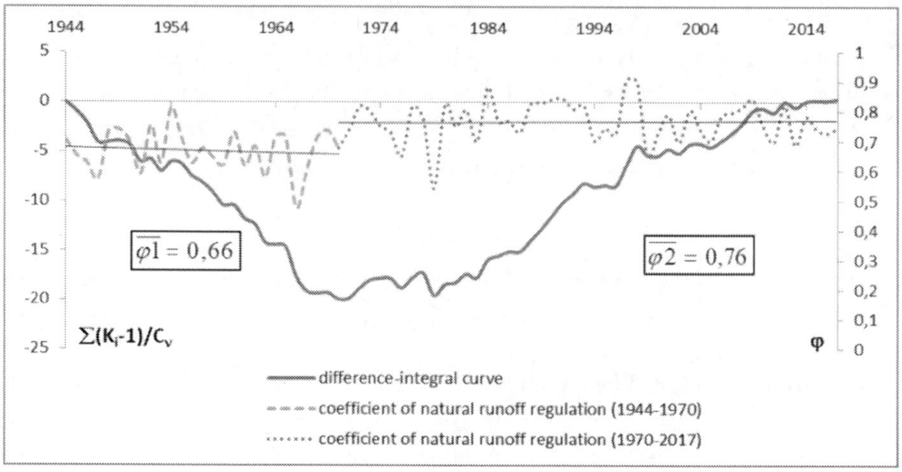

a

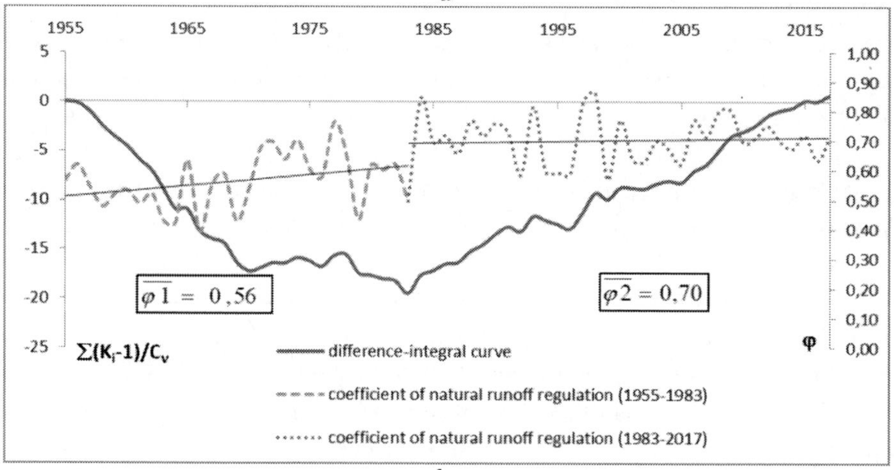

b

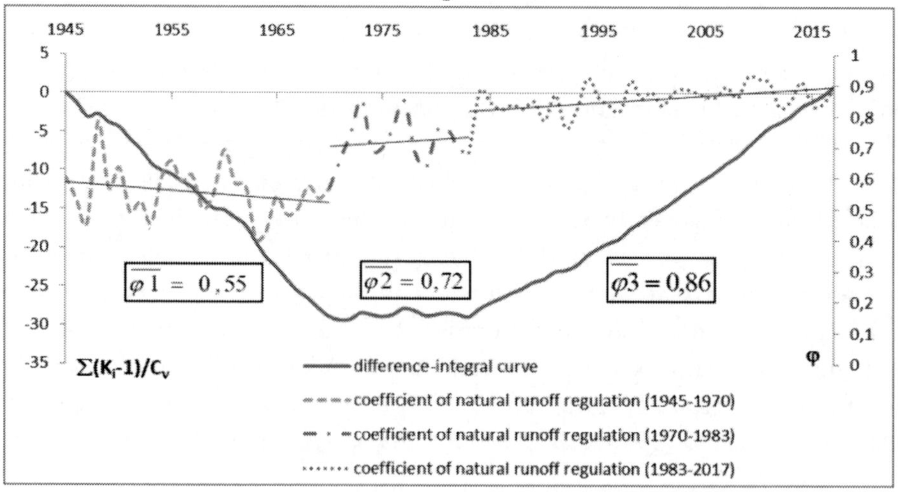

c

Fig. 4. Difference-integral curves and chronological graphs of the coefficient of natural runoff regulation (a – the Pripyat River at the Mozyr station; b – the Tsna River at the Dyatlovichi station; c – the Yaselda River at the Bereza station)

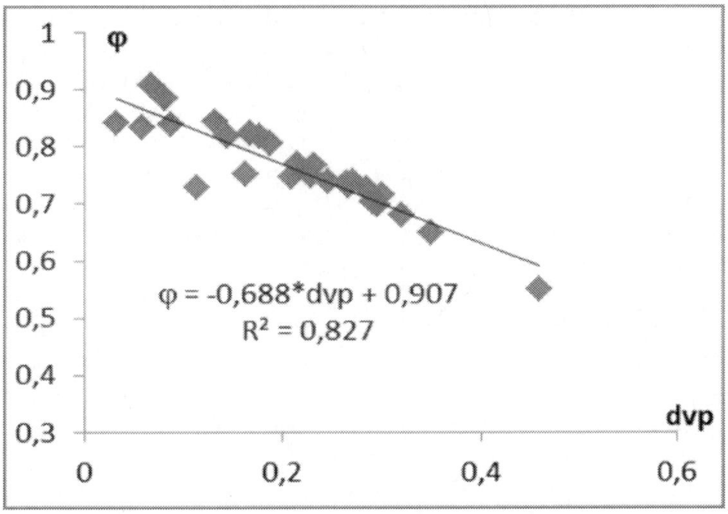

Fig. 5. Relationship between the share of spring flood runoff *dvp* and the coefficient of natural runoff regulation φ

Fig. 5 shows that the share of spring flooding in annual runoff is closely related to its intra-annual unevenness (R^2=0.83). Moreover, as the share of spring flood runoff decreases this relationship weakens. Using the established relationship it is possible to obtain dvp indicators for the entire period under consideration (1944–2017). Fig. 6 shows changes in the φ-coefficients and *dvp* for the Pripyat River at the Mozyr station.

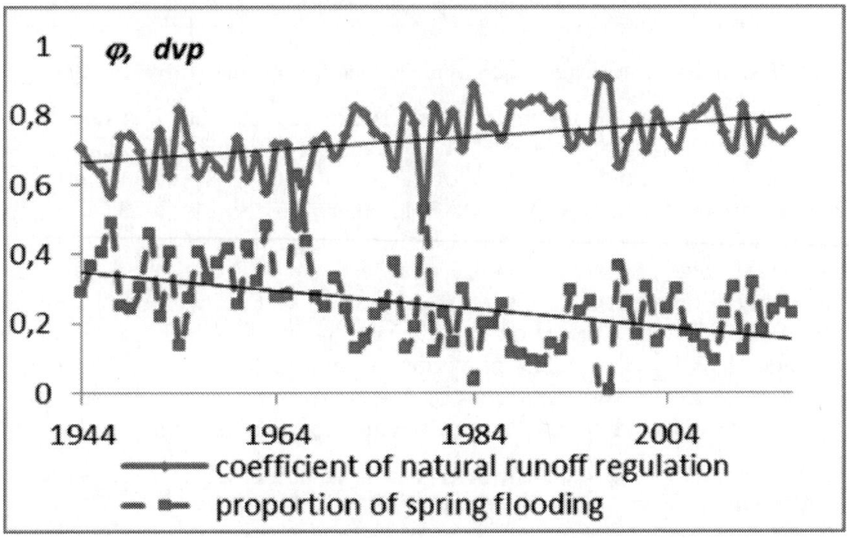

Fig. 6. Change in the share of spring flood runoff *dvp* and the coefficient of natural runoff regulation φ over a multi-year period

Changes in the natural runoff regulation are reflected in the form of hydrographs. There is a change in some phases of the water regime in the last decade when averaging hydrographs over decades (Fig. 7).

The changes in the average monthly runoff of the Pripyat River at the Mozyr station by decades for the period 1948–2017 in the form of a cross table were calculated for a more detailed study of changes in hydrographs. The cross-table values were the percentage of changes in runoff between the periods under consideration and the correlation coefficient between the periods.

Fig. 7 and calculated values show decrease in runoff in January in the last ten-year period (2008–2017) compared to the 4 previous ten-year periods. In March the changes in runoff in 2008–2017 were positive compared to all previous ten-year periods. However, it should be noted that these changes are gradually decreasing and in the current decade the changes in runoff in March compared to the previous decade will be negative, i.e. runoff in March will decrease. The same situation is observed in February. For this month, there has already been a decrease in runoff in the last decade by 5.53 % compared to the period 1998–2007.

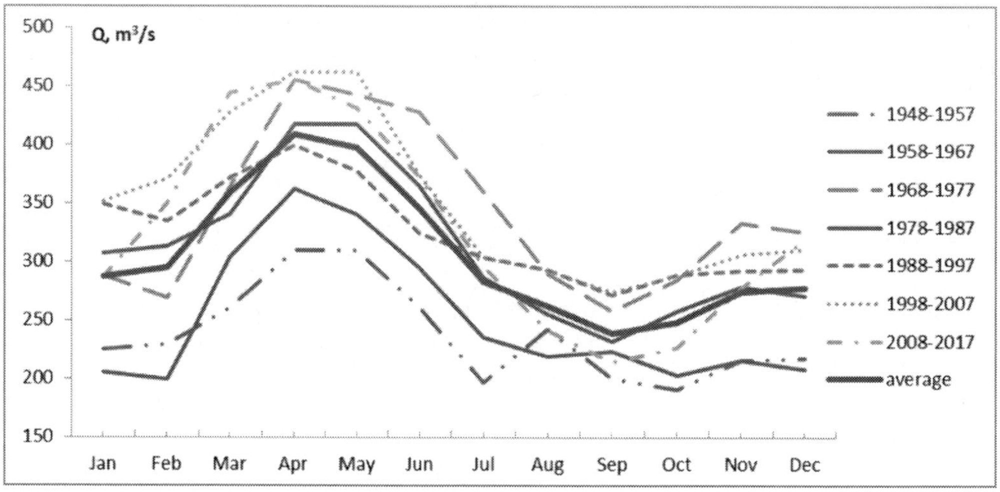

Fig. 7. Averaged by decades hydrographs of the Pripyat River at the Mozyr station

A significant decrease in runoff over the period 2008–2017 is observed in August, September, October, and November. The percentage of reduction reaches 21.88 %. When comparing the data on runoff changes for the period 2008–2017 with the average multi-year values for months for the period 1948–2017, there was positive trend in changes in most cases except August, September, and October. During these months runoff has decreased by an average value of 9 % over the past decade.

Intra-annual unevenness of river runoff can act as a major factor of hydro-ecological safety of the territory. The number of years (k) with exceeding the maximum water consumption by 15% was selected as a criterion for assessing the hydro-ecological safety of the territory. Table 1 shows the values of the φ-coefficient averaged over two periods and the calculated values of k for the Pripyat River at the Mozyr station.

Table 1. Averaged values of the φ-coefficient and the number of years with exceeding the maximum runoff of 15% security

Period	Averaged values of the φ-coefficient	Calculated values of *k*
1944–1980	0.67	9
1981-2017	0.78	2

Table 1 shows that during the period 1981–2017 the number of years with exceeding the maximum runoff of 15% security has significantly decreased and the φ-coefficient has increased by 16 % compared to the previous period.

4 Conclusions

The article considers the features of intra-annual runoff distribution of rivers in the Pripyat River basin. Changes in the nature of natural runoff regulation of rivers are shown based on data from 10 gauging-stations for the period 1945–2017. The relationship between the share of spring flooding and the coefficient of natural runoff regulation was established. There was an increase in the natural runoff regulation coefficient after 1970–1983 by 15–56% and a decrease in the runoff irregularity for all rivers of the Pripyat River basin. The coefficient of natural runoff regulation of rivers can serve as an important factor of hydro-ecological safety of the territory. The number of years with the maximum runoff exceeding 15% safety decreases when the φ-coefficient increases for the rivers of the Pripyat River basin.

References

1. P. Choudhury, B. Sil, J of Hydrol., **385**, 313 (2010)
2. R. Harding, G. Weedon, H. van Lanen, D. J of Hydrol., **518**, 186 (2014)
3. S. Huang, F. Hattermann, V. Krysanova, A. Bronstert, Climate Change, **116**, 631 (2013)
4. V. Mikhailov, Water Resources, **37**, 145 (2010)
5. Y. Sang, D. Wang, J. Wu, Q. Zhu, L. Water Sci. and Eng., **4**, 36 (2011)
6. A. Volchak, S. Parfomuk, Limnol. Rev., **4**, 185 (2018)
7. N. Bosch, J. Allan, D. Dolan, H. Han, R. Richards, J of Great Lakes Res., **37**, 268 (2011)
8. J. Makarewicz, T. Lewis, E. Rea, M. Winslow, D. Pettenski, J of Great Lakes Res., **41**, 123 (2014)
9. E. Schiefer, E. Petticrew, R. Immell, M. Hassab, D. Sonderegger, Anthropocene, **3**, 61 (2013)
10. V. Krysanova, D. Müller-Wohlfeil, A. Becker, Ecol. Model., **106**, 261 (1998)
11. A. Lobanova, H. Koch, S. Liersch, F. Hattermann, Hydrol. Proc., **30**, 5039 (2016)
12. J. Stagl, F. Hattermann, Water, **7**, 6139 (2015)
13. M. Wortmann, V. Krysanova, Z. Kundzewicz, B. Su, X. Li, Hydrol. Proc., **28**, 6337 (2014)
14. S. Shaghaghi, H. Bonakdari, A. Gholami, I. Ebtehaj, M. Zeinolabedini, Appl. Math. and Computation, **313**, 271 (2017)
15. W. Wang, K. Chau, C. Cheng, L. Qiu, J Hydrology, **374**, 294 (2009)
16. A. Zaji, H. Bonakdari, B. Gharabaghi, Acta Geophysica, **66**, 717 (2018)
17. M. Zeynoddin, H. Bonakdari, A. Azari, I. Ebtehaj, B. Gharabaghi, H. Madavar, J Env. Man., **222**, 190 (2018)
18. A. Volchak, I. Kirvel, S. Parfomuk, R. Makhambetova, Limnol. Rev., **4**, 221 (2013)
19. I. Kirvel, A. Volchak, S. Parfomuk, Quaest. Geogr., **35**, 57 (2016)
20. A. Furat, M. Al-Faraj, M. Scholz, Wat. Res. Man., **28**, 5477 (2014)
21. D. Hannah, B. Smith, A. Gurnell, G. McGregor, Hydrol. Proc., **14**, 317 (2000)
22. L. Liucci, D. Valigi, S. Casadei, Wat. Res. Man., **28**, 881 (2014)
23. X. Tu, V. Singh, X. Chen, L. Chen, Q. Zhang, Y. Zhao, Wat. Res. Man., **29**, 2677 (2015)

24. A. Volchak, O. Natarova, *Scientific notes of RSMU*, **52**, 67 (2018)
25. An. Volchek, *Scientific and technical problems of water management and energy complex in modern conditions of Belarus*, 109 (2011)
26. V. Evstigneev, D. Magritsky, *River flow and hydrological calculations* (Moscow, 2013)
27. A. Komlev, G. Proskurina, Meteorology and Hydrology, **10**, 95 (1977)
28. A. Volchak, S. Parfomuk, S. Sidak, *Natural environment of Polesie: features and prospects of development*, **11**, 29 (2018)
29. N. Frolova, D. Nesterenko, N. Schoenberg, Bul. Mosc. Un. Geography, **6**, 8 (2010)

Availability of hydro-chemical indicators of river water

Alexander Volchek[1], *Maya* Taratenkova[1]*
[1]UO BrSTU, Brest, Republic of Belarus

Abstract. The article contains studies to identify the law of distribution of availability for hydrochemical quality indicators such as dissolved oxygen, BOD5, COD, suspended solids, pH and mineralization. Time series analysis was carried out using data from the Mukhavets River. The revealed law of distribution allows considering the issues of design and [1]construction of water facilities more reasonably.
Keywords: hydro-chemical quality, availability, river water, dissolved oxygen, BOD5, COD, suspended solids, pH, mineralization.

Introduction

The intensification of anthropogenic impact on water resources requires changes in the regulation and methodological approach in determining quantitative characteristics. The authors in [7] propose a new approach for sustainable water resources management and achieving long-term water security for people and the environment. It consists in the perception of water resources not only as an integral object but also in understanding the functioning of separate components of the system. The hydrochemical regime of river waters is a complex system that cannot be considered without hydrological parameters as evidenced by the following studies [8]. The authors in the article [9] propose to consider the hydrochemical regime of the catchment through the prism of hydrological and climatic parameters. The results of studies [11, 21] testify to the correlation of hydrological and hydrochemical characteristics. When studying the dynamics of river water quality indicators such as total alkalinity, electrical conductivity, dissolved oxygen and others seasonal changes were revealed. This has been confirmed in studies [16]. In work [10], the main sources and factors of changes in the quality of surface waters were identified which include geological, climatic and anthropogenic. The influence of the latter factor on the ecological system of the Republic of Belarus is also discussed in [17]. As stated in article [19], the concentration of dissolved substances such as chlorides, sulfates, etc. also increases with an increase in the degree of urbanization. The mechanisms of interaction between surface and ground waters are described in [12]. The analysis was carried out for the main components of natural waters such as cation-anion composition and electrical conductivity. Another important factor in the formation of the hydrochemical regime of surface waters is the type of feeding as evidenced by the results of studies [18].

* Corresponding author: taratenkava@mail.ru

In most cases, the study of the hydrochemical regime is carried out according to the average concentrations of quality indicators. The use of such hydrochemical parameters does not often make it possible to adequately assess the state of aquatic ecosystems.

When designing water facilities, building and road structures, the criterion of the availability of hydrological quantities is widely used. This indicator reflects the probability of exceeding the characteristic among its possible values. Most often, it is required to determine the hydrological characteristics of a given supply which is easily solved if the distribution law of this quantity is known. However, in practice, the distribution law is often not known. As for the determination of the availability of hydrochemical values, then mention in the literature, does not occur. Therefore, the study of this issue is of practical interest for engineering calculations in water management design.

Literature sources mention the distribution of errors in the data of hydrological and hydrochemical observations while the complexity and multicomponent nature of the error in observing water quality is noted [6]. The law of distribution of probabilities of various indicators of surface water quality is discussed in works [13, 14]. The derivation of the statistical distribution is carried out through the balance equation for impurity and water at the water intake. The fact of seasonal variations in surface water quality indicators was also established. The article [15] discusses a number of statistical methods that can be used to describe the dynamics of hydrochemical indicators which indicates the widespread use of statistical approaches in hydrochemistry.

The purpose of this work is to establish the distribution law for the availability of hydrochemical indicators.

Method

Data of the State Water Cadastre (SWC) of the Republic of Belarus were used in the course of the research in the period from 2010 to 2019. The starting material used for the time series on the Mukhavets River in the alignment above the city of Brest according to the following hydrochemical parameters: BOD5, suspended solids, total mineralization, COD, dissolved oxygen.

Empirical security was determined by the formula:

$$p = \frac{m}{(n+1)} \cdot 100\% \,; \qquad (1)$$

where m – is the ordinal number x in the ranked row; n – is the length of the row.

Based on the above hydrochemical indicators, empirical supply curves were constructed, according to which the law of supply distribution was selected. The empirical law of distribution of the availability of hydrochemical parameters was carried out for a normal distribution according to the criterion of consent χ^2.

The Mukhavets River is a typical Polesye river experiencing anthropogenic pressure from the urbanized territory. The hydrochemical regime of the river was investigated earlier in our work [1] and regularities were found between such parameters as total mineralization and specific conductivity from discharge. According to the results of the study [2, 3], it was revealed that the probability density distribution of some water quality indicators obey the two-parameter lognormal law. The synchronicity of the distribution of hydrochemical parameters on the Mukhavets River at sections 1.7 km below Kobrin and 0.8 km above Brest was studied in work [4].

Results and their discussion

A hypothesis is put forward in this paper that a random variable which is characterized by a certain distribution law can be a mathematical model for describing the dynamics of hydrochemical processes. This approach has found wide application in hydrology, climatology and other natural sciences. There is no information in the literature on the application of these methods for hydrochemical parameters.

As mentioned above, the random variable (x) is characterized by a distribution law. Analytically, the law is expressed by the distribution function which is described as a parametric expression. With such a record for each value of the random variable, the corresponding value of the distribution function is uniquely determined by the parameters of the analytical expression [5].

When determining the law of distribution of hydrochemical quantities the choice of the type of distribution function should be made taking into account the range of variation of its argument. Changes in the value of hydrochemical parameters usually exceed zero and are theoretically not limited by the upper limit.

When considering the distribution of some hydrochemical parameters (Fig. 1) a certain asymmetry is observed which justifies the assumption of a lognormal distribution.

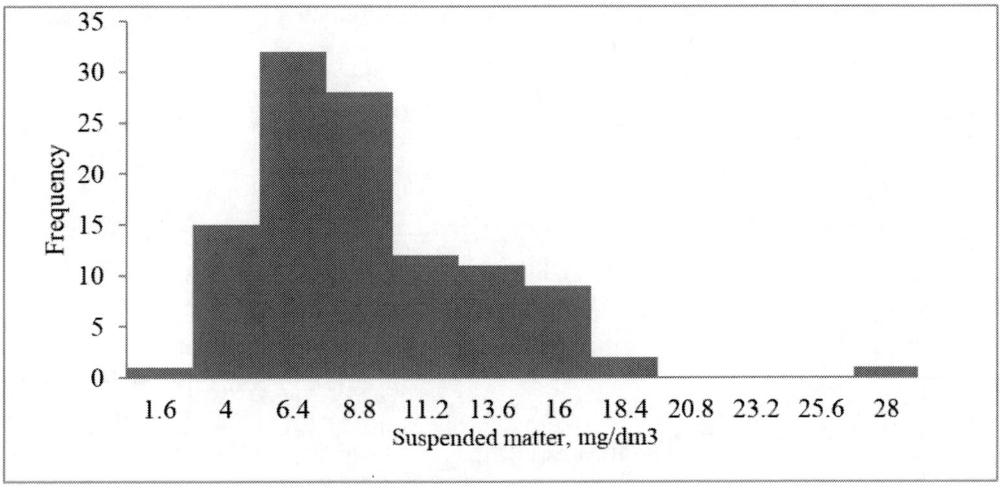

Fig. 1. Histogram of the distribution of suspended solids

Lognormal distribution assumes that the logarithm of the random variable (z = lnx) is distributed according to the normal law. The cumulative distribution function in this case has the form:

$$F(x) = \begin{cases} 0, \ x \le 0 \\ F(u) = \frac{1}{\sqrt{2\pi}} \int_{-\infty}^{u} exp \frac{-s^2}{2} ds, \ x > 0 \end{cases} \quad (2)$$

$$u = \frac{(z - m_z)}{\sigma_z} \quad (3)$$

where s – is an integrated variable;
mz – mathematical expectation of a random variable Z;
σz – is the standard deviation of the random variable Z;

Differentiating expression (2), we obtain the probability density function of the lognormal distribution:

$$f(x) = \begin{cases} 0, \ x \le 0 \\ \frac{1}{\sigma_z x} f(u) = \frac{1}{\sigma_z x \sqrt{2\pi}} exp\left(\frac{-u^2}{2}\right), \ x > 0 \end{cases} \quad (4)$$

From expressions (2) and (3) it can be seen that the distribution law depends on two parameters: the mathematical expectation and the standard deviation of a random variable.

The variance, standard deviation and mathematical expectation of a normal distribution of a random variable and a lognormal distribution are related by the following relations [5]:

$$D_z = \sigma_z^2 = \ln(\frac{\sigma_x^2}{m_x^2} + 1) \quad (5)$$

$$m_z = \ln(m_x) - \frac{\sigma_z^2}{2} \quad (6)$$

The series of observations of hydrochemical parameters are, as a rule, of limited length which does not allow determining the values of the maximum or higher supply. An analytical distribution curve was constructed to solve this problem, the curve logically determines the value of the parameter of any security. Distribution curves of security are constructed for such parameters as BOD_5, suspended solids, dissolved oxygen, COD, total mineralization. As an example (Figs. 2 and 3), distribution curves for the parameters suspended matter and COD are shown.

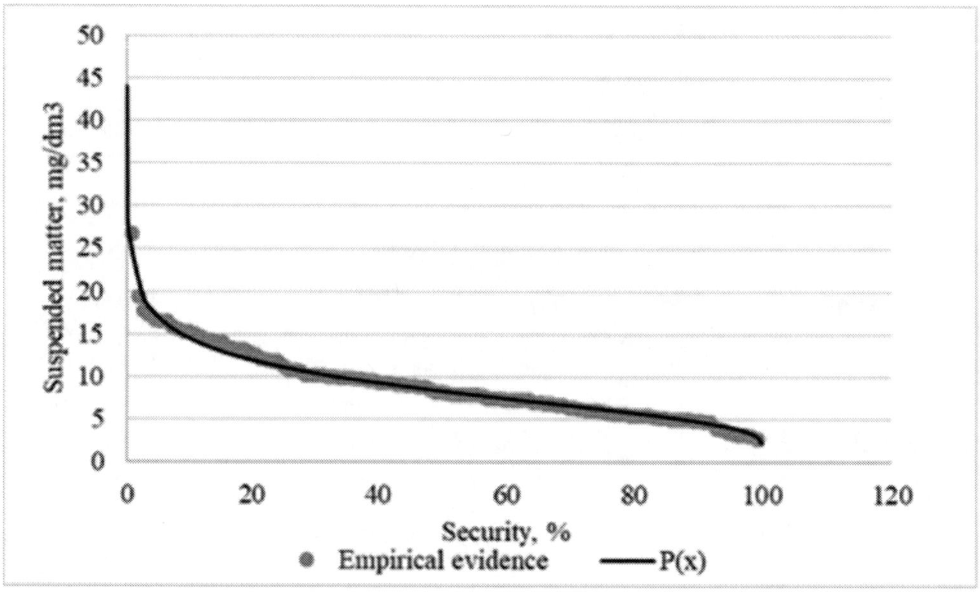

Fig. 2. Schedule of suspended solids supply

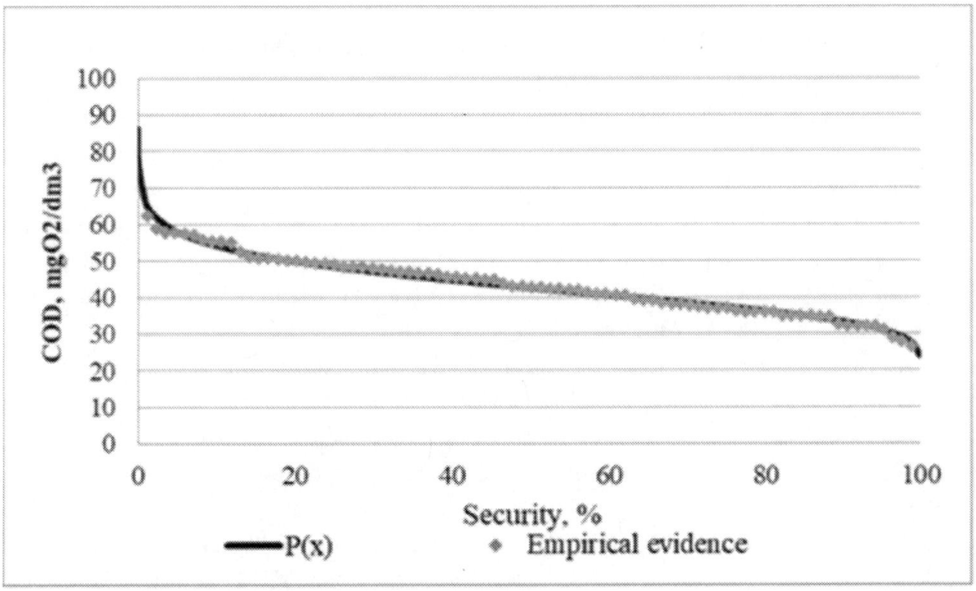

Fig. 3. Schedule of COD supply

Table 1 shows the probability density equation for the distribution of hydrochemical parameters as well as the parameters of the lognormal distribution.

Table 1. Parameters of the lognormal distribution

Indicator	Density equation of probabilities of a lognormal distribution	Lognormal distribution parameters
BOD_5	$f(x)=\frac{1}{0,73}\exp(-\frac{(lnx-0,29)^2}{0,87})$	m_z =0,29 σ_z =0,66
Suspended matters	$f(x)=\frac{1}{1,12}\exp(-\frac{(lnx-2,10)^2}{0,41})$	m_z =2,10 σ_z =0,45
Mineralization	$f(x)=\frac{1}{0,28}\exp(-\frac{(lnx-5,82)^2}{0,02})$	m_z =5,82 σ_z =0,11
Dissolved oxygen	$f(x)=\frac{1}{0,55}\exp(-\frac{(lnx-2,14)^2}{0,10})$	m_z =2.14 σ_z =0,22
pH	$f(x)=\frac{1}{0,06}\exp(-\frac{(lnx-2,04)^2}{0,01})$	m_z =2.04 σ_z =0,03
COD	$f(x)=\frac{1}{0,37}\exp(-\frac{(lnx-3,75)^2}{0,06})$	m_z =3,75 σ_z =0,18

Conclusion

As a result of studying the time series of quality indicators of the Mukhavets River, it was revealed that the distribution of the availability of hydrochemical quantities, such as BOD_5, suspended solids, dissolved oxygen, COD, total mineralization and calcium-ion obey the lognormal distribution law. This makes it possible to determine these parameters with a given security. What can help in water facilities construction. The law of distribution of the availability of the hydrochemical indicator makes it possible to design water treatment facilities with the required degree of reliability and validity.

References

1. A.A. Volchek, M.A. Taratenkova, Mat. of the intern. conf. ded. to the 145th anniv. of UE "Minskvodokanal" in two parts, 32-36, (2019).

2. A.A. Volchek, M.A. Taratenkova, Coll. of mat. of the IV Inter. scientific-practical conf., timed to the 1000th anniversary of the city of Brest "Actual problems of geosciences research of transboundary regions" in two parts, 6-10, (2019).

3. A.A. Volchek, M.A. Taratenkova, Coll. of mat. of the regional scientific and practical conf., timed to the 50th anniversary of the Department of Water Supply, Wastewater Disposal and Protection of Water Resources "Advanced Methods of Natural and Waste Water Treatment", 18–23, (2019).

4. A.A. Volchek, M.A. Taratenkova, Agrar. Land., their stability and development features: collection of scientific papers based on materials from Mezhdunar. scientific. eco. conf., 401–404, (2020).

5. A.V. Sikan, Methods of statistical processing of hydrometeorological information, 279 p., (2007).

6. H. Mcmillan, J. Freer, T. Krueger, Hydrological Processes 4078-4111. (2012).

7. Th. Wagener, M .Sivapalan, P.A Troch, B.L. McGlynn, C. J. Harman, H.V. Gupta, P. Kumar, P.S.C. Rao, N.B. Basu, J.S. Wilson. Water Resources Research, 46 (5), 1-10 (2010).

8. A. Hodson, M. Tranter, A. Gurnell, M. Clark, J.O Hagen, 2002, 1-4, 91-114.

9. C. Soulsby, C. Gibbins, A. J.Wade, R. Smart, R. Helliwell, Science of the Total Environment, 1-3, 73-94, (2002).

10. R. Bouza-Deaño, M. Ternero-Rodríguez, A. J. Fernández-Espinosa, Journal of Hydrology, 631 (3-4), 227-239, (2008).

11. A. R.C. Ovalle, C. F. Silva, C. E. Rezende, C. E.N. Gatts, M. S. Suzuki, R. O Figueiredo, Journal of Hydrology, 191-203 (2013).

12. M. R. Guggenmos, C. J. Daughney, B. M. Jackson, U. Morgenstern, Hydrology and Earth System Sciences, 15, 3383-3398, (2011).

13. B.M. Dolgonosov, K.A. Korchagin, Water Resources, 1, 39-48, (2014).

14. B.M. Dolgonosov, K.A. Korchagin, Water Resources, 3, 345-358, (2011).

15. C. E.M. Lloyd, J. E. Freer, A. L. Collins, P. J. Johnes, J. I. Jones, Journal of Hydrology, 514, 297-312, (2014).

16. V.M. Shulkin, N.N. Bogdanova, L.V. Perepelyatnikov Water Resources, volume 36, 4, 428-439, (2009).

17. V.S. Khomich, Nature management, 22, 26-32 (2012).

18. C. Neal, J. Wilkinson, M. Neal, M. Harrow, H. Wickham, L. Hill, C. Morfitt, Hydrology and Earth System Sciences, 1, 583-617, (1997).

19. S. Rose, Journal of Hydrology,341(1-2):42-54, (2007).

20. C. Soulsby, C. Gibbins, A. J. Wade, R. Smart, R. Helliwell, Science of the Total Environment, 256 (1-3), 51-63, (2002).

High-performance electric arc source of ultraviolet radiation to ozone generate

Andrey Veremeychik[1,*], *Evgeny* Dmukhaylo[1], *Sergey* Onysko[1], *Mikhail* Sazonov[1], *Vitaly* Khvisevich[1]

[1]Brest State Technical University, Moskovskaya str. 267, Brest, 224017, Belarus

Abstract. Studies of electric arc combustion in the swirling flow of a cylindrical discharge chamber of a plasma torch at oxygen consumption of 10-160 g/s are carried out. It is established that the electric field strength is constant along the channel at the initial section of the arc, and the length of this section is determined. On the basis of the similarity theory in a wide range of changes in the parameters of the plasma torch, methods for determining the electric field strength of the arc and the length of the initial section are obtained. The length of the plasma torch is calculated, which mainly includes the length of the initial section, where the voltage is constant, and the current-voltage characteristics. The design of the plasma torch was created. Theoretical studies of near-electrode processes at the cathode and anode were carried out, on the basis of which electrode materials were selected and cathode and anode structures were created that provide high thermal efficiency and service life of the plasma torch. The developed plasma torch can be used as a highly efficient arc source of ultraviolet radiation for generating ozone, for heating oxygen, as well as in metallurgy, chemical industry, for example, the production of titanium dioxide and ozonation.

Keywords: plasma torch, electric arc, ozone, ultraviolet radiation, current-voltage characteristics, voltage.

1 Introduction

The field of ozone usage has expanded significantly and new research in this direction is being intensively carried out all over the world in recent years. The rapid development of technologies using ozone is facilitated by its ecological purity. Ozone forms saturated oxides during the reactions unlike other known oxidants. In this case, unused ozone is decomposed into atomic and molecular oxygen. All these products as a rule do not pollute the environment and do not lead to the formation of carcinogenic substances as for example when oxidized with chlorine or fluorine.

The most economically profitable way of producing ozone on an industrial scale is currently considered to be its synthesis in electric discharges of oxygen or oxygen-containing gas mixtures. A number of articles have been published concerning water purification using gas electric discharges over the past (10 - 15) years. Most of this research

*Corresponding author: vai_mrtm@bstu.by

is related to the oxidative degradation (destruction) of organic pollutants using non-equilibrium plasma. Improved oxidation processes which include the formation of highly reactive chemical oxidants (such as O, OH, O_3, H_2O_2) have become an important step in accelerating the oxidation process and hence the breakdown of a wide range of organic pollutants in water.

Traditional approaches and technical solutions in the field of water treatment are often ineffective for the treatment and disposal of wastewater containing toxic, biodegradable and difficult to oxidize organic pollutants. In this regard, water purification technologies based on the use of strong oxidants are promising.

Due to the constant complication of the composition of wastewater, in particular due to biologically intractable and toxic organic compounds, technologies based on the use of various oxidants are promising.

In terms of its oxidizing capacity, ozone (redox potential 2,07 V) ranks third among the known oxidants, inferior to fluorine (oxidizing potential 2,41 V) and oxygen fluoride while chlorine (oxidizing potential 1,73 V) is in the eighth place and ordinary oxygen (O_2) is only in the thirteenth place. The reactions of ozone with aromatic compounds formed the basis of modern technologies for deodorization of various environments, rooms, drinking and waste waters. The advantages of using ozone in the processes of purification and disinfection of water, air and other liquid and gaseous areas are generally recognized. Currently, about 95% of drinking water in Europe is ozone treated. In the USA, the process of transferring water supply systems from chlorination to ozonation is underway [1].

2 Ozone and its usage

Currently, ozonation is the only universal method of water treatment that allows to effectively affect a large number of various pollutants of artificial and natural origin with the simultaneous disinfection of water [1]. Long-term experience in the use of ozone and the operation of ozonation installations convinces that this method is highly effective. Further improvement of the ozonation technique will eliminate the disadvantages inherent in the method and it will be widely used.

In [2] are described methods for measuring odor at treatment facilities, the main methods used to reduce emissions of foul-smelling substances from wastewater disposal facilities are described.

Wastewater treatment of industrial enterprises using ozonation is one of the most promising and modern methods [3]. Wastewater ozonation can achieve the following goals:

- oxidation of organic and inorganic compounds;
- discoloration of water;
- deodorization of water containing substances of biological origin and contaminated with organic substances;
- treatment of effluents containing oil, oil products and phenols;
- treatment of effluents containing nitro compounds, carcinogenic and other substances;
- wastewater treatment after biological treatment;
- purification of acetone-containing and hydrogen sulfide-containing waters.

Ozone destroys all known microorganisms: viruses, bacteria, fungi, algae, their spores, cysts, etc. and acts very quickly - within a few minutes. Ozone removes unpleasant odors and does not form toxic by-products. Ozone destroys microorganisms 300 to 3000 times faster than any other disinfectant.

In crop production, ozone is currently used in the following areas:

- stimulation of plant cultivation by reducing microbial contamination of the plants themselves, soil and air as well as increasing the synthesis and accumulation of nutrients;

- pre-sowing treatment of agricultural crops in order to increase their sowing qualities and yield properties, as well as resistance to adverse influences;
- controlling plant pests and diseases and reducing the use of herbicides and pesticides;
- the use of an ozone-air mixture as a drying agent;
- conservation and preservation of agricultural products, including in a wet state, reducing energy consumption during drying, reducing losses during storage;
- reducing the toxicity of harmful substances contained in agricultural products.

Crop products are often heavily contaminated with various infections, mold as well as small insects it can lead to significant losses during storage and processing and is especially dangerous, contributes to the spread of infectious diseases (staphylococcus, E. coli, salmonella, etc.).

The use of ozone-air mixtures allows:

- accelerate the processing of grain, reduce its energy intensity, ensure an increase in sowing quality and prevent spoilage;
- to increase the terms of safe storage of grain (1,5 - 2) times, to increase seed quality by 15%;
- to reduce the duration of low-temperature drying and energy consumption for drying grain by (15 - 20)%;
- to reduce the viability of mold and leads to the destruction of toxins.

According to [3], one of the main ways to increase the yield of agricultural crops is to protect plants from diseases, in particular, from those phytopathogens the spores of which are localized on the surface of seeds. The most harmful of these are pathogens of hard smut and root rot. Losses of grain crops from these diseases can reach (20 - 35)%. One of the ways to solve this problem is to use ozone. It is known that at an ozone concentration of about 2 mg/m^3 and a 30-minute treatment, corn, wheat, barley and oats are sterilized from all types of microorganisms. When treating barley seeds with an ozone-air mixture with ozone concentrations of 6 mg/m^3 and 10 mg/m^3, the yield was 12,0% and 17,5% higher than when treated with pesticides. Also, according to field trials, an increase in yield was recorded for wheat - 22%, peas - 11%, buckwheat - 31%.

In [4], the dynamics of the state of beetles and insects in grain and grain products after their treatment with ozone was studied. It is shown that when using an ozone-air mixture with an ozone concentration of 1 g/m^3 on the first day after treatment, a large number of paralyzed individuals are noted, which are not able to restore their state to normal and always die on the following days.

Recently, scientific researches have appeared on the effectiveness of the use of ozone in the fight against coronavirus. More and more clinical data is emerging on the successful application of "ozone therapy for coronavirus". It turned out that physiotherapy treatment with ozone can prevent the transition of COVID-19 to a severe form and reduce the need for potent medications. The pioneers in the application of this method of physiotherapy for COVID-19 were the countries that took the especially powerful blow of the new viral infection (USA, Spain, Italy, China, etc.). US Doctors R. Rowan and H. Robins report on the effectiveness and economic benefits of ozone therapy for the treatment of COVID-19. There is evidence of successful experience in the use of ozone against the SARS-CoV-2 virus in Romania, Georgia, Peru, Chile, Ecuador, Turkey and other countries [5].

The official expert opinion of the International Scientific Committee for Ozone Therapy ISCO3 on the use of ozone in the fight against ARVI infection with the COVID-19 strain is given in [6]. It provides a detailed explanation on whether ozone is effective in combating coronavirus. The International Scientific Committee on Ozone Therapy ISCO3 together with WHO (World Health Organization) officially declare that "there are currently no vaccines or special pharmaceutical treatments for COVID-19," therefore alternative methods of limiting the spread of the virus are being developed and considered. It is known that the virus lives on surfaces for a long time (furniture, door handles, elevator buttons, handrails, etc.) room ozonation allows you to sterilize the room with high efficiency and destroy all strains of the virus and infections. Ozonation is much more effective than sprays and disinfectants since it has a deep penetration rate into surfaces and crevices and, unlike chemicals, does not leave behind odors and harmful traces as it is natural gas. The room becomes completely safe for a person. At this difficult time, all public places that are allowed to be visited and these are medical institutions, educational institutions, shops, courtrooms, hairdressers, beauty salons, office and industrial premises need treatment (disinfection).

3 Ozone synthesis

International Scientific Committee on Ozone is formed in a gas environment containing oxygen if conditions arise in which oxygen dissociates into atoms. This is possible in all forms of electric discharge: glow, arc, spark, corona, surface, barrier, electrodeless, etc. The main reason for dissociation is the collision of molecular oxygen with electrons accelerated in an electric field. In addition to the discharge, the dissociation of oxygen is caused by ultraviolet radiation with a wavelength of less than 240 nm and various high-energy particles: α-, β-, γ-particles, X-rays, etc.

There is a group of reactions as a result of which ozone decomposes in almost all sources of ozone formation. These are thermal decomposition in the volume and on the walls of the reactor, its reactions with radicals and excited particles, reactions with additives and impurities that can come into contact with oxygen and ozone.

The main methods for the synthesis of ozone:

- Photochemical method isthe main share of ozone produced on Earth in nature is formed by this method. The main area of its use is the production of medium and low ozone concentrations. In practice, mercury and excimer xenon lamps are used to produce ozone by this method.
- Barrier discharge is a discharge that occurs between two dielectrics or a dielectric and a metal. Due to the fact that the electrical circuit is broken by a dielectric, power is supplied only with alternating current. At low capacities, the ozonizerneed not be cooled, since the released heat is carried away with the flow of oxygen and ozone.Ozone is also synthesized in arc ozonizers (plasmatrons), in glow ozone generators (lasers) and surface discharge in industrial production.
- Corona discharge - this method of producing ozone is the most reliable and efficient of all known and therefore is most widely used in industrial conditions. It features an optimal ratio of energy consumption to the concentration of generated ozone. A corona discharge occurs in a gas in a highly inhomogeneous electric field between two electrodes - high-voltage and grounded, separated by a gap (discharge gap) and a dielectric. Ozone is formed as a result of the dissociation of an oxygen molecule as a result of the action of the energy of electrons moving between the electrodes through the discharge gap. The ozone concentration depends on the voltage, its frequency, dielectric thickness, dielectric constant, as well as on the oxygen concentration in the working gas, which is determined by the type of working gas - dried or undried air, oxygen, as well as the working gas pressure in the discharge gap. A known method for the synthesis of ozone [7] in the corona discharge of direct and alternating current. It was established in [7] that in all cases of a corona discharge, there is a dependence of the formation of ozone on the material of the electrodes, as well as on the operating time of the ozonizer, which is explained by corrosion and sputtering of metal electrodes.
- Electrolytic synthesis method - usually aqueous solutions of perchloric or sulfuric acid are used as an electrolyte, electrodes are made of platinum. The use of some acids has

shown that they do not give up their oxygen during the formation of ozone. Therefore, the gross scheme should take into account only the decomposition of water ($H_2O + O_2 \rightarrow O_3 + 2H + \ + e^-$) with possible intermediate formation of ions or radicals.

- The formation of ozone under the influence of ionizing radiation - ozone is formed in a number of processes accompanied by the excitation of an oxygen molecule either by light or by an electric field. When oxygen is irradiated by ionizing radiation, excited molecules can also arise and ozone is observed to form.
- Ozone formation in a microwave field - when a jet of oxygen was passed through a microwave field, ozone formation was observed. This process is poorly understood, although generators based on this phenomenon are often used in laboratory practice.

According to [8], the most economically profitable way of producing ozone on an industrial scale is currently considered to be its synthesis in electric discharges of oxygen or oxygen-containing gas mixtures. It should be noted that ozone electrosynthesis as a rule is a rather energy-intensive process.

A significant number of articles are devoted to the synthesis of ozone in electrical discharges [9–23]. The article [9] proposes a method for the synthesis of ozone in a combined discharge, in which the surface and barrier discharges are connected in series in an electric circuit, experimental studies of the characteristics of the synthesis of ozone from oxygen at the atmospheric pressure of this discharge are carried out depending on the parameters of the discharge circuit, voltage and gas flow rate. An ozonizer built on the basis of a combined discharge allows one to obtain high ozone concentrations up to 100 g/m³ at an energy consumption of about 12 kWh/kg. In [10], the influence of the source gas (dried air and oxygen) on the process of ozone synthesis in a barrier discharge, on the structure and parameters of ozonation stations, on the operating conditions of ozonation equipment, and on the economic indicators of the ozone production process is considered. The authors of [11] propose to use the synthesis of ozone in a corona discharge between coaxial electrodes under the influence of pulses up to 27 kV with a duration of 25 ns with a repetition rate of 130 Hz, while the efficiency of O_3 formation reached 130 g/kWh.

In [12], a set of conditions and requirements for the discharge circuit necessary for the generation of ozone from air and oxygen with minimal specific energy consumption is considered, the possibility of generating ozone with an energy efficiency close to the thermochemical limit of 0.85 kWh/kg O_3 is shown, the possibility of water disinfection is shown using a corona flare discharge with energy consumption at the level of 1 kWh/m.

4 High performance plasma generator for wastewater ozonation process

Electric arc gas heaters (plasmatrons) are widely used for stationary irradiation to the state of low-temperature plasma. Their use in industrial technologies has provided a variety of design solutions for plasma devices. The energy and electrophysical characteristics of plasmatronsare mainly determined by the geometric and gas-dynamic parameters of the discharge chamber [24–28]. Oxygen plasma generators are used in metallurgy, chemical industry, for example, titanium dioxide production and ozonation, and can also be used to produce ozone in various fields.

To obtain initial data for theoretical research, modeling and development of plasmatrons intended for the production of ozone, measurements of the electric field strength of an arc burning in oxygen were carried out. The nature of combustion is determined by the swirling flow and the speed of the gas flowing around the arc. The modes of formation of the boundary layer near the channel wall and flow around the arc strongly affect the processes of converting electrical energy into heat.

Arc studies were carried out on an experimental plasmatron (Fig. 1) with a fixed arc length and geometrically similar electric discharge chambers. The inner electrode was a zirconium cathode 2 mm in diameter, sealed flush into a copper cage.

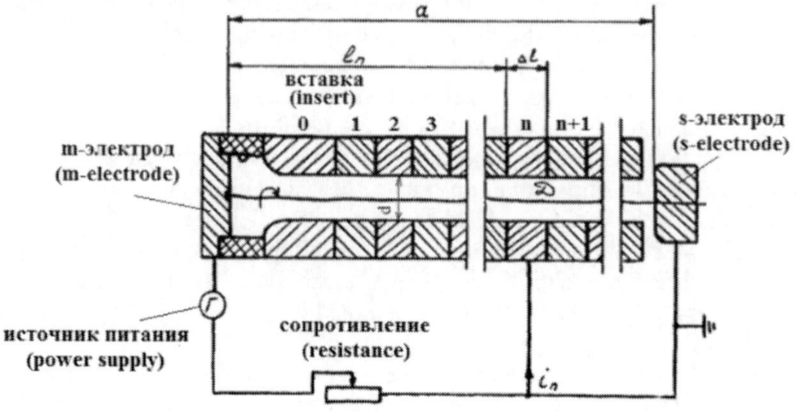

Fig. 1. Diagram of the experimental device.

A cylindrical discharge chamber is formed in this plasmatron with the help of sections in order to change the arc length and to measure the arc radiation, the development of the boundary layer and the length of its initial section where there is no convective heat exchange between the arc and the gas flow. This makes it possible to determine the length of the discharge chamber where the gas is exposed only to ultraviolet radiation.

The electric field strength along the arc E was determined by measuring the arc length, measuring the arc voltage [27]. The electric field strength of the arc was determined from the slope of the potential distribution curve along the *z* axis. The ranges of parameter variation when investigating the arc in oxygen are as follows: diameter of the discharge chamber (0,01 – 0,03) mm, distance between electrodes (0.15 - 0.36) m, arc current (30 - 180) A, oxygen consumption (2 - 4) g/s. A typical dependence of the electric field strength of the arc on the current I at a gas flow rate $G = 4$ g/s is shown in Fig.2.

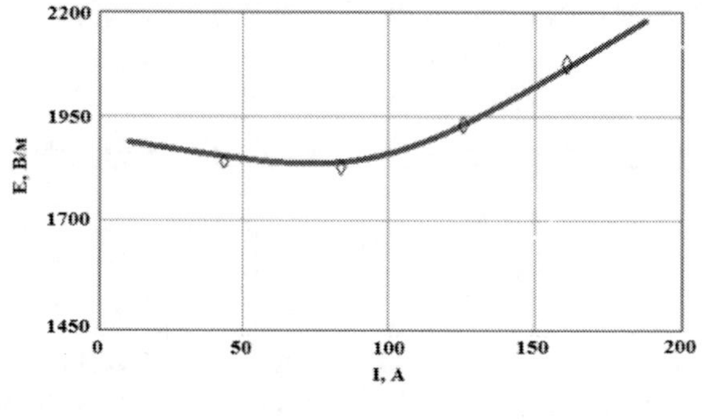

Fig. 2. The electric field strength of the arc in the oxygen flow at d = 1 cm.

A method for calculating plasmatrons has been developed on the basis of the similarity theory. The following criterion formula was obtained for calculating the E arc in an oxygen flow by processing the experimental results and generalizing them:

$$E = 3,17\times10^{-2}\left(\frac{G}{d}\right)^{0,15}(pd)^{0,13}\left(347 - 1\times10^{-2}\frac{I}{d} + 9,26\times10^{-2}\frac{I^2}{d^2}\right).$$

 https://doi.org/10.1051/e3sconf/202021201018

Dimensions of quantities: E - V/m, G - kg/s, I - A, d - m, p - Pa.

Current-voltage characteristics of the arc were studied [27, 28] to compare the electric field along the arc in vortex flows of various gases.Plasmatron was developed and constructed the diagram of which is shown in Fig. 3 with the help of the developed technique using the criterion formula. The main elements of the plasmatron are the zirconium cathode, the ignition electrode and the anode. The ignition electrode is made of copper in the form of a washer section with an inner diameter of d_1 = 16 mm. The tested copper anode had a diameter d_2 = 16 mm. The anode length l varied from 90 mm to 150 mm. The cathode, ignition electrode, anode and solenoid are intensively cooled with chemically purified water. Water is supplied to the plasma torch cooling system under pressure (10 - 15) atm. To calculate the heat fluxes into the electrodes, the temperature of the water at the inlet and outlet of the plasmatron was determined using chromel-copel thermocouples.

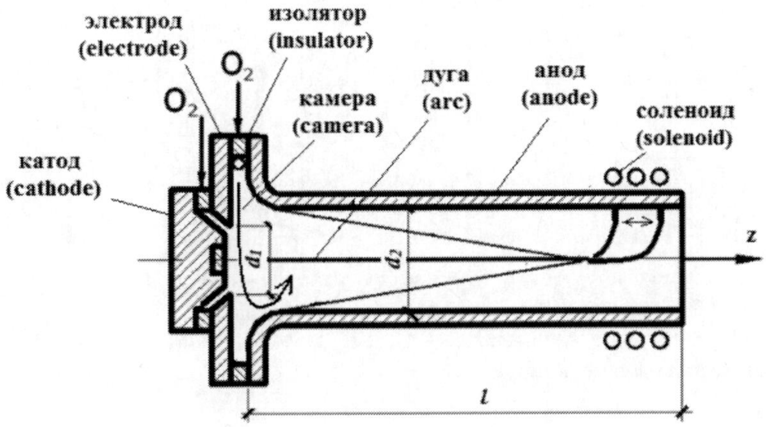

Fig. 3. Plasmatron diagram.

The oxygen pressure in front of the flow meters at the plasma torch inlet was (4 - 8) × 10^5 Pa. Oxygen was fed into the gap between the cathode and the ignition electrode through two tangential holes 1.2 mm in diameter located in a twist ring with an inner diameter of 50 mm, oxygen was supplied at a rate of (0,1 – 0,2) g/s. The second swirl chamber located between the ignition electrode and the anode was supplied with gas through four tangential holes 2.4 mm in diameter located in a swirl ring with an inner diameter of 64 mm. The total oxygen consumption varied from 0.5 g/s up to 2.4 g/s. The plasmatron was powered from an APR-404 type power source with a rated no-load current of 600 A and a voltage of 900 V. Because its external electrical characteristic is rigid and current-voltage characteristic of the arc is falling due to unstable arc burning; a stepwise ballast resistance is included in the power circuit in series with the arc in order to ensure its stable combustion. Plasmatronwas ignited using a high-voltage high-frequency oscillator with a voltage applied to the igniting electrode. A resistance (10 - 20) Ohm is connected between the ignition electrode and the anode via a contactor.

Current-voltage characteristic of the plasmatronwere studied. In addition, measurements of the arc radiation were carried out in the plasmatron section, cathode, anode and ignition electrode. This made it possible to calculate the efficiency of plasmatron, enthalpy and average mass temperature T of the gas depending on the magnitude of the arc current and the length of the anode at different gas flow rates. It was found that the efficiency is equal to (0,6 - 0,74). The diagram of the plasmatronis shown in Fig. 4. This plasmatron is a radiation generator. A reactor with tangential holes is installed at the outlet of the plasmatron through which waste water is supplied for quenching and mixing with O_3.

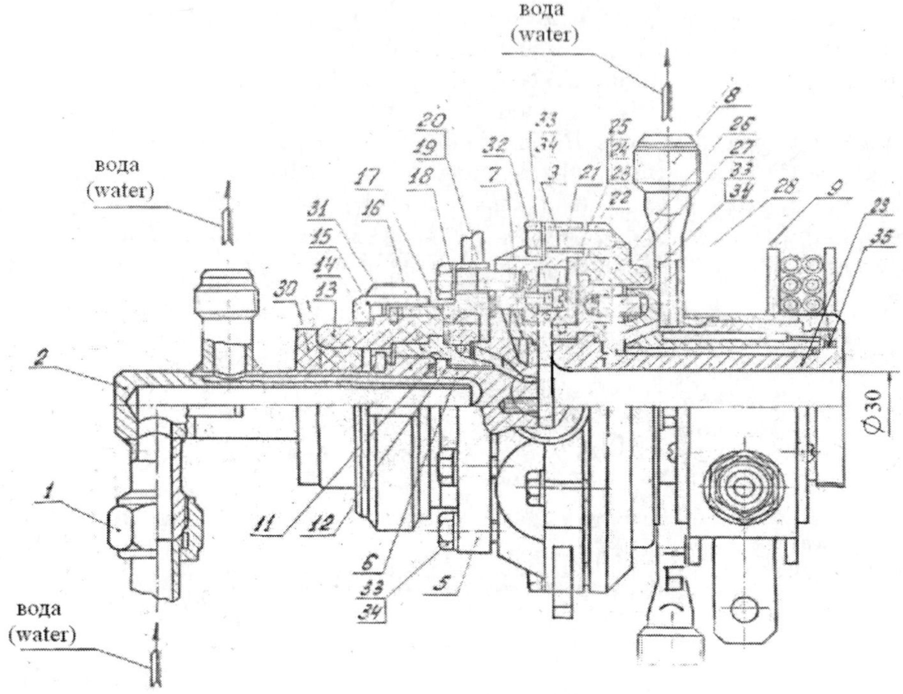

Fig. 4. General view of the plasmatron.

Conclusion

An experimental plasma installation was created, the current-voltage characteristics of the arc in the flow of oxygen and air at different arc lengths, the electric field strength along the arc and the length of its initial section were studied, and it was found that E_n=const. The radiation along the arc is measured, which is constant at different distances from the cathode in this section of the arc.

Conducted theoretical research on the basis of similarity theory and research results are summarized as a dependency for electric field strength of the arc, the current-voltage characteristics of the arc length of the initial section, which allows to count the torches of the given power.

A radiation generator with a power of up to 0.5 MW has been developed and created. The plasma torch was tested and the erosion characteristics of the electrodes were studied, which confirmed the high service life of the plasma torch.

References

1. V.A. Orlov. Ozonation of water. M.: Stroyizdat, 1984. 88 p.

2. S.V. Svitskov, D.A. Danilovich, V.N. Azarov. Wastewater treatment plants as a source of unpleasant odor: causes, characteristics and methods of control // Water supply and sanitary engineering. No. 7. 2016. P. 24–43.
3. http://www.kaufmanntec.ru
4. G.L. Zakladnoy, M.A. Osman. Biological assessment of ozone as a means of combating pests of grain and grain products // Storage and processing of agricultural raw materials. No. 5, 2011. P. 8–9.
5. https://www.secret-dolgolet.ru/ozonoterapiya-ot-koronavirusa
6. Adriana Schwartz, Gregorio Martínez-Sánchez. Potential use of ozone in SARS-CoV-2 / COVID-19. Official Expert Opinion of the International Scientific Committee of Ozone Therapy (ISCO3). ISCO3/EPI/00/04 (March 14, 2020). Approved by ISCO3 on 13.03.2020. P. 15.
7. Yu.V. Filippov and others. Electrosynthesis of ozone // Publishing house of Moscow State University, 1987. P. 21.
8. E.M. Silkin Synthesis of ozone in electrical discharges and increasing its efficiency // Components and technologies. No. 6. 2008. P. 136–143.
9. A.V. Tokarev, A.Yu. Makhkambaev. Synthesis of ozone in a combined discharge with a plasma induction electrode. Bulletin of KRSU. 2006. Volume 6. No. 5. P. 16–19.
10. V.G. Samoilovich, L.Yu. Abramovich Arguments "for" and "against" the use of air or oxygen for the industrial production of ozone // First All-Russian Conference "Ozone and other environmentally friendly oxidants" Materials conference, M. 2005. S. 144.
11. R.H. Amirov, E.I. Asinovsky, I.S. Samoilov, A.V. Shepelin. The characteristics of ozone generation from air by nanosecond corona discharge // 20 Int. Conf. Phenom. Ionized Gases, II Ciocco, 8th–12th July, 1991: Contrib. Pap. 1. Pisa, 1991. P. 279–280.
12. A.V. Tokarev Determination of the characteristics of a corona flare discharge as a source of ozone production, abstract of Ph.D. dis. ... Cand. physical-mat. Sciences: 01.04.14. Bishkek, 2000. 146 p.
13. W.J.M. Samaranayake, Y.Miyahara, T.Namihira, S.Katsuki, T.Sakugawal, R.Hackad, H.Akiyama. Pulsed Streamer Discharge Characteristics of Ozone Production in Dry Air // ZEEE Traitsactions on Dielectrics and Electrical lnsulafion. Vol. 7. No. 2, April 2000. P. 254–260.
14. A.A. Mynka, N.P. Polyakov. Method for the synthesis of ozone and a device for its implementation / Patent No. RU2220093C2.
15. Yu.V. Filippov, V.A. Voblikova, V.I. Panteleev Electrosynthesis of ozone. M.: Publishing house of Moscow State University, 1987.
16. V.V. Lunin, N.V. Karyagin, S.N. Tkachenko Methods for producing ozone and modern design of ozonizers: Textbook. M.: Max Press, 2008.
17. P.2346886RF, MKI S01V13 / 11. Ozone generator / E.M. Silkin. Publ. B.I. 2009. No. 5.
18. V.A. Voblikova, E.A. Shabrova, E.V. Shabrova. Optimization of ozone synthesis using special dielectric materials and increased current frequency // Abstracts of reports. Russian conference dedicated to ozone and other environmentally friendly oxidants, science and technology. Moscow. June 7–9, 2005.
19. V.V. Lunin, V.I. Gibalov, I.S. Tkachenko Ozone synthesis in a surface barrier discharge in oxygen // Abstracts of reports. Russian seminar dedicated to ozone and other environmentally friendly oxidants. Moscow. June 18-19, 2008.
20. Dingkun Yuan, Can Ding, Yong He, Zhihua Wang, Sunel Kumar, Yanqun Zhu &Kefa Cen. Characteristics of Dielectric Barrier Discharge Ozone Synthesis for Different Pulse Modes // Plasma Chemistry and Plasma Processing Vol. 37, P. 1165–1173 (2017).

21. I.S. Tkachenko. Modeling the synthesis of ozone in a surface barrier discharge in oxygen / Author's abstract of the dissertation for the degree of candidate of physical and mathematical sciences 02.00.04. Physical chemistry. Moscow, 2010. 23 p.
22. V.V. Andreev, Yu.P. Pichugin Effect of voltage polarity on ozone synthesis in a dielectric barrier discharge. - Applied Physics. 2017 (3). S. 47–51.
23. V.V. Andreev, A.N. Matyunin, Yu.P. Pichugin, V.G. Telegin, G.G. Telegin. Investigation of the efficiency of plasma-chemical ozone generators based on a barrier discharge in air at atmospheric pressure. VANT. Ser. Thermonuclear fusion. 2011, No. 4. P. 112–117.
24. M.F. Zhukov, A.S. Koroteev, B.A. Uryukov. Applied dynamics of thermal plasm / Novosibirsk: "Science" SB AN. 1975. 299 p.
25. G.Yu. Dautov, M.I. Sazonov. Electric field strength in a vortex-stabilized arc. PMTF, No. 4, 1967. S. 127–131.
26. A.F. Bublievsky. Criteria dependencies for non-consumable electric arcs in the channel // IFZh. 1997. T. 70, No. 1. P. 99–104.
27. A.I. Veremeychik, M.I. Sazonov, D.L. Tsyganov. Study of an electric arc in a helium flow in a high-current plasmatron // Instruments and experimental techniques. 2006. No. 5. P. 99–102.
28. V. V. Batrak, A. I. Veremeychik, M. I. Sazonov, V. M. Khvisevich. Investigation of an electric arc in an argon flow / // Bulletin of Brest. state tech. university. 2008. No. 4: Mechanical engineering. S. 26–28.

Effect of calcium sulfoaluminate additive on linear deformation at different humidity and strength of cement mortars

Kadhim Abdulwahid Al-Musawi[1], *Kottayil Bindhu Abraham*[2]*, *Tatsiana Potses*[1], *Leonovich Sergey*[1], *Beulah M*[2], *and Jerzy Obolewich*[3].

[1]Belarusian National Technical University, Minsk, Belarus.

[2]Christ (Deemed to be University), Bangalore, India.

[3]Politechnika Bialostocka, Poland.

Abstract. The effect of calcium Sulfoaluminate additives (CSA) on the compression and bending strength of mortar, as well as linear deformation of prism samples at different environmental humidity was studied. Test results indicate that bending strength of mortars with CSA and the referent at the age of 28 days are practically equal. Compressive strength of mortars with CSA reduced by 20 ... 23% for all dosages of CSA. Relative linear deformations depend on the humidity of the environment. At a humidity of 100%, the relative linear deformations are positive and the expansion increases with increasing dosage of the expanding additive. When hardening in dry air at a humidity of 55%, the greatest shrinkage deformations were observed for mortars with CSA. We can conclude that the expanding effect of CSA is fully manifested at high humidity, i.e. under construction conditions, this means very high-quality moisture care for concrete structures.

Keywords: Shrinkage, cracks, calcium Sulfoaluminate additives, expanding additives.

1 Introduction

The issue of concrete durability is one of the main issues in modern construction. It is well known that one of the most significant factors affecting the durability and security of monolithic and prefabricated structures is the appearance and development of various kinds of cracks. Different types of cracks and their causes are described in detail in the literature [1,2,3,4,5,8,9].

Generally, all cracks in concrete can be divided into 2 categories: cracks that appear in fresh concrete and cracks that appear after hardening of concrete.

Cracks in fresh concrete can appear as a result of plastic settlement and/or plastic shrinkage of concrete, as well as the result of both the movement of the formwork, foundation or other technological reasons. There are many reasons for cracking in hardened concrete.

*Corresponding author: bindhuabraham.23@gmail.com

It can be chemical shrinkage (volume reduction due to cement hydration), autogenous shrinkage (typical for concrete with a low water/cement ratio), drying shrinkage (water loss of hardened concrete in an environment with low humidity). Shrinkage cracks can also be caused by temperature deformations of massive structures, by errors in structural design/manufacturing/ operation, corrosion of reinforcing bars (wedge effect of corrosion products), carbonization, etc.

It is obvious that depending on the composition, the conditions of hardening and care, the age of concrete, different types of shrinkage deformations will prevail. However, for ordinary concrete (W/C >0.4) the most significant contribution to the shrinkage deformation of the structure is the drying shrinkage [2].

Three most effective ways to combat drying shrinkage can be [4]:

- Implementation of high-quality moisture care of the concrete structure;
- Compensation of shrinkage with the help of expanding mineral additives;
- Reduction of shrinkage due to Shrinkage Reducing Admixtures (SRA) based on glycols.

The use of the first method is of the most preferable, but it is not always feasible in construction.

Thus the most reliable and widespread way is the use of expanding mineral additives, that are used for concrete, mortar dry mixes, grouting etc. Expanding additives act due to the formation of ettringite (Sulfoaluminate additives) or calcium hydroxide (additives based on CaO). Sulfoaluminate additives, which are safer for humans than CaO-based additives, are most prevalent in the world and in Belarus. CSA-expanding additives have been used since the 80s of the last century [6]. Many years of experience in their use have revealed some of the disadvantages of using these modifiers. They are supplied only in dry form and have high dosages (usually 10% by weight of cement). There is also literature date that concrete with CSA expanding additives is very sensitive to the quality of moisture care. In this regard, this research is related to its study of the effect of humidity on the expanding potential of CSA-expanding additives.

2 Experimental Investigation

Materials

The properties of the materials used in the experiment are shown in the tables 1-5.

Table 1 – Cement

Type, producer	Mineralogical composition, %				Specific Surface, m^2/kg	Density kg/m^3	Standard Consistency, %	Strength of Cement, 28 days MPa
	C_3A	C_4AF	C_3S	C_2S				
CEM I 42.5N Belorussian cement plant, Costukovichy	7.3	13.5	53.7	21.1	330	3150	31	52.1

Table 2 – Fine aggregate

Type, producer	Modulus of fineness	Density, kg/m^3	Bulk density, kg/m^3	Water absorption, %	Specific Surface, m^2/kg
Quarry sand, Crapugino	3.25	2650	1660	0.66	8.9

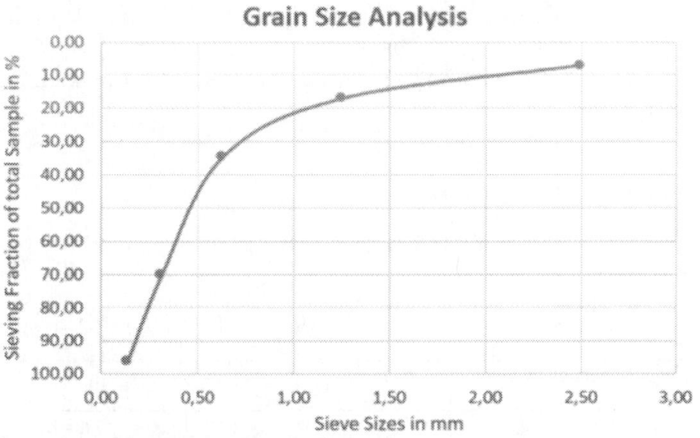

Figure 1 – Grain size distribution of used fine aggregate

Table 3 – Calcium Sulfoaluminate additive (CSA)

Type, producer	Humidity,%	Density, kg/m^3	Specific Surface, m^2/kg	Content, %		
				Al_2O_3	SO_3	Cl
Expanding Sulfoaluminate Modifier «ESAM»,	< 0.1	2850	450	10	30	< 0.1

Table 4 – Polycarboxylate-based admixture Relamix PK

Type, producer	pH	Density, kg/m^3	Dry content,%
Polycarboxylate-based admixture Relamix PK, Polyplast	8.2	1080	30

Table 5 – Water

Content, mg/l,			
Soluble salts	SO_4^{-2}	Cl^{-1}	Suspended particles
< 3000	< 2000	< 600	< 200

3 Experimental Methods

Prism samples for determining compressive and bending strength, linear deformation were made in accordance with the requirements of Self-stressing cement - technical conditions - 1335 [7]. In accordance with Self-stressing cement - technical conditions - 1335, the ratio of cement/sand=1. All compositions had the same water/binder ratio and workability, which was achieved due to the adding of the polycarboxylate-based admixture Relamix PK into the mortar mixture.

3.1 Experimental Program

Table 6 - Experimental program

№	Modifier	W / B	Dosage CSA,% by weight of Binder	Storage
1.1	Referent (without CSA)	0.3		Storage at humidity 55% ± 5
2.1	CSA	0.3	5	
2.2			10	
2.3			15	
10.1	Referent (without CSA)	0.3		Storage in the water
20.1	CSA	0.3	5	
20.2		0.3	10	
20.3		0.3	15	

3.2 Experimental Results

The results of experimental studies are shown in tables 7 - 11 and Figures 2-5.

Table 7 – Results of Cone Spread

NO		Cement (gm)	Sand (gm)	CSA(%)	Water (gm)	W/Binder ratio	Cone Spread (mm)
1.1	10.1	1000	1000	-	150	0.3	113
2.1	20.1	1000	1000	5	150	0.3	111
2.2	20.2	1000	1000	10	150	0.3	110
2.3	20.3	1000	1000	15	150	0.3	108

Table 8 – Flexural strength

No	Flexural strength, MPa at age, days		
	1	3	28
Referent	4.8	6.52	9.1
5 % CSA	5.6	6.7	8.9
10 % CSA	5.5	6.9	8.2
15 % CSA	5.9	7	9.2

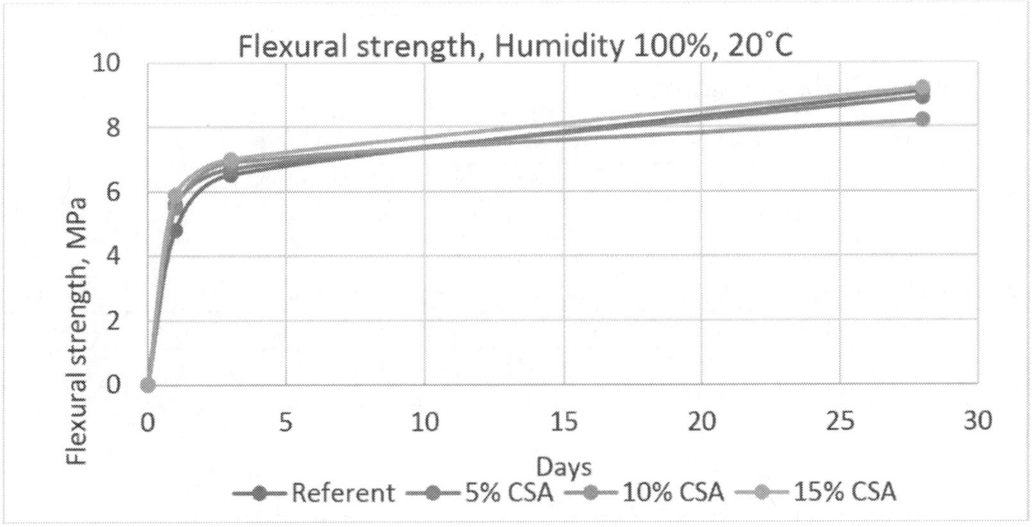

Figure 2 – Flexural strength results

Table 9 – Compressive strength

No	Compressive strength, MPa at age, days		
	1	3	28
Referent	33	54.1	65.5
5 % CSA	32.2	50	52.1
10 % CSA	29.5	46.8	51.8
15 % CSA	28.5	42.8	49.9

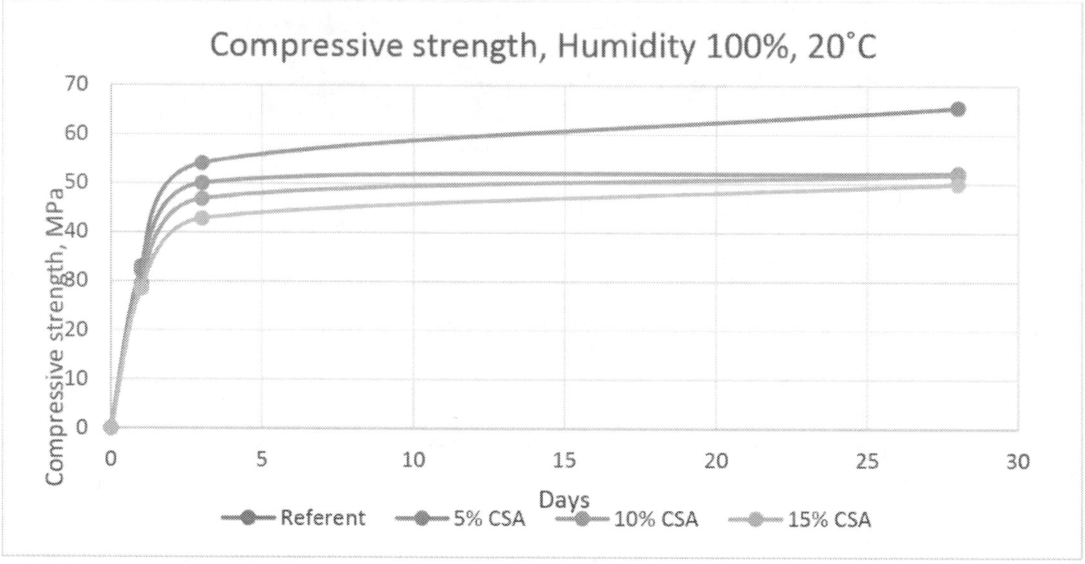

Figure 3 – Compressive strength results

Table 10 – Relative Linear Deformation Humidity 55 %

No	Relative Linear Deformation, %						
	1	2	3	7	14	20	28
Referent	0	0.006	-0.03	-0.045	-0.04	-0.03	-0.02
5 % CSA	0	0.01	-0.065	-0.09	-0.065	-0.065	-0.062
10 % CSA	0	0.012	-0.066	-0.093	-0.049	-0.054	-0.051
15 % CSA	0	0.015	-0.085	-0.12	-0.1	-0.07	-0.07
Referent	0	0.01	-0.065	-0.09	-0.065	-0.065	-0.06

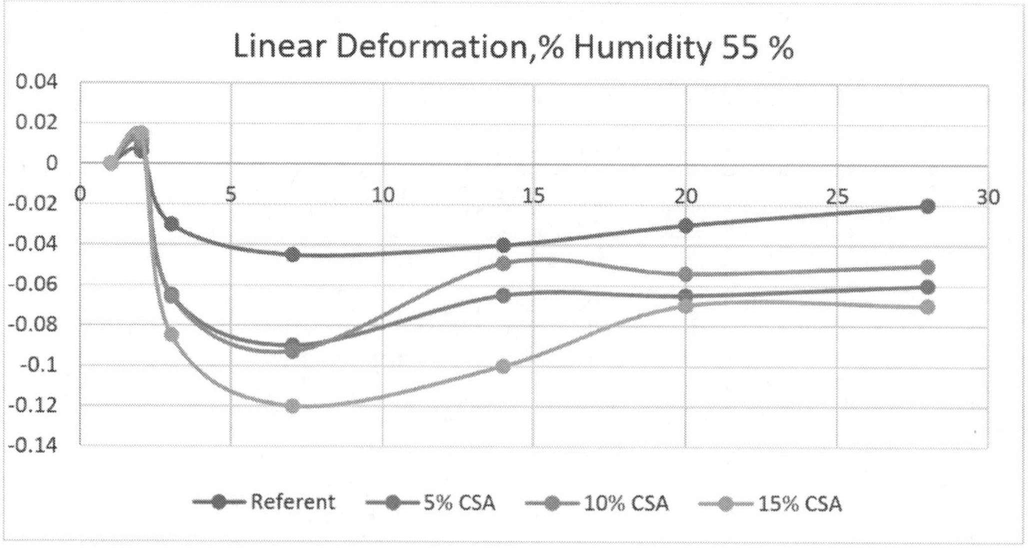

Figure 4 – Relative Linear Deformation at Humidity 55 %

Table 11 – Relative Linear Deformation Humidity 100 %

No	Relative Linear Deformation, %						
	1	2	3	7	14	20	28
Referent	0	2	3	7	10	14	28
5 % CSA	0	0.0889	0.1	0.12	0.05	0.03	0.01
10 % CSA	0	0.13	0.19	0.27	0.27	0.25	0.2
15 % CSA	0	0.17	0.19	0.32	0.32	0.27	0.23

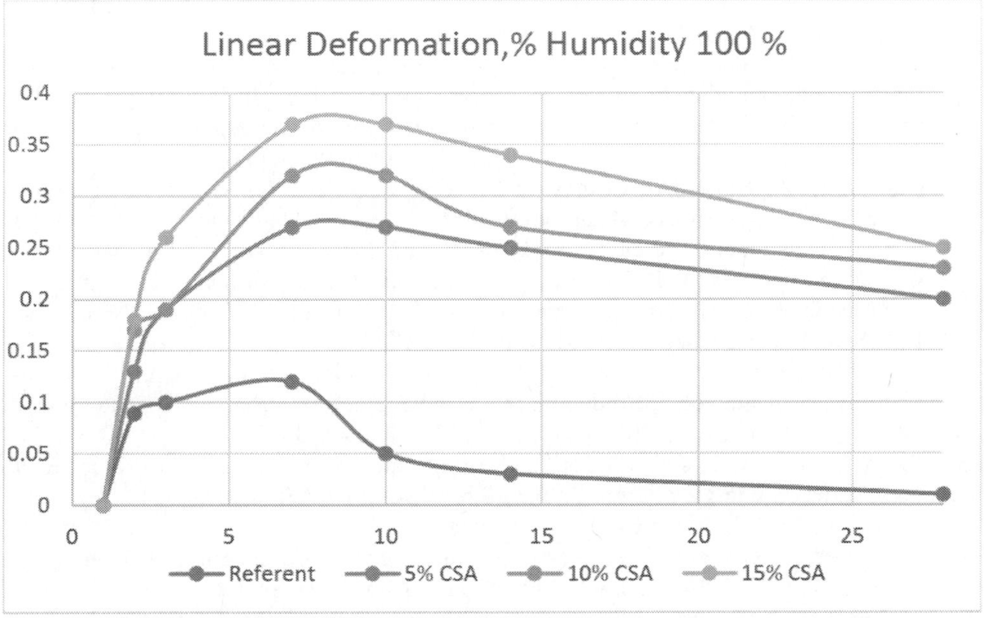

Figure 5 – Relative Linear Deformation at Humidity 100 %

4 Conclusions

1. The initial stage of the study on the influence of expanding additives on the crack resistance of concrete is to determine the effect of Sulfoaluminate based additives on the main properties of a mortar - compression and bending strength, as well as linear deformation of prism samples at different environmental humidity.

2. As a result of the work, the following patterns were identified:

- Bending strength of mortars with CSA increased at the age of 1-3 days by 12.5...20% compared with referent mortar. Moreover, the strengths at the age of 28 days of mortars with CSA and the referent are practically equal.
- Compressive strength of mortars with CSA reduced by 20...23% for all dosages of CSA.
- Relative linear deformations depend on the humidity of the environment. At a humidity of 100%, the relative linear deformations are positive and the expansion increases with increasing dosage of the expanding additive. Peak expansion occurs on 5...10 days of hardening.

- When hardening in dry air at a humidity of 55%, the greatest shrinkage deformations are characteristic for mortars with CSA. Moreover, with an increase in CSA dosage, shrinkage increases.

- Based on the results of shrinkage deformations, we can conclude that the expanding effect of CSA is fully manifested at high humidity, i.e. under construction conditions, this means very high-quality moisture care for structures.

In the case of hardening of structures in low humidity, i.e. when it is impossible to provide high-quality moisture care, for example, for vertical structures, the use of expanding additives is impractical and harmful.

References

1. ACI committee 224 " Control of cracking in concrete structures" ACI- 224-R American concrete institute, Farmington Hills MI, 2011.
2. Physical and mechanical properties of heavy self-compacting concrete: dissertation for the degree of candidate of technical sciences: specialty 05.23.05 Building materials and products / Kotov Dmitry Svyatoslavovich; scientific adviser Bleshchik NP; Research Republican Unitary Enterprise for Construction "Institute BelNIIS".
3. Collepardi M., Borsoi A., Collepardi S., Olagot J.J.O., Troli R. Effects of shrinkage reducing admixture in shrinkage compensating concrete under non-wet curing conditions // Cement and Concrete Composites. Vol.27, Issue 6. – 2005. – Pp. 704-708.
4. 4. N.N. Kalinovskaya, D.S. Kotov, E.V. Shcherbitskaya Shrinkage deformations of modified concrete. Causes and remedies // Bulletin of Polotsk State University. Series F. Construction. Applied Sciences. - 2018.No. 8. - S. 43-48.
5. 5. N.N. Kalinovskaya, D.S. Kotov, E.A. Ivanova Durability of concrete. Analysis of the causes and methods of reducing the shrinkage deformations of modified concrete. Tekhnologii betonov. - 2017. - No. 11-12 (136-137). - S. 14-17.
6. Ramachandran V.S., Feldman R.F. Concrete admixtures handbook. Properties, Science, and Technology // Noyes Publication. USA. 1984, 575 p.
7. STB 1335-2002 Stressing cement. Specifications.
8. Obolewicz J.: Evolution of quality in technique, Safety Engineering of Anthropogenic Objects ISSN 2450-1859 e-ISSN 2450-8721 No. 2 (2017) pp. 8-15.
9. Obolewicz J. and Dąbrowski A.: An application of the Pareto Method in Surveys to Diagnose the Managers and Workers Perception of Occupational Safety and Health on Selected Polish Construction Sites, International Journal of Occupational Safety and Ergonomics (JOSE) Vol. 23, no 3 (2017), s.43.

Structural Mechanics Aspect of Strength of Composite Reinforcement

Vladimir Barsukov[1], *Alla* Volik[1,*], *Svetlana* Sazon[1]

[1] Yanka Kupala State University of Grodno, 230023 Grodno, 22 Ozheshko str., Belarus
* a.volick@grsu.by

Abstract. When operating in wet and corrosive conditions, composite reinforcement is in some cases an effective substitute for the suitable steel reinforcement. However, many questions related to its design, production and operation are insufficiently studied. The aim of the work is to create a methodology and carry out a comparative structural and mechanical analysis of strength indexes when working under load of coils and a bar of composite construction reinforcement. A method has been developed for analyzing the loading capacity of composite reinforcement according to the criteria of strength at rupture of a bar, as well as shear strength and bearing strength of its braid coils. Analytical and numerical modeling of conditions of equal strength of structural elements of composite construction reinforcement was carried out. Theoretically substantiated the assumption of a relatively low loading capacity of the braid coils in comparison with the loading capacity of the reinforcement bar. A predictive estimate of the conditions for ensuring the uniform strength of the bar for breaking and braid coils for shear and crushing by the method of lower estimation (for the minimum values of strength) has been carried out using the example of composite reinforcing bars produced by the industry of Republic of Belarus. In conclusion, results of work are formulated and it is noted that the results can be used by manufacturers and consumers of construction composite reinforcement, also in the educational process in the training of engineering personnel for the construction profile.
Keywords: composite reinforcement, coils, bar, strength, shear, breaking force, crushing.

Introduction.

Steel reinforcement traditionally used in the production of reinforced concrete has a number of disadvantages, the most important of which are low corrosion resistance, as well as the relatively high specific gravity and thermal conductivity of the metal. In the Russian Federation [1–4], in the Republic of Belarus [5–10] and many scientific and technical centers of foreign countries [11–13], research and development is being carried out aimed at creating composite reinforcing materials based on high-strength fibers and polymer matrixes. Research and development are carried out both in relation to composites based on glass and basalt fibers that have already become traditional [14–16], and in the direction of creating hybrid composites that simultaneously include several types of reinforcing fibers (glass + carbon, glass + aramid) [11, 17–23].

Taking into account the economic side of the problem, polymer composites based on glass and basalt fibers are the most promising in the construction, transport and household spheres. However, the widespread use of such materials is hindered by a number of factors, among which one can single out [8, 9] the lack of study of the work under load both a whole composite construction reinforcement, and its individual elements (bar, coils, braids), as well as the lack of a regulatory framework for assigning mechanical characteristics, control and testing methods.

For example, it is known that in steel reinforcement the ribs on the surface of the bars provide high- quality adhesion of the reinforcement of a helical, crescent-shaped or other periodic profile to concrete and contribute to an increase in the reinforcing effect of the bars [24]. In composite reinforcement, braid coils can act as an analogue of corrugations. However, their work under load has not been adequately studied, there are no **methods for comparative predictive estimate of strength indexes**, and scientifically grounded recommendations on the appointment of winding parameters have not been developed. All this makes it difficult to predict the performance in terms of the strength of building structures containing composite reinforcement.

The aim of the research is to create a methodology and carry out a comparative analysis of strength indexes working under a load of coils and a bar of composite construction reinforcement.

Methodology of predictive estimate of the braid coils strength indexes.

Coils of composite reinforcement work under longitudinal loading (tension of a bar surrounded by hardened concrete) on a shear at the matrix (in the zone of adhesion to the bar) and on crushing of the side surfaces. To draw up the design scheme, we will perform a sweeping of a cylindrical part of the bar with coils. As a result, we get a rectangle with a base equal to the circumference (L = πd), on which inclined stripes are applied, characterizing the sweep of the spiral braid (Fig. 1).

a)

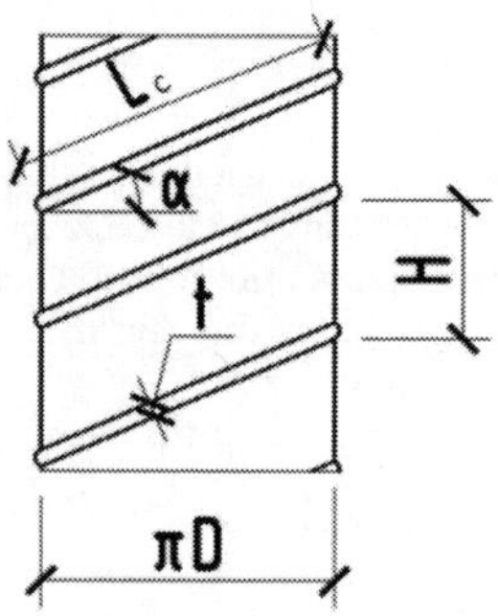

b)

Fig. 1. General view (a) and the surface sweep (b) of the composite reinforcement.

It can be seen from the surface sweep that the length L_c of a coil (spiral line) for a composite reinforcement bar with a core diameter d, on which the braid is wound at an angle α with a winding step H, can be determined by the following formula:

$$L_c = \frac{\pi d}{\cos\alpha} = \sqrt{(\pi d)^2 + H^2} \quad (1)$$

The area A_c of the coil base with the bar connecting is equal to the product of the coil spiral line length L_c by the coil width t at the base:

$$A_c = \frac{\pi d}{\cos\alpha} t = t\sqrt{(\pi d)^2 + H^2} \ . \tag{2}$$

The one coil shear force F_s can be found as the product of the resin shear strength limit τ_s by the shear area which is equal to the cross- sectional area of the coil at the base A_c:

$$F_s = \tau_s A_c = \tau_s t \frac{\pi d}{\cos\alpha} = \tau_s t\sqrt{(\pi d)^2 + H^2} \ . \tag{3}$$

Whence after removing the rod diameter d from under the radical we obtain:

$$F_s = \tau_s t d\sqrt{\pi^2 + \left(\frac{H}{d}\right)^2} \ . \tag{4}$$

Shear force F_{sn} for n_s coils is:

$$F_{s_n} = n_s \tau_s t d\sqrt{\pi^2 + \left(\frac{H}{d}\right)^2} \ . \tag{5}$$

The breaking force of the bar F_p is equal the product of own cross-sectional area on the bar material strength limit σ_b at break:

$$F_p = \sigma_p \frac{\pi d^2}{4} . \tag{6}$$

As a result of equating the shearing force of the braid coil and the breaking force of the reinforcement bar we obtain:

$$\tau_s t d n_s\sqrt{\pi^2 + \left(\frac{H}{d}\right)^2} = \sigma_p \frac{\pi d^2}{4} .$$

Whence the required number of winding coils ns from the position of equal strength of the bar at breaking and coils per shearing:

$$n_s = \frac{\pi d}{4t\sqrt{\pi^2 + \left(\frac{H}{d}\right)^2}} \frac{\sigma_p}{\tau_s} . \tag{7}$$

In addition to cutting off the coils at the base, the second reason for the loss of strength is the destruction of the coil surface due to its compressive breaking. The crushing force F_{b1} of one coil can be found as the product of the resin compressive strength σ_b and the crushing area which is equal to the product of the coil length L_c and the coil cross section height h:

$$F_{b1} = \sigma_b L_c = \sigma_b hd\sqrt{\pi^2 + \left(\frac{H}{d}\right)^2}. \tag{8}$$

The crushing force F_b for n_b coils is found as the product of the corresponding crushing force for one coil determined by formula (8) by the number of coils n_b:

$$F_b = n_b F_{b1} = \sigma_b hdn_b\sqrt{\pi^2 + \left(\frac{H}{d}\right)^2}. \tag{9}$$

As a result of equating the coil lateral surface crushing force and the breaking force of the composite reinforcement bar we obtain the required number of bars ensuring for the condition of equal strength of the bar breaking and coils crushing:

$$n_b = \frac{\pi d}{4h\sqrt{\pi^2 + \left(\frac{H}{d}\right)^2}} \frac{\sigma_p}{\sigma_b}. \tag{10}$$

Methodology for predictive estimate of the ultimate tensile strength for a composite structural reinforcement bar.

As can be seen from formulas (7) and (10), one of the most important factors determining the minimum required number of coils, which ensures the uniform strength of the elements of composite reinforcement to rupture of the bar, as well as shear and crushing of the braid coils , is the strength of the bar. Despite the seeming simplicity of the analysis, the problem of predicting the strength of unidirectional composites has so far been solved only in an approximate estimated formulation.

In a simplified theoretical analysis of the physical and mechanical properties of unidirectional composites, formulas based on the so-called "rule of mixtures" are usually used which for the elastic modulus and density of the composite bar have the following form [13]:

$$E_k = E_f\theta_f + E_m\theta_m = E_f\theta_f + E_m\left(1-\theta_f\right) \tag{11}$$

$$\rho_k = \rho_f\theta_f + \rho_m\theta_m = \rho_f\theta_f + \rho_m\left(1-\theta_f\right) \tag{12}$$

where E_k, E_f, E_m – elastic moduli of the composite reinforcing fiber and polymer binder (matrix) respectively; θ_f, θ_m – volumetric content of reinforcing fiber and binder respectively; ρ_k, ρ_f, ρ_m – respectively, density of the composite, fibers and binder.

In this case $\theta_m = 1 - \theta_f$.

Formulas (11) and (12) make it possible to calculate the volumetric content of fibers in two-component systems if the elastic modulus or density of the composite is previously determined. It is easy to make sure that the corresponding dependence will have the following form:

$$\theta_f = \frac{E_k - E_m}{E_f - E_m} = \frac{\rho_k - \rho_m}{\rho_f - \rho_m}$$

Relationship between mass C_f and volumetric θ_f content of fibers.

Since the two-component composite volume V_k is equal to the sum of the fibers V_f and the polymer binder V_m volumes, and the volume of each component is equal to the ratio of its mass to density, then to calculate the volumetric content of fibers we obtain the following formula

$$\theta_f = \frac{C_f}{C_f + (1 - C_f)\frac{\rho_f}{\rho_m}} = \frac{1 - C_m}{1 - C_m + C_m\frac{\rho_f}{\rho_m}}$$

Numerous experiments have confirmed that formulas (11) and (12) for nonporous unidirectional composites are in good agreement with experimental data [13].

With regard to predicting the longitudinal strength for unidirectional composites the "rule of mixtures" gives the following calculated dependence

$$\sigma_t = \sigma_f\theta_f + \sigma_m(1 - \theta_f) \tag{13}$$

where σ_t – ultimate tensile strength of the unidirectional composite; σ_f – ultimate tensile strength of fibers; σ_m – stresses in the matrix during deformation on fiber destruction; θ_f – volumetric fiber content.

In this case it is assumed to obtain the maximum strength of the fiber composite it is necessary a high content of fibers θ_f and that the destruction was happening due to the breaking of the fibers but not the matrix. The last condition is fulfilled if the volume fraction of fibers is higher than a certain minimum value approximately 0,1 [13]. As a rule, the fiber volumetric content in composite reinforcement is θ_f = 0,6 – 0,7, i.e. this condition is guaranteed.

Experiments show that the calculation according to formula (13) gives overestimated values of strength in comparison with those observed in practice. Let analyze this aspect. It is known that the brittle materials strength has a significant scatter which can be described by the Weibull's distribution function [13, 24]. The strength distribution for a lot fiber can be described by this function too. As the main reason for the phenomena consideration the fluctuations of their length arising during the production of fibers, due to the scattering (within the manufacturing tolerance) of the spinneret dimensions, as well as local fluctuations the melt temperature and its rheological properties during the drawing of the fibers should be taken. All this in combination contributes to decreasing the strength of products made of unidirectional fiber composites in comparison with the theoretical strength predicted by equation (11c) for ideally laid fibers and its ideally identical properties.

So the fiber bundle strength σ_p is less than the elementary fibers average strength σ_f n general and the ratio $\eta = \sigma_p/\sigma_f$ is depend on the dispersion (spread degree) of the fibers strength also the inevitable bending of some fibers in the bundle (roving) during their production, impregnation with a polymer binder, and squeezing out excess of this binder before curing it. As a rule, brittle and bending- sensitive fibers of the carbon type have bundle strength in the range 50 – 65% of their average strength for individual fibers [13]. For somewhat less brittle glass fibers, the strength of the bundle ranges from 65% to 80% of the fibers average strength [13] and higher values are typical for bundles with a minimum properties spread and the absence of fiber bending in the bundles.

Taking into account the foregoing, when predicting the strength of unidirectional composites, it is proposed to modify the equation (11) presenting this dependence in the following form

$$\sigma_t = \eta \sigma_f \theta_f + \sigma_m \left(1-\theta_f\right) \tag{14}$$

where η – the realizable strength factor for the fibers bundle which is equal to 0,5 ... 0,65 for carbon fibers and 0,65 ... 0,80 for glass fibers.

Matrix stresses σ_m during deformation of fiber destruction can be determined using the hypothesis of plane sections and the condition of equality the relative deformations ε of fibers and binder. From Hooke's law we can write

$$\varepsilon = \frac{\sigma_f}{E_f} = \frac{\sigma_m}{E_m}$$

Whence for stresses in the matrix (binder) we obtain

$$\sigma_m = \sigma_f \frac{E_m}{E_f}$$

Then the formula for calculating the predicted theoretical strength of the composite takes the form

$$\sigma_t = \eta \sigma_f \left[\theta_f + \frac{E_m}{E_f}\left(1-\theta_f\right) \right] \tag{15}$$

As a rule the second term can be neglected in composite reinforcement due to the small ratio of the binder elastic modulus to the fiber elastic modulus and a calculated predictive designed estimate of the strength can be made using the following simplified formula

$$\sigma_t \approx \eta \sigma_f \theta_f \tag{16}$$

The above dependences (7), (10), (15) and (16) allow us to make a comparative structural and mechanical calculating estimate the of the coils and the of bar strength for composite reinforcement if the quantities included in these equations.

Research results and their discussion.

The available in the modern scientific and technical literature data on the construction composite reinforcement strength indexes are characterized by a significant scatter of digital values. This may be due to the discrepancy between the volumetric content of reinforcing fibers in the tested samples of the composite, differences in the composite components properties, also the imperfection of test methods [9]. The clearest evidence of such a data scatter is the dependence of the ultimate strength and elastic modulus of the composite on the diameter of the tested fiberglass (FRP) and basalt-plastic (ABP) structure reinforcement, noted in a number of experiments [1, 2]. The mechanism of this dependence has not yet been clarified in a rigorous formulation. Apparently the stress concentration effect in the reinforcement bar near the fastening half couplings has an result [9]. The reinforcement diameter influence on the tensile strength can also be explained as follows. Composite reinforcement is a material consisting the fibers bonded with a polymeric binder the deformability of which is several times higher than that of fibers. When testing the samples, the force that squeezes the bar in the grippers is taken up by the fibers located on the surface of the bar, and then, through the binder interlayers, are transmitted to the fibers in its core.

In connection with the foregoing in this work under consideration, the predicted property indicators are determined by the lower estimate method based on the minimum values of the reinforcing fibers strength and the minimal value of the realizing factor of this strength in the beam.

Typically used for the production of fiberglass reinforcement the glass fibers of type E have the following physical and mechanical properties: density 2540 ... 2560 kg/m^3; tensile elastic modulus 73 ... 77 GPa; ultimate tensile strength 3100 ... 3700 MPa [3, 25] depending on the technological features of production. Cured polyester or epoxy resins used in the production of fiberglass reinforcement depending on the grade of resin, have a density of 1100 – 1200 kg/m^3 and an elastic modulus of 2.75 ... 3 GPa [3, 25]. Basalt fiber has an elastic modulus of 70 – 90 GPa and a tensile strength of 2700 – 3500 MPa [1, 2].

The shear resistance τ_s of a cured resin depend on its type, polymerization degree, type and content of hardener, also on the test method. In particular, according to [10], the shear resistance τ_s, determined by pulling out the fiberglass reinforcement bars Ø7.5 mm embedded in concrete cylinders, is 10,14 ... 10,82 MPa, depending on the thickness of the walls of the concrete cylinder. Similar results were also obtained in a number of works for composites based on carbon fibers [2]. In connection with the above, for estimated technical calculations, it is possible to take $\tau_s \approx 10$ MPa.

The crushing (compressive) strength of cured epoxy or polyester resins is 37 – 42 MPa [24]. We take for estimated calculations the average value $\sigma_b = 40$ MPa.

Since no unambiguous recommendations for the purpose for the coils braid step of the composite construction reinforcement have been developed, and the range of its change is quite wide, then for the estimated calculations we will take the height of the turn h half its width at the base, and the remaining parameters are as follows: $d/t=4$; $d/h=8$; $H/d=1$.

The experimental values of the density and mass content of the binder for a number of standard sizes of fiberglass reinforcement produced in the Republic of Belarus are shown in Table 1.

Table 1. Experimental values of density and mass content of binder in industrial fiberglass reinforcement.

Type of reinforcement	Nominal diameter, mm	Density, kg/m^3	Binder content, mass.%
AKC 6	6	2018	19,3
AKC 10	10	1980	18,63
AKC 12	12	1944	19,25

Note: the diameter of the AKC 6 reinforcement bar excluding the braiding coil is 5,2 mm, taking into account the coil is 6,2 mm.

Table 2 shows the calculated values of the predicted fiber volumetric content indexes, also the strength of the composite reinforcement and the required number of braid coils from the standpoint of ensuring equal shear and crushing strength. In the calculations, the following values of the physical and mechanical properties indexes for fibers and polymer binder were taken ($\sigma_f = 3100$ MPa; $E_f = 73$ GPa; $E_m = 3$GPa).

Table 2. Calculated values of the binder volumetric content, predicted the composite reinforcement strength indexes and the required number of braid coils from the standpoint of ensuring equal strength with the bar.

Type of reinforcement	Calculated volumetric content of binder,%	Predicted minimal strength, MPa	Calculated number of coils from the point of view of ensuring equal strength with the bar	
			for shear	for crush
AKC 6	66,2	1361	130	65
AKC 10	67,2	1354	129	64
AKC 12	66,3	1364	130	65

The analysis shows that the calculated estimate values of the minimal predicted strength 1354 – 1364 MPa are in good agreement with those specified in the STB 1103-98 "Fiberglass reinforcement" standard (taking into account the amendments). The standard stipulates ultimate strength σ_c for bars with a diameter of 6 mm is at least 1300 MPa and a standard resistance R_n is at least 1200 MPa. This indexes are at the correctness for the application of the modified notation of Eq. (14) to predict the longitudinal strength of unidirectional composites.

From the values presented in Table 2, it can be seen that the coils of the reinforcement braid work for shear and crush not efficiently and provide mainly technological strength, preventing the bar from delamination at the stage of reinforcement manufacturing. It is advisable for increasing the efficiency to make in accordance with the accumulated experience, a rough surface with the application of silicate particles, which contribute to strengthening the setting of concrete with the surface of the reinforcement (Fig. 2.)

Fig. 2. Samples of composite construction reinforcement with a layer of mineral particles applied to the surface.

In this case, the step of the braid does not have a determining value and can be assigned from the conditions of the convenience of winding, i.e. may be large in comparison with the one assumed in the calculations in this article.

Conclusion.

It is made a calculating estimate the conditions for ensuring the uniform strength of rupture and braid coils of construction composite reinforcement bar for shear and crushing. Theoretically substantiated the assumption of a relatively low loading capacity of the braid coils in comparison with the loading capacity of the reinforcement bar. The results obtained can be used by manufacturers and consumers of construction composite reinforcement, also in the educational process at the training of engineering personnel for the construction profile.

References

1. V.F. Stepanova, A.Yu. Stepanov, E.P. Zhirkov. Composite polymer reinforcement. Moscow: Bumazhnik, 2013. – 200 p.
2. V.F. Stepanova, D.A. Ilyin, A.V. Buchkin. Hybrid composite reinforcement with increased modulus of elasticity. Natural and technical sciences, 2014, No. 9-10, p. 435-437
3. N. P. Frolov. Fiberglass reinforcement and fiberglass concrete structures. Moscow.: Stroyizdat, 1980.– 104 p.
4. Ya.A. Pronozin, V.F. Bai, Yu.V. Zazulya, R.V. Melnikov. Application of nonmetallic materials as the main reinforcement of concrete bending elements. //Industrial and civil construction. - 2009. - No. 7. – S. 60 – 61.
5. M.Yu. Oshchepkov. Union of Science and Production in the Belarusian Composites Industry // Composite World-2018.-№2, pp. 34 – 41.
6. E. Ya. Sadin. Anchoring in concrete of fiberglass reinforcement produced in the Republic of Belarus // Architecture and construction, 2016-№3, p.68-71
7. E. Ya. Sadin. Estimation of anchoring in concrete of fiberglass reinforcement produced in the Republic of Belarus // Science and technology, T.15, No. 4., 2016. – PP. 308 – 314.
8. V.G. Barsukov, A.R. Volik, A.A. Maruk. Influence of reinforcement made of composite materials on the position of the neutral layer during bending of asymmetrically reinforced concrete beams // Bulletin of the Yanka Kupala Grodno State University. Series 6. Technique, Volume 9 No. 1, 2019. – PP.117 – 125.
9. A.R. Volik, S.A. Sazon, K.Yu. Churilo. Peculiarities of tensile tests of composite reinforcement // Bulletin of the Yanka Kupala Grodno State University. Series 6- Technique, Volume 10, No. 1, 2020. – PP. 110 – 118.
10. V.V. Taletsky. Increasing the elastic modulus of fiberglass reinforcement // Issues of introducing design norms and standards of the European Union in the field of construction. Materials of the international scientific and methodological seminar, Minsk, BNTU, 22-23.05.2013. p. 130-136
11. Y. Qiu, P. Schwartz (1993) Micromechanical behavior of Kevlar-149 / S-glass hybrid seven-fiber micro composites I: tensile strength of the hybrid composite. Compos Sci. Technol. 24: PP. 19 – 32.
12. Yihua Cui, Moe M. S. Cheung, Bahman Noruziaan, Stephen Lee, Jie Tao Development of ductile composite reinforcement bars for concrete structures // Materials and Structures (2008) 41: PP. 1509 – 1518.
13. F. Matthews. Composite Materials: Mechanics and Technology. - M .: Technosphere, 2004. – 406 p.
14. O. Chaallal, B. Benmokrane. (1996) Fiber-reinforced plastic rebars for concrete applications. Composites B 27 (B): PP. 245 – 252.
15. Lither land KL, Oakley DR, Proctor BA (1981) The use of accelerated aging procedures to predict the long term strength of GRC composites. Cement Concrete Res 11: PP. 455 – 466.
16. F. Micelli, A. Nanni (2004) Durability of FRP rods for concrete structures. Constr. Build Mater 18: PP. 491 – 503.
17. M.G. Bader, P.W. Manders (1981) The strength of hybrid glass / carbon fiber composites, part 1: failure strain enhancement and failure mode. J Mater Sci 16: PP. 2233 – 2245.
18. C.E. Bakis, A. Nanni, J.A. Terosky (2001) Self-monitoring, pseudo-ductile, hybrid FRP reinforcement rods for concrete applications. Compos SciTechnol 61: PP. 815 – 823.
19. A.R. Bunsell , B. Harris (1974) Hybrid carbon and glass fiber composites. Composites 5: PP. 157 – 164.
20. G. Kretsis (1987) A review of the tensile, compressive, flexural, and shear properties of hybrid fiber-reinforced plastics. Composites 18: PP. 13 – 23.

21. A. Nanni A, Nenniger JS, Ash KD, Liu J (1997) Experimental bond behavior of hybrid rods for concrete reinforcement. StructEngMech 5 (4): PP. 339 – 353.
22. A. Nanni, J. Liu (1997) Modeling of bond behavior of hybrid rods for concrete reinforcement. Struct. Eng. Work 5 (4): PP. 355 – 368.
23. V.V. Tur. Reinforced Concrete Structure. Basis of Design. / V.V. Tur, V.S. Semianiuk, A.V. Tur - Brest, BrSTU Publishing House, 2018. – 402 p.
24. V.P. Stavrov. Mechanics of Composite Materials – Minsk: BSTU, 1996. –165 p.
25. V.M. Kenko. Nonmetallic materials and methods of their processing. – Minsk .: Design PRO, 1998. – 240 p.

The temperature distribution in the load-bearing designs using LSTC

Roman Borisevich[1*], *Aleksey* Dragan[1]

[1]Brest State Technical University, 224017 Moskovskaya 267, Belarus

Abstract. The article discusses wall enclosing structures based on light steel thin-walled structures using mineral wool as a heater. The aim of this work is to determine the most energy-efficient building envelope. For this, two types of steel profile were chosen: thermal profile and thin-walled steel profile. Based on heat engineering calculations, the values of their resistance to heat transfer were obtained. The analysis of the calculations revealed the most energy-efficient design.
Key words: steel profile, lightweight steel thin-walled structures, heat transfer resistance, temperature, cold bridge.

1 Introduction

Construction is one of the fastest growing industries in the modern world. In almost any country, significant attention is paid to the issues of construction and architecture. One of the problems that has always attracted the attention of specialists is the reduction in the construction time and the cost of building operation. One of the directions for solving this problem is the introduction of the results of modern scientific research into the design and construction of buildings, the use of more advanced building materials that have improved characteristics for reliability and durability. The most important factor affecting the cost of operating a building is its insulation and the cost of heating the finished buildings. Warming is produced around the perimeter of the building, so even in a small area of the building insulation may be hundreds of square meters, which leads to significant costs. In this regard, the task of optimizing architectural and planning decisions of buildings seems quite relevant.

Fig. 1. The supporting frame of a residential building from LSTC

*Corresponding author: borisevich.roma@gmail.org

Recently, the technology of building using lightweight steel thin-walled structures has become increasingly popular (LSTC) (Fig.1) [1-2]. Steel structures are successfully used both for the construction of industrial facilities, and residential and public buildings. Meanwhile, this material has both advantages and disadvantages. LSTC is a wonderful and promising direction in architecture. After all, metal is one of the most suitable materials for the embodiment of the most incredible ideas. The modern technology of prefabricated construction of LSTC, widely used in the world, makes it possible to significantly save resources and reduce construction time.

Fig.2. Steel thin-walled structures (STC)

Fig. 3.Thermal profile

Due to the fact that the assembly and installation of structures are performed lower than from prefabricated reinforced concrete structures of factory manufacture. A variety of materials are used to decorate building facades. It can be constructions made of metal-plastic, aluminum, siding and corrugated board, special ventilated facade panels (cassettes) or outdoor aquapanel plates. Aquapanel [3] cement-mineral slabs are sheet products consisting of a core based on fine-grained lightweight concrete, all of whose planes (front, back, longitudinal edges), except for the end edges, are reinforced with fiberglass. However, the low heat engineering characteristics of the steel profile (fig. 2) (λ = 58 W/(m·°C)), despite all the advantages of LSTC, limited the scope of their mass application in the construction of non-residential buildings and structures.

It should be noted that during the construction of a house from LSTC, a large number of so-called "cold bridges" cannot be avoided.

Thanks to them, the building heat loss increases, which leads not only to a lack of comfort indoors in cold weather travel, but also significantly reduces the service life. In fact, "bridges" are in every building, regardless of building materials. In houses of gas silicate blocks, they are in the seam area. Brick in itself refers to "cold" materials and it is usually insulated with polystyrene foam or mineral wool. In the case of LSTC reduce the influence of "cold joints" using special thermal profiles (fig. 3).

The development of perforated steel profiles (thermal profiles) allowed us to solve the problem of reducing the thermal conductivity of the metal. According to studies conducted in Finland and Sweden, steel profiles with perforation along the entire length from 4 to 8 rows are most effective from the point of view of energy saving. Their use in building envelopes allows to reduce thermal conductivity along the profile by 70–80% compared with STC without perforation.

2 Methods

Let us calculate the LSTC design shown in fig.4 as a uniform structure. The calculation was carried out in accordance with [4] by (1).

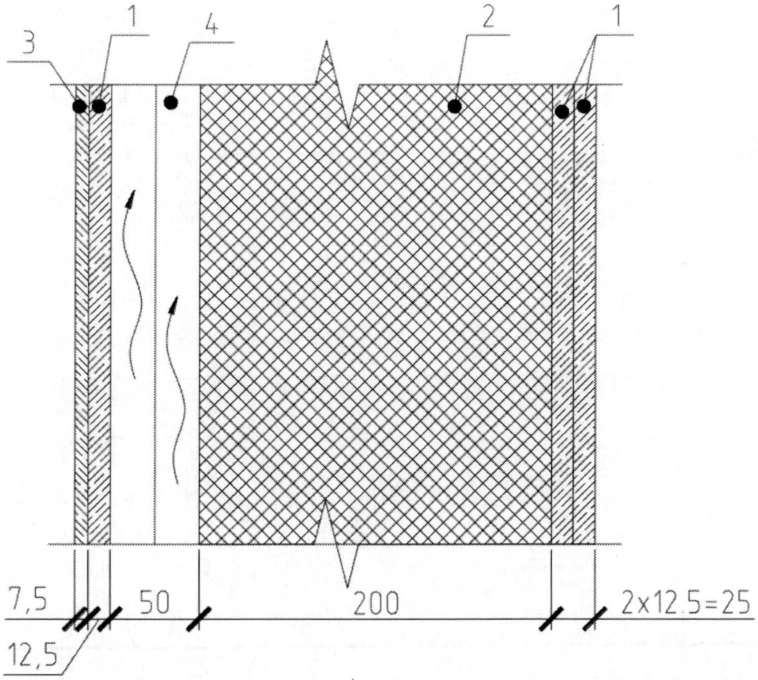

Fig.4. The technical solution of the frame-sheathing external wall includes a steel frame, the outer cladding of Aquapanel Exterior (1 - reinforced cement-mineral slab Aquapanel; 2 - mineral wool ROCKWOOL; 3– finishing layer; 4– air gap)

Table 1. Characteristics of the materials used

Name of material	Density, kg/m^3	Thermal conductivity, W/m·°C	Heat capacity, J/kg·°C
Reinforced cement-mineral slab Aquapanel	1100	0,32	950
Mineral wool plate ROCKWOOL	40	0,042	1050
Steel profile	7900	58	434

$$R=1/\alpha_{in}+\delta_1/\lambda_1+\delta_2/\lambda_2+1/\alpha_o \tag{1}$$

$$R=1/8.7+0.025/0.32+0.2/0.042+1/12=5.038\ m^2\cdot{}^\circ C/W \tag{2}$$

Since two wall options, in one of which STC is used, and in the other a thermal profile, are calculated without taking into account the profile passing through the thickness of the heat-insulating layer, we get the same value of thermal resistance in both cases.

To take into account the reduction in thermal resistance in the presence of profiles, we determine the required heat transfer resistance on the inner surface of the wall using the COMSOL [5] software package. Conditionally accept wall 1.2x1m. fragment [6-7], since the frame consists of walls LSTC rack profiles set at 600 mm. Let us identical for the two variants of the boundary values of temperature and heat transfer coefficients for the inner and outer surfaces (table 2), as well as the thickness and structure (fig. 4) and the proposed materials (table 1).

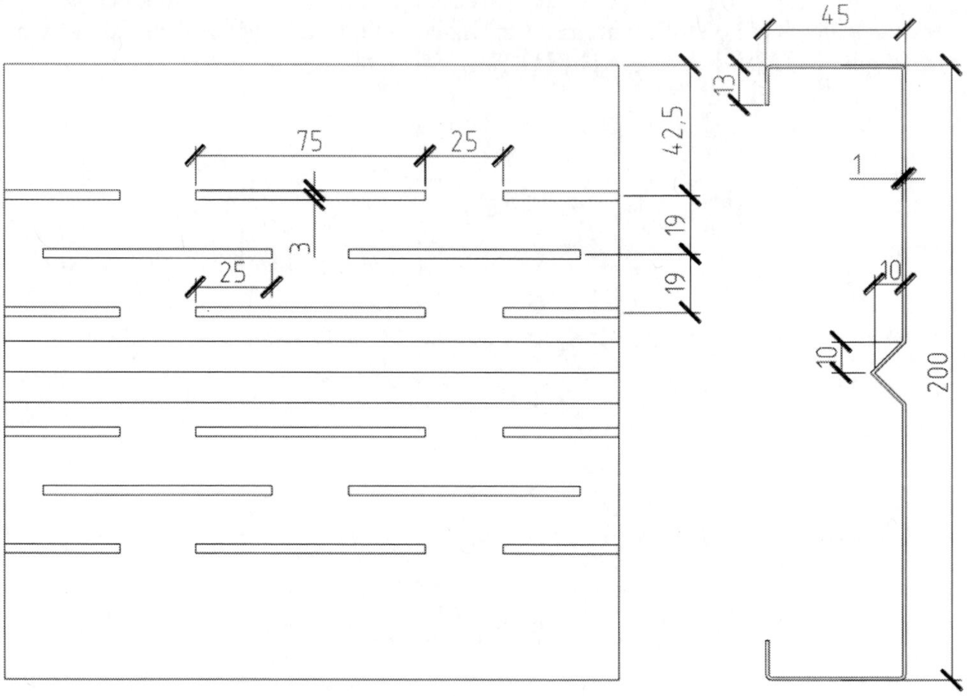

Fig.5. Thermal profile geometrical characteristics

Required thermal resistance, $m^2 \cdot °C/W$, determined by the formula:

$$R= n(t_i-t_o)/(\alpha_{in}\ \Delta t_{in}) \tag{3}$$

t_i – calculated temperature of indoor air, ° C, taken in accordance with the process design;

t_o – calculated winter outdoor temperature, °C, taken according to table 4.3 [4], taking into account the thermal inertia of the building envelope D (except for filling openings) according to table 5.2 [4];

n – coefficient taking into account the position of the outer surface of the building with respect to the outside air, taken according to table 5.4 [4];

α_{in} – heat transfer coefficient of the inner surface of the building envelope, $W/(m^2 \cdot °C)$, taken according to table 5.5[4];

Δt_{in} – the calculated difference between the temperature of indoor air and the temperature of the inner surface of the enclosing structure, °C, taken according to table 5.5[4].

3 Results and Discussion

Table 2. Result

Profile	n	α_{in}, W/(m^2·°C)	t_{in}/ t_o, °C	t_{in}-t_o, °C	$\Delta t_{in.s}$, °C	Δt_{in}, °C	q, W/m^2	R, m^2·°C /W
STC	1	8,7	+18/ -21	39	15,96	2.04	17,79	2.19
Thermal profile			+18/ -21	39	16,69	1.31	11,44	3.41

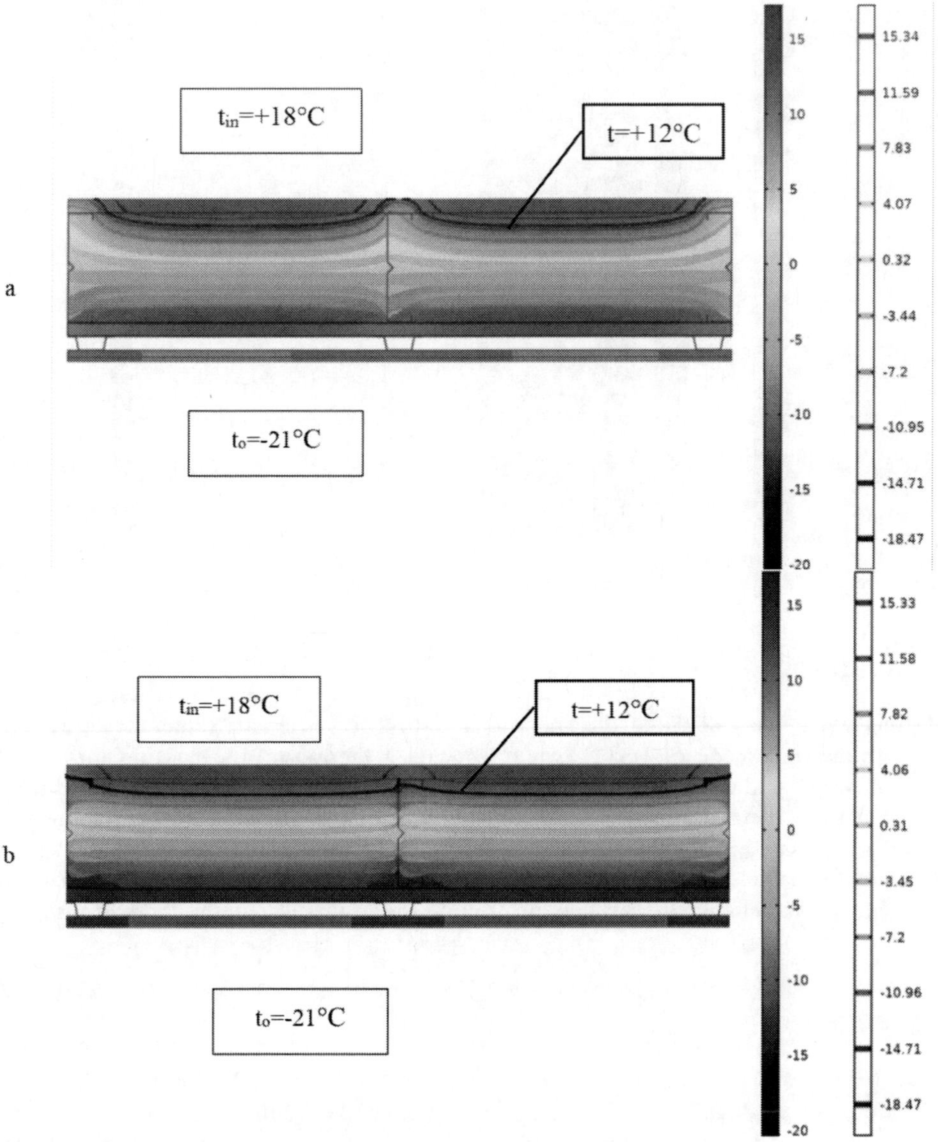

Fig.6. Distribution on a heater temperature: (a) – STC, (b) – Thermal profile

The result of calculation obtained data allow to conclude that the perforations in the more smoothly thermal profiles help distribute the heat loss of insulant (fig. 6b) and reduce the heat losses in the profile (fig. 7). At the same time, in Figure 6a, it is seen that at the wall using STC, the calculated difference between the temperature of the internal air and the temperature of the internal surface is more than 6 °C, which causes freezing. Also, according to the results STC profile when adopted to the wall thickness is not appropriate, since the thermal resistance of less standard, which is 3,2 $m^2 \cdot °C /W$.

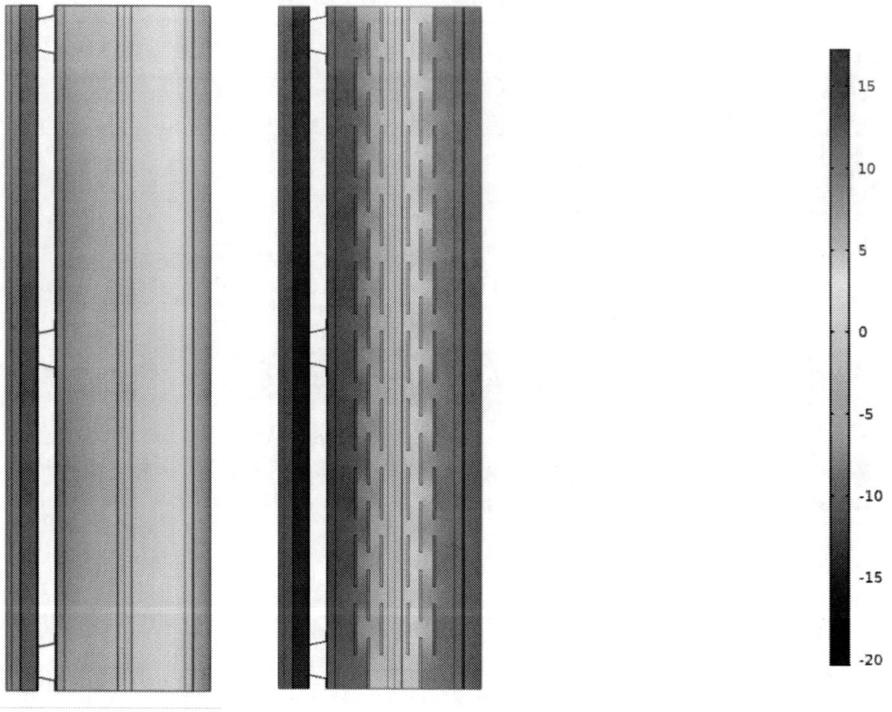

Fig. 7. Distribution of temperature profile (left – STC, right– thermal profile)

4 Conclusion

In this work, we calculated the specific heat loss for external frame-sheathing walls using a framework made of LSTC.Structures with a thermo-profile passing through the insulation were studied. A great influence on the structural design characteristics of perforation of a thermal profile is noted. Conventional profiles in such structures are practically unsuitable for use due to large heat losses. To further improve the characteristics of the structures under consideration, it is necessary to increase the perforation of the thermal profile and change their shape, or provide for an increase in the thermal insulation layer.

References

1. M. Veljkovich, B. Johansson, Thin -Wall Str **44**, pp. 1272–1279 (2006)
2. B. Johansson, Magazine of Civil Engineering **6**, pp. 2–3 (2009)
3. The album of technical solutions "KNAUF AQUAPANEL® External wall. External load-bearing frame-sheathing walls with a frame made of steel thin-walled cold-formed galvanized profiles using KNAUF materials "(2018)
4. TCP 45-2.04-43-2006. *Construction heat engineering* (Minsk, 2006)

https://doi.org/10.1051/e3sconf/202021202003

5. R.W. Pryor, *Multiphysics modeling using COMSOL: a first principle approach* (Jones & Bartlett Publishers, Inc, 2011)
6. P 1.04.115.13. *Recommendations for calculating the reduced heat transfer resistance of building envelopes and calculating heat loss through the enclosure* (Minsk, 2013)
7. GOST P 58774–2019. *The external walls are frame-sheathed self-supporting and non-bearing with a frame made of cold-rolled steel galvanized profiles. General specifications.* (Moscow, Standardinform, 2020.)

A cataluminescence sensor for the detection of trichloroethylene based on PEG200/ZnO nanocomposite

Xinli Cai[a], *Fei* Liu[a], *Fakang* Pan[a], *Bai* Sun*[,a], *Fangwen* Xu[a], *Zhuo* Tang[a], *Jie* Zhang[a], *Yunming* Cheng[a], *Jinyun* Liu[b], *Fanli* Meng[c], *Shuguang* Zhu*[,a]

a Key Laboratory of Water Pollution Control and Wastewater Resource of Anhui Province,College of Environment and Energy Engineering, Anhui Jianzhu University, Hefei 230601, PR China

b Key Laboratory of Functional Molecular Solids, Ministry of Education, College of Chemistry and Materials Science, Anhui Normal University,Wuhu 241002, China;

c State Key Laboratory of Synthetical Automation for Process Industries, Northeastern University, Shenyang 110819, PR China

Abstract: The content of volatile organic compounds (VOCs) in the atmosphere will endanger the health and safety of human beings which makes it very important to develop a simple and rapid method for the determination of gas pollutants. Based on this, a new type of gas sensor was proposed for the detection of trichloroethylene in air. PEG200/ZnO nanocomposite were prepared by hydrothermal method. The materials were characterized by scanning electron microscope, X-ray energy spectrum and fourier infrared spectrum. The high selectivity of the materials was verified by using the cataluminescence (CTL) intensity of 9 kinds of VOCs on the surface of the materials as a reference. The results show that trichloroethylene can produce CTL response on the surface of PEG200/ZnO nanocomposite. Temperature, air flow rate and detector concentration all have certain effects on the CTL intensity. By comparing the CTL intensity under different reaction conditions, it is found that the suitable temperature and air flow rate are 120 °C, 180 mL/min and there is a good linear relationship between the relative CTL intensity and the concentration of the detected substance ($y = 28.588\,x - 285.56$, R=0.9593). The gas sensor has the advantage of rapid response, and trichloroethylene can produce the maximum CTL on the surface of the material within 3 ~ 5 s.

Key words: VOCs; gas sensor; detection; cataluminescence

1 Introduction

With the development of coal chemical industry, petrochemical industry and solvent manufacturing industry, as well as the increase of industrial exhaust gas and automobile exhaust gas, the content of volatile organic compounds (VOCs) in air is increasing year by year[1,2]. As an important precursor of haze and photochemical smoke, VOCs has become the main culprit of air pollution. The final result of air pollution is to endanger the health and safety of human beings [3], so it is particularly important to find a rapid detection method of VOCs in order to ensure the health of human beings.

At present, VOCs detection technology mainly includes spectrophotometry, gas chromatography, high performance liquid chromatography, etc. [4]. However, there is a disadvantage in these methods that rapid monitoring cannot be achieved. In view of this situation, we urgently need to develop a portable method which can quickly determine VOCs at very low concentration level [5]. Currently, cataluminescence (CTL)-based sensors showed significant advantages. Yu et al [6]. found that NiO with various morphologies showed different CTL capabilities to detect H_2S. Tang et al[7]. found that $NaYF_4$:Er had a good CTL effect on ketones. Jiao et al [8]. have studied a new alcohols sensor based on CTL on nano-CdS. These and numerous studies have shown that a method of measuring one or more VOCs based on a CTL sensor is possible [9-12].

Trichloroethylene is the most common VOCs in the air. It not only has potential carcinogenic risk to the human body [13,14], but also has great damage to the eyes and the central nervous system. Therefore, in combination with the advantages of the rapid detection method based on the CTL, it is necessary to study a method to quickly detect the trichloroethylene.

A large number of studies have shown that nanometer ZnO has excellent photocatalytic and photoluminescence properties [10,15]. Xu et al.[16] found that PEG200 can be used to improve the CTL properties of materials. In view of this, nanometer ZnO modified materials were used as sensing elements to design and construct CTL - based sensors for rapid detection of VOCs in this paper. The PEG200/ZnO nanocomposite were prepared by hydrothermal. Taking nine typical VOCs pollutants including trichloroethylene as the research object, the high CTL sensitivity nanomaterials for the detection of trichloroethylene were found.

2 Experiment

2.1 Experimental materials

All the reagents were of analytical grade and used as received without further purification. The PEG200, $ZnCl_2$, Na_2CO_3, carbon tetrachloride, n-hexane, chlorobenzene, ammonia, butanone, ethanol, xylene, trichloroethylene, and acetone are all purchased from Shanghai Group Chemical Reagent Co., Ltd.

2.2 Preparation of PEG200/ZnO nanocomposite

Surfactants can affect the morphology and structure of substances and PEG200 can be used to improve the CTL properties of materials [16-18], so PEG200 was selected to modify ZnO. The preparation of materials refers to the relevant literature [19]. 2 mL PEG200 was added to 20 mL 1.0 $mol \cdot L^{-1}$ Na_2CO_3 solution and stirred for 30 min, and then the mixed solution was added to 20 mL 0.5 $mol \cdot L^{-1}$ $ZnCl_2$ solution (30 °C). Secondly, stir the reaction mixture for 30 min, and then leave it for another 30 min to completely precipitate the product. After filtered, washed and dryed at 70 °C for 24 h, the obtained product was calcined at 550 °C for 3 h as resulting materials.

2.3 Characterization

The photos of field-emission scanning electron microscope (SEM) were obtained by using field-emission SEM (ARURIGA) to study the morphology of the material, and the elemental analysis of the samples was carried out using the accompanying X-ray energy (EDS) spectrum. The fourier infrared (FT-IR) spectrum of the sample was determined by the NEXUS-670 infrared spectrometer of NICOLET.

2.4 Cataluminescence sensing measurement

The ceramic rods coated with nanomaterials were placed in the quartz tube and the upper cover of the reaction chamber was covered to keep the reaction chamber in an impermeable state. The appropriate flow rate and temperature of reaction can be adjusted by controlling the flow pump and adjustable transformer. After the reaction conditions were stable, a certain amount of VOCs to be detected was injected into the sample inlet by using a microsyringe, and the CTL signal was detected by ultra-weak chemiluminescence analyzer (BPCL-1-TIC).

3 Results and discussions

3.1 Characterization of PEG200/ZnO nanocomposite

The SEM images of PEG200/ZnO nanocomposite are shown in Fig. 1a and Fig. 1b. As can be seen from Fig. 1a and Fig. 1b, the sample shows a wrapped bulk structure. Further, the detection multiple is further enlarged, and the surface of the material is found to be in a loose floc shape, which increases the specific surface area of the material, thereby facilitating the detection of the contact between the gas and the material.

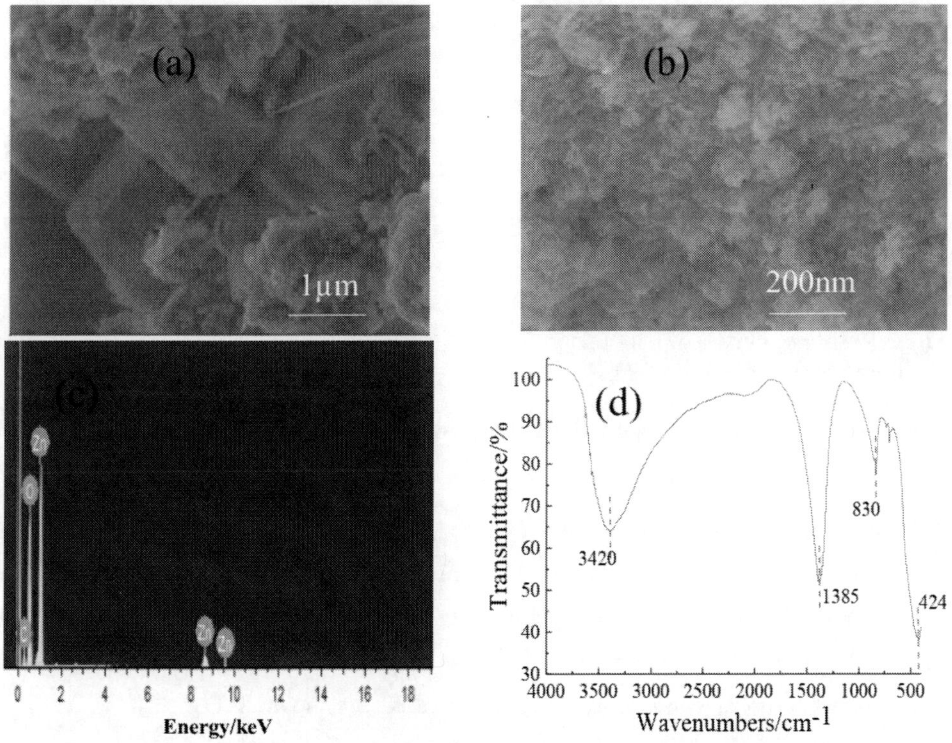

Fig. 1 The SEM images, EDS and FT-IR spectrum of PEG200/ZnO nanocomposite

Combined with Fig. 1c, the sample contains not only Zn and O elements, but also C elements in PEG 200. Fig. 1d shows the FT-IR spectrum of the PEG200/ZnO nanocomposite. Where in 424 cm^{-1} is a characteristic absorption band of ZnO, and 830 cm^{-1} correspond to the O-H feature peaks in the PEG200. While 1385 cm^{-1} and 3420 cm^{-1} corresponds to the absorption band of the ZnO surface O-H [20,21]. So the PEG 200 and ZnO were bonded together to form PEG200/ZnO nanocomposite.

3.2 CTL response of trichloroethylene to nanometer ZnO and PEG200/ZnO nanocomposite

The CTL response of trichloroethylene to nanometer ZnO and PEG200/ZnO nanocomposite was determined under suitable reaction conditions. After injecting the detected material, the signal was monitored by BPCL-1-TIC. The response results are shown in Fig. 2. Fig. 2a shows that trichloroethylene has no obvious CTL response on the surface of nanometer ZnO, but has obvious response effect on the surface of PEG200/ZnO composite nanomaterials. This indicates that the CTL properties of nanometer ZnO have changed after adding PEG200 [16]. The material has a stable response peak in four parallel assays within 400 s, so this method has good reproducibility.

 https://doi.org/10.1051/e3sconf/202021202004

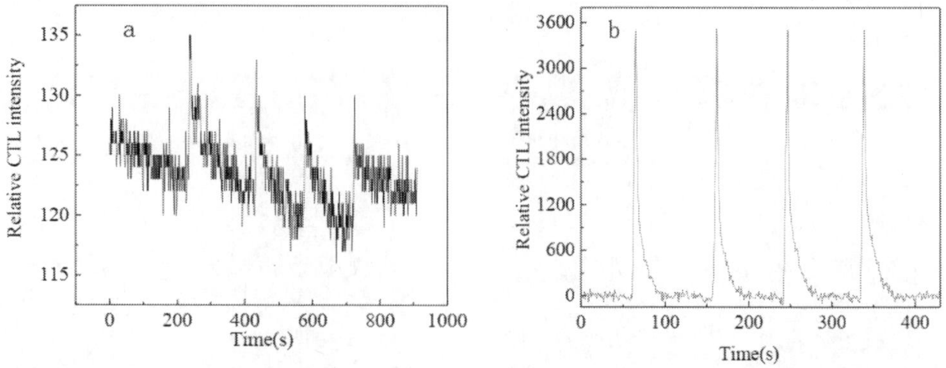

Fig. 2 Response curves of nanometer ZnO (a) and PEG200/ZnO nanocomposite (b) for trichloroethylene based on CTL.

3.3 Effect of working temperature, air flow rate and analyte concentration on CTL intensity

A large number of studies have shown that temperature and air flow rate have a great influence on the intensity of CTL [22-27]. Therefore, under the condition of keeping other conditions unchanged, we investigated the CTL intensity of trichloroethylene on the surface of PEG200/ZnO nanocomposite at different temperatures and air flow rate in order to obtain better response conditions, the specific results are shown in Fig. 3.

As can be seen from Fig. 3a, the CTL signal increases with the increase of temperature. However, under the same time background, the noise is also increasing, and finally the signal-to-noise ratio (S/N) increases first and then decreases. Using this as a reference, the suitable working temperature of the sensor can be determined to be 120 °C, which indicates that the sensor has good CTL activity at lower temperature. Fig. 3b shows that with the increase of air flow rate, CTL signal and signal-to-noise ratio (S/N) increase at first and then decrease. The reason for this result may be that when the air flow rate reaches a certain value, the contact time between the detector and the sensor is not enough, which reduces the sensitivity of the detection system[26,27]. Using this as a reference, the suitable carrier air flow rate of the sensor can be determined to be 180 mL / min.

To investigate the effect of the concentration of the analyte on the intensity of the CTL, a series of concentrations of trichloroethylene were injected into the sensor. The results show that the CTL intensity of trichloroethylene on the surface of nanomaterials increases with the increase of the concentration of trichloroethylene. And there is a good linear relationship between the relative CTL intensity and the concentration of the detected substance (y = 28.588 x - 285.56, R=0.9593). As can be obtained from Fig. 3c, the relative CTL intensity introduced into the trichloroethylene 3 to 5 s reaches the maximum, which indicates that the reactor has the advantage of rapid response. Besides, the limit of detection defined as 3 N/S (N refers to the noise and S refers to the slope of the calibration curve) is about 8.5 μg/mL.

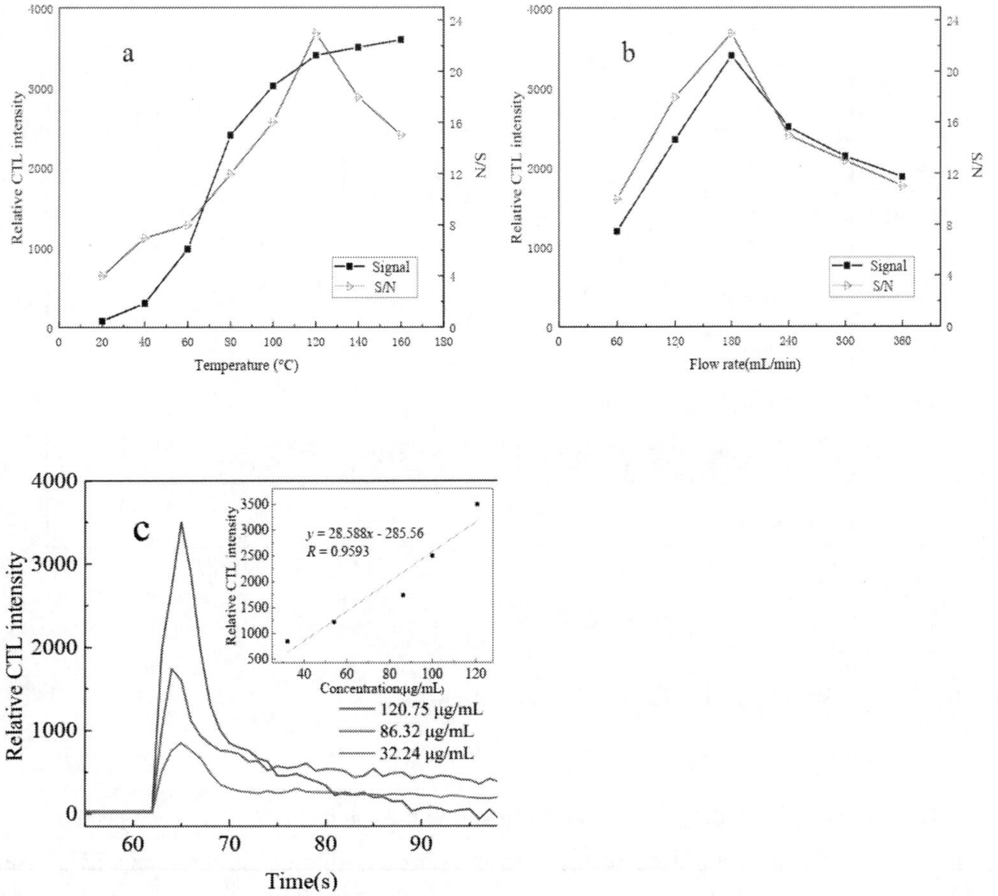

Fig. 3 Effects of operating temperature (a), air flow rate (b) and analyte concentration (c) on the relative CTL intensity. (a) air flow rate: 180 mL/min, trichloroethylene concentration: 120.75 μg/mL; (b) operating temperature: 120 °C, trichloroethylene concentration: 120.75 μg/mL; (c) operating temperature: 120 °C, air flow rate: 180 mL/min.

3.4 CTL selectivity of PEG200/ZnO nanocomposite

During the study, it could be found that the PEG200/ZnO nanocomposite as sensing material showed an excellent selectivity toward trichloroethylene by testing another eight sorts of common possible VOCs, including carbon tetrachloride, n-hexane, chlorobenzene, ammonia, butanone, ethanol, xylene and acetone under the same investigating conditions. At the concentration of 120.75 μg/ mL, under the conditions of the air flow rate of 180 mL/min, and the temperature of 120 °C, the object to be examined was injected into the CTL-based sensors of the PEG200/ZnO nanocomposite respectively, and the relative CTL intensity the surface of the material was detected. The results show that trichloroethylene has a strong rapid CTL reaction on the surface of PEG200/ZnO nanocomposite, while other eight kinds of VOCs on the surface of PEG200/ZnO nanocomposite has no CTL response, which indicates that the material has good CTL selectivity for trichloroethylene (Fig. 4).

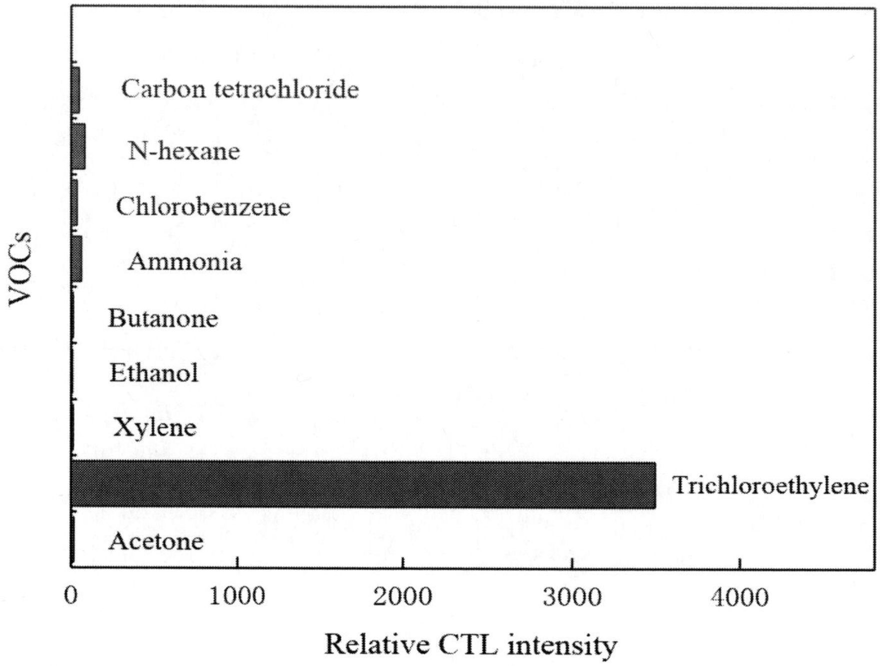

Fig. 4 Gas selectivity of PEG200/ZnO nanocomposite

3.5 Possible mechanisms

According to the widely accepted CTL reaction theory, the CTL response is generated in the process of converting the excited state intermediate into the ground state [28]. Thus, when the trichloroethylene is in contact with the PEG200/ZnO nanocomposite, the reactants are converted from the ground state to the excited state intermediate and a CTL is generated when returning to the ground state again.

The main active species in the CTL reaction are superoxide radical ($\cdot O_2^-$) and hydroxyl radical (O-H) [28]. Combined with Fig. 3, the composite surface contains O-H and the reason for the CTL signal of trichloroethylene on the surface of PEG200/ZnO nanocomposite is that the surface active agent PEG200 provides the O-H for the CTL reaction. In addition, PEG200 plays the role of adhesive in the composite material, and makes the surface of the material appear loose floc structure, so the adhesion and the contact effect of the PEG200/ZnO nanocomposite to the detection object are better than that of the nanometer ZnO, which caused CTL ability of PEG200/ZnO nanocomposite is naturally stronger than that of nanometer ZnO.

4 Conclusion

The trichloroethylene sensor prepared by using PEG200/ZnO nanocomposite as CTL sensitive material has the advantages of fast response and strong selectivity. Through condition optimization, the suitable reaction temperature and air flow rate of the sensor are determined to be 120 °C and 180 mL/min., respectively. Besides, there is a good linear correlation between the concentration of the detected substance and the relative CTL intensity ($y = 28.588\, x - 285.56$, R=0.9593), and the reactor has good reproducibility for the detection of trichloroethylene.

Acknowledgements

This work was supported by the Project of National Key Research and Development Program(2019YFC0408503), the Natural Science Major Research Projects of Anhui Education Department (KJ2017ZD40), the National Natural Science Foundation of China (61873003, 61673367), the Science and Technology Major projects of Anhui Province (18030801106, 16030801118), the Key Research and Development Plan of Anhui Province (201904a07020070), the Scientific Research Start-up Foundation for Introduction of Talent, Anhui Jianzhu University (2016QD113), the General project of Anhui Provincial Department of Education (No.KJ2015JD06).

References

[1] Yuan HW, Yin HL, Huang Y, Qian S, Xie ZW, Su QK, Qin Q. Study on the pollution characteristic of VOCs in ambient air of Chengdu. Adv Mater Res. 2012; 610-613: 1889-1894
[2] Zhang YL, Yang WQ, Simpson I, Huang XY, Yu JZ, Huang ZH, Wang ZY, Zhang Z, Liu D, Huang ZZ, Wang YJ, Pei CL, Shao M, Blake DR, Zheng JY, Huang ZJ, Wang XM. Decadal changes in emissions of volatile organic compounds (VOCs) from on-road vehicles with intensified automobile pollution control: Case study in a busy urban tunnel in south. China Environ Pollut. 2018; 233: 806-81
[3] Janke K, Propper C, Henderson J. Do current levels of air pollution kill? The impact of air pollution on population mortality in England. Health Econ. 2009; 18: 1031-1055
[4] Li MN, Zhang HZ, Zhang SZ, Ma J, Jia WK. Research progress in detection methods of volatile organic compounds in the environment. Chinese Journal of Analysis Laboratory. 2018; 37: 730-739 (in Chinese)
[5] Sha W, Ni S, Zheng C. Development of cataluminescence sensor system for benzene and toluene determination. Sensor Actuat B-Chem. 2015; 209: 297-305
[6] Yu KL, Hua JX, Li XH, Zhang LC, Lv Y. Camellia-like NiO: A novel cataluminescence sensing material for H_2S. Sensor Actuat B-Chem. 2019; 288: 243–250

[7] Tang J, Song HJ, Zeng BR, Zhang LC, Lv Y. Cataluminescence gas sensor for ketones based on nanosized $NaYF_4$:Er. Sensor Actuat B-Chem. 2016; 222: 300-306

[8] Jiao X, Zhang LC, Lv Y, Su Y. A new alcohols sensor based on cataluminescence on nano-CdS. Sensor Actuat B-Chem. 2013; 186: 750-754

[9] Li B, Liu JF, Shi GL, Liu JH. A research on detection and identification of volatile organic compounds utilizing cataluminescence-based sensor array. Sensor Actuat B-Chem 2013; 177: 1167-1172

[10] Marcello M, Sara G, Flavio DP, Corrado DN, Sinazo Q, Emmanuel I, Paola P, Dario C. Peptide modified ZnO nanoparticles as gas sensors array for volatile organic compounds (VOCs). Front Chem. 2018; 6: 105

[11] Zhang RK, Cao XA, Liu YH, Chang XY. Development of a simple cataluminescence sensor system for detecting and discriminating volatile organic compounds at different concentrations. Anal Chem. 2013; 85: 3802-3806

[12] Luo M, Shao K, Long Z, Wang LX. A paper-based plasma-assisted cataluminescence sensor for ethylene detection. Sensor Actuat B-Chem. 2017; 240: 132-141

[13] Tachachartvanich P, Sangsuwan R, Ruiz HS, Sanchez SS, Durkin KA, Zhang LP, Smith MT. Assessment of the endocrine-disrupting effects of trichloroethylene and its metabolites using in vitro and in silico approaches. Environ Sci Tec. 2018; 52: 1542-1550

[14] Zhang C, Yu Y, Yu JF, Li BD, Zhou CF, Yang XD, Wang X, Wu CH, Shen T, Zhu QX. Viral mimic polyinosine-polycytidylic acid potentiates liver injury in trichloroethylene-sensitized mice-Viral-chemical interaction as a novel mechanism. Ecotox Environ Safe. 2018; 155:101-108

[15] Xia H, Zhou RH, Zheng CB, Wu P, Tian YF, Hou XD. Solution-free, in situ preparation of nano/micro CuO/ZnO in dielectric barrier discharge for sensitive cataluminescence sensing of acetic acid. Analyst. 2013; 138: 3687-3691

[16] Xu HL, Li QY, Zhang LC, Zeng BR, Deng DY, Lv Y. Transient cataluminescence on flowerlike MgO for discrimination and detection of volatile organic compounds. Anal Chem. 2016; 88: 8137-8144

[17] Yang T, Xia DG, Chen G, Chen Y. Influence of the surfactant and temperature on the morphology and physico-chemical properties of hydrothermally synthesized composite oxide $BiVO_4$. Mater Chem Phys. 2009; 114: 69-72

[18] Sabet M, Jahangiri H. The effects of surfactant on the structure of $ZnCr_2O_4$ dendrimer like nanostructures used in degradation of eriochrome black T. Mater Res Express. 2017; 5

[19] Xu HL, Li QY, Zhang LC, Zeng BR, Deng DY, Lv Y. Transient Cataluminescence on Flowerlike MgO for Discrimination and Detection of Volatile Organic Compounds. Anal Chem. 2016; 88: 8137-8144

[20] Li F, Xia L, Liu XD, Yang QL, Cheng KJ, Huang QJ. Preparation and photocatalytic properties visible-light-driven ZnO nanostructures. Journal of Shenyang University. 2017; 29: 179-183 (in Chinese)

[21] Wu ZF, Li SJ. Infrared spectral characteristics of zinc hydroxide and zinc oxide. Spectral Laboratory. 2012; 29: 2172-2175 (in Chinese)

[22] Liu YH, Tang F, Kang CJ, Cao XA. Detection of hydrogen sulphide using cataluminescence sensors based on alkaline-earth metal salts. Luminescence. 2012; 27: 274-278

[23] Xu YS, Zheng W, Liu XH, Zhang LQ, Zheng LL, Yang C, Pinna N, Zhang J. Platinum single atoms on tin oxide ultrathin film for extremely sensitive gas detection. Mater Horiz. 2020; 7: 1519-1527

[24] Xu YH, Zheng LL, Yang C, Zheng W, Liu XH, Zhang J. Oxygen vacancies enabled porous SnO_2 thin films for highly sensitive detection of triethylamine at room temperature. ACS Appl Mater Inter. 2020; 12: 20704-20713

[25] Zheng W, Xu YS, Zheng LL, Yang C, Pinna N, Liu XH, Zhang J.. MoS_2 van der waals p-n junctions enabling highly selective room-temperature NO_2 sensor. Adv Funct Mater. 2020; 30: 2000435

[26] Wang NJ, Cao XA, He RW, Liu YH, Huang YJ. A cataluminescence-based sensor for detecting benzene, toluene and xylene vapors utilizing the catalytic reduction on the surface of nanosized Al_2O_3/Pt. Adv Mater Res. 2013; 663: 335-342

[27] Zheng JZ, Zhang WX, Cao J, Su XH, Li SF, Hu SR, Li SX, Rao ZM. A novel and highly sensitive gaseous n-hexane sensor based on thermal desorption/cataluminescence. RSC Adv. 2014; 4: 21644-21649

[28] Zhang RK, Cao XA, Liu YH, Chang XY. A new method for identifying compounds by luminescent response profiles on a cataluminescence based sensor. Anal Chem. 2011; 83: 8975-8983

Flexible prefabricated system moravamont 2000

Živko P. Cuckič[*1]

[1]Serbia and European Union, 21000, Ul. Radnička 26, Novi Sad, Sebija

Abstract. At the end of a decade-long research work at the Moravamont plant in Gnjilane, a new completely prefabricated building system was created from reinforced concrete and prestressed precast elements on the track, which was called Moravamont 2000. Presented in paper final results demonstrates that the construction is well and rationally designed, that the construction behaviour for the maximum expected earthquake effects with a return period of 500 years, according to the criterion of regulation, is resistant and resistant to an earthquake without major damage. The idea is to present through new system moravamont 2000 its flexible side along with the possible development using 3D technology.
Keywords: prestressed precast element, seismic action, structural system, 3D

1 Introduction

Moravamont Factory from Gnjilane (Figure 1) is one of the four factories in the Vemont system in the former Yugoslavia. In addition to the Vemont system, the factory Adrijamont and the IMS system of professor Branko Žeželj also have a factory. The factory owns its raw material base – quarry, concrete base, as well as modern mold hydraulics (flares), reinforcement, internal transport and all necessary contents, which can be seen from the attached documentation. In addition to adhesion pre stressing, it also has the possibility of expanding the basic product range by applying universal pre-precision on the track, as well as innovation within the system, as well as possible combinations. This idea of the combination was supported in 1982 by my professor Momir Krastavcevic in postgraduate studies in Nis.

The factory's position allows expansion of production capacities, which the author started in 1989, at the time he was at the head of Binačka Morava, but, unfortunately, the built structure was not put into operation, although the feasibility studies are still good for all new products today. The complex itself enables the creation of a universal concept of construction according to the system: concrete, metal, wood, which practically provides a leading position in the construction industry in the Balkans.

*Corresponding author: dadge_serbia@yahoo.com

Fig. 1. *Moravamont complex in Gnjilane*

2 Methods

The idea to make Moravamont 2000 a flexible one came as a need to resolve construction projects on different locations, both in constructing houses and apartments, business halls and other public buildings. The solution lies in merging elements, their easy transport and quick creation of technical line for constructing elements for each construction site. The fundamental models for the construction of apartments and halls have been presented for this occasion (fig. 2), along with the fundamental model for the transport of all elements (fig. 3), as well as all the elements separately (fig. 4).

3 Result and Discussion

The results coming out of this flexible system Moravamont 2000 do not fall behind in terms of quality with fabricated ones, what is more, they allow better construction organization as well as the possibility to mix with local materials. The idea resulted from the project of building around 10 000 apartments in Africa, which was postponed for 2021 due to the pandemic caused by corona virus. In this case, the energy reduces the price for 15 per cent.

a) b)

Fig. 2. Housing program (a) , Hall program (b)

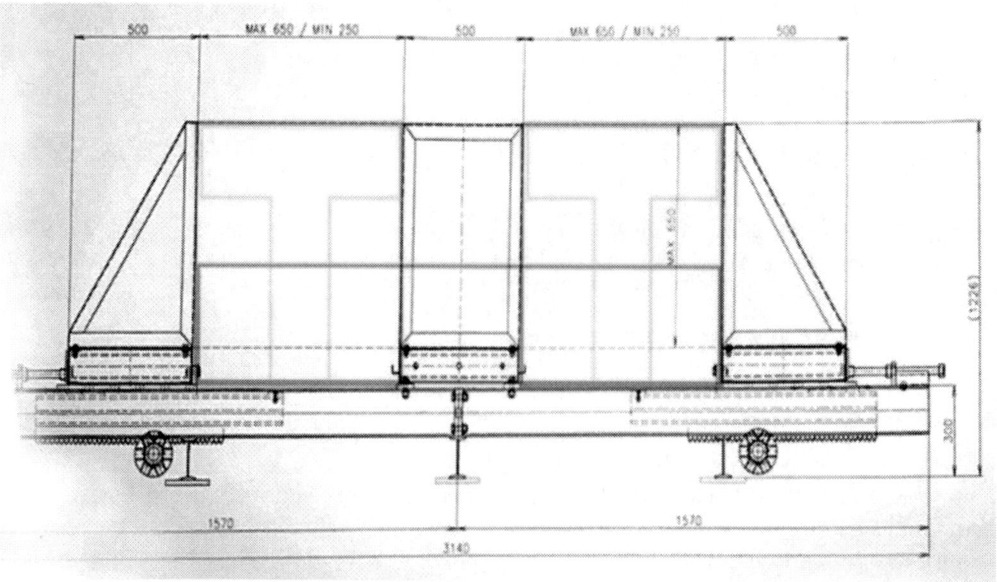

Fig.3 *- Complete transport technology (2,8 x 12)m*

Fig.4*. Elements: Panel plate(a), T-carrier (b), Pillars (c), Prestressing plate (d), Basic rate and casica (e)*

4 Conclusions

Besides achieving good results when it comes to the strength and resistance of the structure as well as its duration in the areas with seismic activities, suggested flexible variation of the system Moravamont 2000 offers the possibility to resolve element layers by using 3D technology (fig. 5). Additionally, it brings the rigidity of the prefabricated structure closer to the monolithic one. All of this provides great possibilities for further research regarding the increase of the structure's span and its optimality.

3D technology starts with resolving the element layers, but it is planned to be used for the elements' construction too. Thanks to that the whole automatization of the process of manufacturing elements will be achieved. In addition, the whole process of construction will be automatazied as well. (the whole process starting with the idea till the realization of the project will be computerized) fig. 6.

Fig.5. *Resolving element layers by using 3D technology*

Fig. 6. *The process of constructing the elements using 3D technology*

https://doi.org/10.1051/e3sconf/202021202005

5 References

1. J. Witzany, *Konstrukce pozemnich staveb 70* (Praha, 2003)
2. Ž.P. Cuckič, J. Witzany, V. Cuckič, International Journal Science and Engineering Investigations **7**, pp. 61-69 (2018)
3. Ž. P. Cuckič, *Improvement of Moravamont mounting System: Seismic active areas* (Lambert, LAP, 2018)

The influence of temperature impact on the strength of adhesion of polyurethane glue-foam with masonry products

Valery Derkach[1*]
[1]Brest State Technical University, 224017, Moskovskaya 267, Brest, Belarus

Abstract. The use of masonry on glued polyurethane foam joints can significantly increase labor productivity in the construction of stone buildings. One of the factors restraining the introduction of polyurethane foam adhesive into the practice of construction of stone buildings is insufficient knowledge of the influence of temperature effects on the adhesion of adhesive-foam to masonry products. In order to obtain data on the adhesion strength of polyurethane joints with masonry products under temperature influences, 28 prototypes of masonry were tested, aged in laboratory conditions and also subjected to heating to 100°C and 200°C. Tests of masonry samples were carried out in accordance with STB EN 1052-5. Based on the experimental studies carried out and new data on the influence of temperature effects on the adhesion strength of polyurethane joints with masonry products were obtained.
Key words: masonry, polyurethane seams, adhesion strength, temperature influences.

Introduction

The introduction of innovative technologies in construction aimed at increasing labor productivity and reducing the cost of products is in line with the priorities, goals and objectives of the Program of socio-economic development of the Republic of Belarus for 2016-2020. Recently, a new technology for the production of masonry works on adhesive polyurethane foam (PPU) joints has been used in the construction of stone walls in the countries of the European Union [1-3]. Research work on the introduction of this technology in construction is currently being carried out in the Republic of Belarus [4-6].

Preliminary studies of the possibility of using the technology of polyurethane foam joints for masonry made of aerated concrete blocks which were carried out in Russia [7-9] show that when using polyurethane masonry mixtures the speed of work has doubled compared to masonry on cement mixtures, at the same time, taking into account the consumption and cost of polyurethane foam glue, the cost of masonry turned out to be lower. According to the results of heat engineering tests, the calculated value of the thermal resistance of a wall

* Corresponding author: v-derkatch@yandex.ru

fragment with a thickness of 375 mm made of blocks of grade D400 density on polyurethane glue with an equilibrium weight moisture content of products in the masonry of 5% Rw1 was 3.44 m^2 °C / W (for comparison, the same thermal resistance for cement joints can be obtained for a wall thickness of 500 mm). The fire resistance limit of a sample of a partition with a thickness of 100 mm made of autoclaved aerated concrete blocks with a density of D400, a compressive strength class B2, on a polyurethane foam glue was EI 150. Thus, preliminary tests confirm the feasibility of introducing the above technology in construction practice.

Polyurethane foam adhesive is an organic material that is subject to pyrolysis at certain temperatures, which can adversely affect the resistance of masonry to force effects during and after fire. Stone structures with polyurethane joints are usually used as floor-by-floor support external walls and internal partitions. Limiting state in such structures under force effects is determined by the shear strength of the masonry and tensile bending which essentially depend on the adhesion strength of mortar joints with masonry products. In this regard, the study of the effect of temperature effects on the adhesion strength of polyurethane foam glue with masonry products is an urgent and important task.

Method

The prototypes consisted of two masonry productsof of dense silicate concrete connected by means of a PPU seam. The width of the masonry products was 70 mm, the length was 240 mm. The supporting surfaces of the masonry products were wiped from dust after which one strip of polyurethane mixture was applied to them, on which the next masonry product was laid. Immediately after fabrication, the prototypes were compressed so that the compressive stresses in the cross section of the specimen were about $2.5 \cdot 10^{-3}$ N / mm^2. In the loaded state, the samples were stored before testing for 7 days in laboratory conditions at a temperature of + 18°C - □ + 20°C and an air humidity of 65%.

4 series of prototypes of masonry were prepared for testing with a total of 28 samples. Series PI and PII included 9 prototypes, series PIII and PIV - 5 samples each. Samples of the PI series were tested after being in laboratory temperature conditions - t° = 20°C. Samples of the PII series were tested after heating to 100°C and subsequent cooling to 20°C and samples of the PIII and PIV series after heating to 200°C. The samples of the PIII series were tested after cooling down to 20°C and the samples of the PIV series were tested in a heated state. After heating, the prototypes were kept at a given temperature for 60 minutes. Heating of prototypes and holding in a heated state were carried out in a muffle furnace. After removing the samples from the furnace, their surface temperature was determined using a laser thermometer.

The tensile strength of the masonry in an unbound section (normal bond strength) was determined by the bending moment method in accordance with STB EN 1052-5 [10] (Figure 1).

The force on the masonry sample was transmitted using a P-10 laboratory test press equipped with an electronic device for recording the force. The loading rate was chosen such that fracture occurred within the range from 2 minutes to 5 minutes from the beginning of the load application.

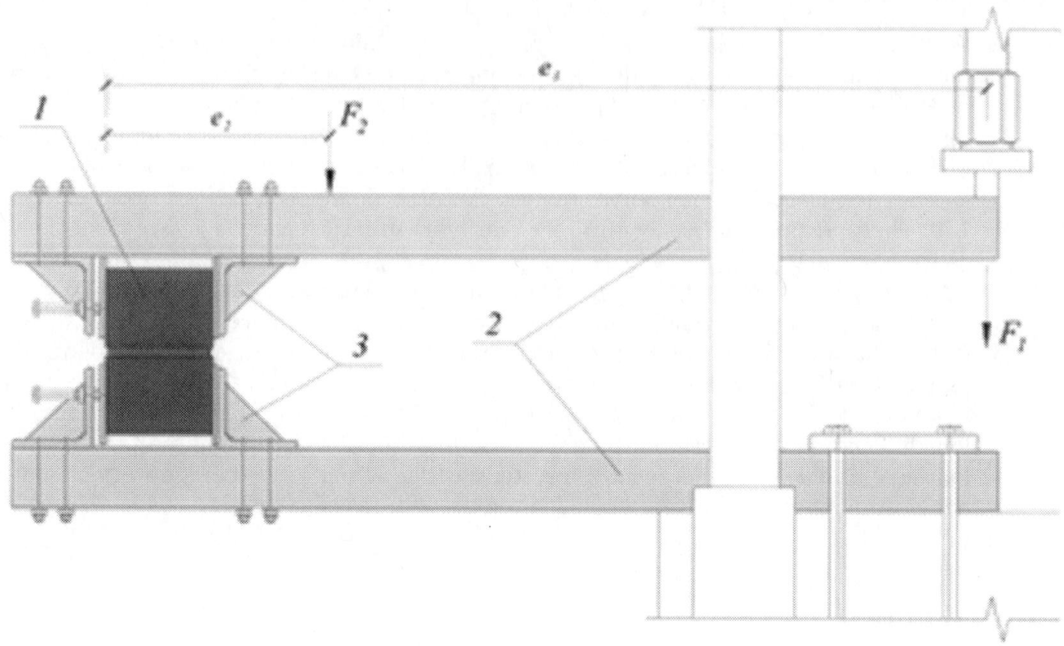

1 – test piece, 2 – levers, 3 – clamps

Figure 1. Scheme of prototype testing

For each case of fracture, the bond strength was calculated with an accuracy of up to 0.01 H / mm² using formula (1).

$$f_{wi} = \frac{F_1 e_1 + F_2 e_2 - \frac{2}{3} d\left(F_1 + F_2 + \frac{W}{4}\right)}{Z} \tag{1}$$

$$Z = \frac{bd^2}{6} \tag{2}$$

where:

- b – average length of the tested horizontal weld in mm;
- d – the average width of the sample in mm;
- $e1$ – distance from the axis of the load application to the stretched face of the specimen in mm;
- $e2$ – the distance from the center of gravity of the lever and clamp to the stretched edge of the sample in mm;
- $F1$ – applied load in H;
- $F2$ – load from the weight of the lever and clamp in H;
- W – load from the weight of the top masonry product.

Results and Discussion

The destruction of most of the prototypes occurred as a result of the rupture of the PPU weld (Figure 2a). In some specimens, the tear off of the upper or lower masonry product along the adhesion boundary was observed (Figure 2b).

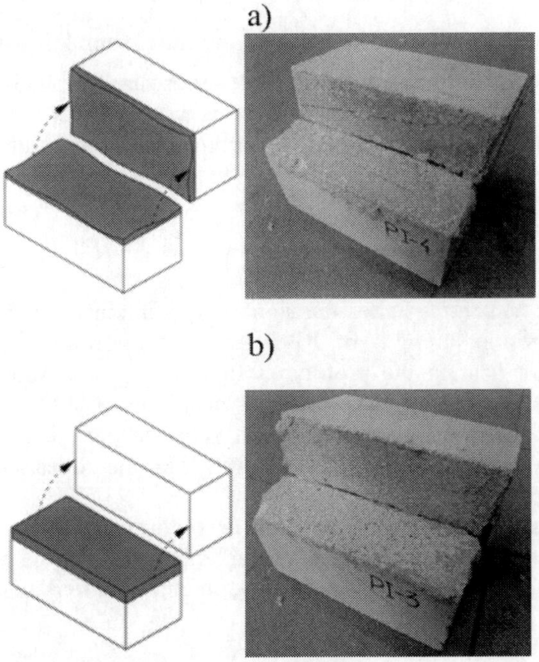

a) rupture of the PPU seam, b) tearing the PPU seam from the masonry product
Figure 2. The nature of the destruction of prototypes

The average value of adhesion strength $\overline{f_w}$ of prototypes of the PI series, cured and tested in laboratory conditions, was 0.82 MPa with a coefficient of variation ν = 20%. The average value $\overline{f_w}$ of the samples of the PII series, heated to 100 ° C and tested in the cooled state to 20 ° C, turned out to be practically the same as for the samples of the PI series. When the prototypes were heated to 200°C and then cooled, the bond strength decreased, the average value of which was 0.68 MPa at ν = 19%. A graphic interpretation of the results obtained is shown in Figure 3.

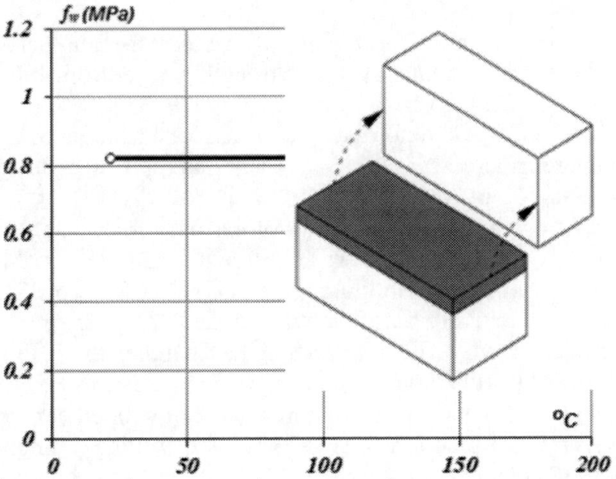

Figure 3 - Graph of the dependence of the strength of normal adhesion of polyurethane foam seam of temperature

The lowest value of adhesion strength $\overline{f_w}$ was obtained when testing prototypes of the PIV series in a state heated to 200°C. The value $\overline{f_w}$ of these samples was 0.3 MPa which is 2.7 times lower than the adhesion strength of the PI series samples. It should be noted that in all cases the adhesion strength of polyurethane seams was no lower than when using thin-layer solutions based on mineral binders (0.3 MPa) which indicates a high degree of fire safety of masonry with the use of polyurethane adhesive foam.

Conclusion

On the basis of experimental studies, new data were obtained on the effect of temperature effects on the adhesion strength of PPU seams with masonry products. It has been established that after heating the prototypes to 200 ° C and their subsequent cooling to 20 ° C, the adhesion strength of PPU seams *fw* decreases. The minimum value of adhesion strength took place when testing prototypes in a state heated to 200 ° C. The value *fw* of these specimens turned out to be 2.7 times lower than the adhesion strength of specimens aged and tested in laboratory conditions.

The results obtained should be considered as preliminary. More complete data on the effect of temperature on the mechanical characteristics of masonry on polyurethane foam joints can be established on the basis of experimental studies of stone walls exposed to temperature effects in a standard fire.

References

1. A. Jäger, Verklebung von Planziegelmauerwerk mit Polyurethanklebern [Joining in masonry with polyurethane adhesives] / A. Jäger, C. Kuhlemann, E. Habian, M. Kasa, S. Lu // Mau-erwerk. – № 15. – pp. 223–231 (2011)
2. E. Aldoghaim, Untersuchungen zur Verbesserung der mechanischen Eigenschaften von Mauerwerk durch Elastomerlager [Studies of the mechanical properties of masonry on elastic joints]/ E. Aldoghaim // Mauerwerk. – № 16. – pp. 93–102 (2012)
3. R. Drobiec, Wplyw rodzaja zaprawy na parametry mechaniczne murow z ABK poddanych sciskaniu [Influence of the type of mortar on the mechanical characteristics of ABK masonry under compression] / R. Drobiec, R. Jasinski // Materialy Budowlane. – №4. – pp. 3–7 (2015)
4. V.N. Derkach, Strength and deformability of masonry from autoclaved aerated concrete blocks on polyurethane seams. Part 1. Strength and deformability in compression / V.N. Derkach // Building materials. – No. 5. – pp. 29–33 (2017)
5. V.N. Derkach, Strength and deformability of masonry from autoclaved aerated concrete blocks on polyurethane seams. Part 2. Tensile strength in bending / V.N. Derkach // Building materials. – No. 7. – pp. 30–34 (2017)
6. V.N. Derkach, Strength and deformability of masonry from autoclaved aerated concrete blocks on polyurethane seams. Part 3. Strength and deformability during shear / V.N. Derkach, IE Demchuk // Building materials. – No. 8. – pp. 32–36 (2017)
7. G.I. Grinfeld, Comparative tests of autoclaved aerated concrete masonry with different designs of masonry seam / G.I. Grinfeld, A.P. Kharchenko // Housing construction. – No. 11. – pp. 30–34 (2013)
8. A.S. Gorshkov, Properties of wall structures made of aerated concrete products of auto-curing on polyurethane glue / A.S. Gorshkov, N.I. Batting // Engineering magazine. – No. 5. pp. 5–19 (2013)

9. A. Glumov Laying on polyurethane compounds: how to eliminate cold bridges / A. Glumov // Building materials, equipment and technologies of the XXI century. – No. 4. – pp. 30–31 (2014)
10. Test methods for masonry. Part 5. Determination of bond strength by the bending moment method: STB EN 1052-5. – Introduction. 02/10/2015 – Minsk: State Standard of the Republic of Belarus – 12 p. (2015)

Environmental safety of construction processes in the Republic of Belarus

Alena Kisel[1,*] and *Liudmila* Sryukina[1]

[1]Brest State Technical University, 224017, Moskovskaya 267, Brest, Belarus

Abstract. In creating or changing real estate property – buildings and structures - a rational use of natural resources and environmental safety of construction activities which result from the actions of all participants at each stage of the construction project implementation are becoming more and more relevant. A construction project should be considered as a subject of investment and construction activities: from its concept to its operation. The theoretical level of environmental safety is laid down at the stage of pre-project studies and design. During the development of the construction site and the execution of construction and installation works, the actual environmental safety is formed. It must be maintained in the future by proper organization of the object's operation. The basis for the development of environmental safety of investment and construction processes is the legal framework and the introduction of environmental management in the activities of participants in the investment and construction cycle. It is possible to achievement of results only under condition of continuous performance of all functions of environmental management by each of the participants in investment and construction activities. These functions must be integrated into the organizational structure of construction organizations, which will allow you to correctly distribute environmental functions within all divisions.
Keywords: investment and construction activities, environmental safety, environmental management, management functions.

Introduction

The object of this research is the processes of forming environmental safety of a project at each stage of investment and construction activities: at the stage of developing a concept of creating or changing a real estate object, choosing a construction site, designing, preparing construction production, performing construction and installation works.

In the investment and construction process, a rational use of natural resources and environmental safety of construction activities which result from the actions of participants at each stage of the construction project implementation are becoming more and more relevant. But the foundation of environmental safety is at the early stages of developing a concept of a construction project and choosing a site for construction. The quality of decisions made at this

* Corresponding author: eleniva2006@mail.ru

stage regarding a choice of materials, structures, engineering and technological equipment depends not only on the initial capital investment, but also on further operating costs, including those related to ensuring a required level of environmental safety of the project throughout its life cycle and creating a sustainable environment for activities.

In creating or changing a real estate property, a large number of participants are involved, each of whom has its own interest in this activity. At the same time, all stages of the construction project must be environmentally friendly and contribute to the creation of a safe facility. To achieve this goal, it is necessary to implement environmental management in the activities of participants in the investment and construction cycle.

Major research in this area is currently focused on integrating the Life Cycle Sustainability Assessment (LCSA) with the building design process and Building Information Modelling (BIM) [1-6], as well as on sustainability issues from a building management perspective.

Key management approaches in construction organizations to move towards sustainable development include: sustainable procurement, integrated project delivery (IPD), environmental management system, building sustainability assessment systems, and post-occupancy management [7].

Integrated project delivery (IPD) is an approach that brings together the capacities and interests of all stakeholders in the investment cycle into a single process to reduce costs and improve efficiency at all stages of planning, design and construction.

Various green rating systems are being developed to assess the sustainability of construction projects: LEED (Leadership in Energy and Environmental Design), BREEAM (Building Research Establishment Assessment Method), CASBEE (Comprehensive Assessment System for Building Environmental Efficiency), Green Star NZ and others. Further research in this area is likely to focus on economic and institutional factors [8] and integration of LCA and LEED methodology [9].

Researchers are working on the environmental safety problems of individual stages of the investment cycle. For example, causal relations of adverse environmental impacts of construction sites are investigated [10], indicators for real-time monitoring of environmental pollutants in construction for the purpose of system planning of work by construction companies and management of pollutants at construction sites are defined [11], various directions of development and improvement of green construction projects are considered: costs, schedule, safety, rework, productivity, risk assessment and mitigation measures, risk management, and leadership [12].

The purpose of this study is to identify the tasks for environmental management at all stages of the investment project in the construction and distribution of environmental responsibilities between offices of the organizations participating in the investment-building cycle to promote environmental safety.

To achieve this goal, it is planned to solve the following tasks: assessment of the impact of individual stages of investment and construction activities on the environment; characteristics of the legal framework for environmental construction in the Republic of Belarus; development of special functions within the framework of environmental management for divisions of organizations participating in the implementation of an investment project in construction.

The study used empirical methods – observation and comparison, theoretical research methods – analysis and synthesis, and mental modeling.

1 Impact of stages of investment and construction activities on the environment

In the existing practice of analysis in the Republic of Belarus, each of the stages of creating (changing) a real estate object is investigated independently, since it is assigned to a separate performer.

Mutual influence is poorly monitored, which means that the strength of management decisions is weakened. This is due to the fact that in general, considering a construction object as a subject of investment and construction activities is very rare. A search for relationships would ensure alternative options for optimizing possible negative impacts on the environment, and would reduce risks throughout an investment and construction cycle as a whole. If all participants in investment and construction processes were guided by environmental management procedures, this would become a basis for formainf an environmental safety management system for finished construction projects. Let's look at the features of each of the following stages:

• Creating a project concept and development of project documentation,

• Site selection for construction,

• Preparation of construction production,

• Construction and installation works.

The concept is formed at the pre-project (pre-investment) stage of the property's life cycle. At this stage, the investment justification is developed, which determines a necessity, technical and economic feasibility of investing in the construction, reconstruction, and restoration of a facility, as well as the environmental impact assessment and calculation of environmental consequences of construction and operational activities. At this stage, the ecological potential of the project is laid for the entire future period of its existence, so the following are extremely important::

- Conscious and competent approach,
- Variant study of architectural solutions, engineering and technological equipment of the project,
- Economic assessment of investment efficiency, taking into account an initial capital investment and upcoming operating costs over the entire life cycle of the project (Table 1).

The main role at this stage belongs to the customer (owner) who makes fundamental decisions on feasibility of developing options for space-planning and structural solutions of the project, as well as solutions related to engineering and technological equipment [13]. It is important for them to understand and accept the concept of "sustainable development" of sustainable planning, designed for a long period of time and aimed at achieving the best result both from the point of view of the owner and from the point of view of the state of the environment, i.e., to create a sustainable environment for activities. Creating a sustainable business environment implies:

- Reducing investment in energy, raw materials, and water while increasing the cost of acquiring knowledge and skills,
- Optimizing a life span of the facility,
- Creating premises with better consumer properties (better living environment),
- Reducing emissions and waste during construction and operation of the facility,
- Creating products with improved environmental parameters (for industrial and engineering infrastructure objects),
- Increasing satisfaction of buyers of products, works, services or users of the facility (for example, tenants),
- Increasing the property owner's income.

The effectiveness of implementation of the first stage is undoubtedly affected by the quality of project documentation development, including technical solutions and measures for rational use of natural resources and prevention of environmental pollution [14, 15].

Table 1. Formation of an environmental potential of a project at the stage of the project concept and design development.

Formation of a project concept		
Spheres of decision-making		
Architectural and planning solutions	Engineering infrastructure of the project	Selection of materials, structures, products, and technological equipment
Emerging potential		
- heat engineering characteristics (energy efficiency), - use of natural renewable sources, - durability, commercial use options, - ability to redevelop and recycle the materials used		
Optimization of the resources to be consumed in the future		

According to the results of the State expertise of project documentation in Belarus, the following typical violations are identified:

- Unsubstantiated or unreliable data on parameters of the project's impact on the environment,
- Incomplete description of the current state of environmental components (before the construction of a project),
- Unreliable or incomplete information on characteristics and condition of green spaces, the number of preserved and removed vegetation; lack of compensatory measures,
- Lack of measures aimed at soil protection,
- Lack of a comprehensive solution to the issues of neutralization, use and disposal of wastes, including the construction ones,
- Non-compliance of the project’s location with the requirements of the established regime of protection and use of the territory, its functional purpose.

The result of the evaluation of the first stage (pre-project stage, design) could be more effective if the choice of a site for construction is more deliberate (Table 2).

Table 2. Influence of the construction site selection process.

Selection of a construction site.				
Spheres of decision-making				
A necessity to withdraw agricultural land from circulation	A necessity to withdraw territories of the forest fund	A necessity to use land near water protection zones	A necessity to use land for recreation areas in localities	A necessity to destroy the habitat of birds and animals
Result of decision making				
Search for new places to grow crops, loss of investment in the development of the territory	Deforestation, ecosystem disruption	Increasing risk of pollution of water and the adjacent territories	Destruction of parks, reduction of green areas in cities	Death, population decline

Location of a construction project significantly affects not only its profitability during its operation and sale cost, but also the environmental friendliness of the construction process and preparation for it (Table 3).

Table 3. Impact of preparation of construction industry.

Preparation of construction industry		
Spheres of decision-making		
Choice of production technology	Planning of logistics processes	Designing a construction site
Result of decision making		
Level of noise, dust, waste, operating time of onstruction machinery and equipment	Level of waste materials, organizational downtime, percentage of defects	Groundwater degradation, landscape change, soil erosion, loss of trees, shrubs, garbage and its sorting

Good quality of preparation for construction production in order to reduce environmental impacts will allow reducing environmental risks at the next stage – the stage of work production (Table 4).

Table 4. Impact of stage of work production.

Construction works during new construction, repair, reconstruction			
Spheres of decision-making			
System of work organization, production practices	Construction waste management system	Upgrading of machinery and transport	Technological discipline
Result of decision making			
*level of resources use *maintenance of construction sites in accordance with regulations *unjustified costs for sites maintenance *unjustified loss of materials *implementation of sanitary standards for household areas *conditions for materials storage	*the amount of waste *compliance with the standard for the volume of construction waste * garbage sorting *handling waste that exceeds the norm	*level of fuel and energy consumed * CO2 emissions *repair costs *downtime *use of the working time fund	*compliance with waste levels *level of waste *compliance with work deadlines

Decisions related to environmental protection in the course of construction and installation works are regulated in Belarus by project documentation - the Construction Organization Project and production documentation – the Work Production Project. These solutions include land reclamation, prevention of loss of natural resources, prevention or treatment of harmful emissions into the soil, reservoirs, and atmosphere, and are reflected in the explanatory notes of these organizational and technological documents and in construction master plans.

It is necessary that environmental protection measures are not just reflected in the project documentation, but also implemented in practice, which is not always the case. Thus, according to the Department of construction control and supervision of the State Committee for standardization of the Republic of Belarus, in 2019, 69% of the inspected construction sites did not meet the established requirements for their maintenance. Construction work began without prior arrangement of the construction site, without installing fences, areas for washing the wheels of vehicles, containers for separate waste collection; untimely removal of construction debris was carried out.

Responsibility for the implementation of measures for environmental protection at the production stage of the work lies on all of the major participants of investment-construction process: the customer (the developer), the project Manager, the technical supervision of the customer, General contractor, and representatives of the project organization.

Consideration of the stages of investment and construction activity reflects the complexity of the impact on the environment. Unfortunately, the negative impact is not reduced to the desired level. Each previous stage has an impact on the next one. Raw materials affect the environmental friendliness of building materials, the choice of material affects the environmental friendliness of finished construction objects, the quality of materials and construction affects the amount of waste generated.

In recent years the legislation of the Republic of Belarus on environmental impact issues has been significantly improved, and control over the creation of construction projects has been strengthened.

2 Legal assessment of environmental construction in the Republic of Belarus

Regardless of types and goals of activities, the state program "Environmental protection and sustainable use of natural resources" for 2016-2020 has been formed. Under the program "the effectiveness of conservation measures significantly increased, providing positive trends in the environment controlled parameters of pollution and has helped to promote the use of natural resources".

Environmental safety in the construction sector is regulated by:

a) The law of the Republic of Belarus of 05.07.2004 No. 300-3 "On architectural, urban planning and construction activities in the Republic of Belarus",

b) The law of the Republic of Belarus of 18.07.2016 No. 399-3 "On state environmental expertise, strategic environmental assessment and environmental impact assessment",

c) The law of the Republic of Belarus of 26.11.1992 No. 1982-XII "On environmental protection",

d) "Regulations on the procedure for commissioning construction projects", approved by resolution of the Council of Ministers of the Republic of Belarus dated 06.06.2011 No. 716,

e) Regulations on the procedure and conditions for conducting state energy efficiency expertise, approved by resolution of the Council of Ministers of the Republic of Belarus No. 216 of 18.03.2016,

f) Resolution of the Council of Ministers of the Republic of Belarus 19.01.2017 N 47 "Regulations on the procedure for conducting an environmental impact assessment, requirements for the composition of an environmental impact assessment report, requirements for specialists performing an environmental impact assessment",

g) Technical code 17.11-10-2014 "environmental protection and nature management. Waste. Rules for handling construction waste".

In the world ranking on the environmental performance index, the Republic of Belarus improved its position and rose from 73rd place in 2005 to 32nd place in 2014. But in 2018 it was ranked 44th. Indicators of the state of the environment and the viability of the ecosystem have decreased. A number of shortcomings that have affected the decline in these indicators can also be noted in the construction sector.

3 Distribution of environment protection functions between divisions of organizations participating in the investment and construction cycle

Development and distribution of special functions within businesses' divisions can ensure managerial functions to get developed, which increases environmental safety of participants of investment-construction cycle (Table 5).

Effectiveness of implementing distributed functions will depend on the quality of organizational structure of an enterprise, which also becomes a basis for sustainable development of organizations participating in the investment and construction cycle [16, 17].

Table 5. Distribution of environment protection functions in an organization.

Name of division	Environment protection functions
Environmental service	Providing control and coordination work within environment protection activities of the construction site.
Production department	Setting and solving tasks related to preventing and minimizing a negative impact of the production activity of the project (participant enterprise) on the environment.
Logistics service	Supply of environmentally friendly materials (causing minimal damage to the environment during transportation, storage, use and recycling) for the production process.
Marketing service	Ensuring that services are clean, safe, and efficient to meet the needs of consumers. Creating a favorable image of the enterprise (project) by forming communications with contact audiences.
Financial department	Financial support for implementing environmental protection measures, projects, and contracts.
Planning and economic department	Creating new and improving existing services for water supply, sanitation and technological processes in order to ensure their environmental safety. Planning of current and capital environmental expenditures. Analysis of economic efficiency of environmental projects and activities.
Accounting	Taking into account actual environmental costs and benefits. Providing financial information to the environmental service.
Personnel department	Preparation of employees of the enterprise (operating organization) to perform environmental tasks at their workplaces; formation of their environmental thinking.
Department of occupational health and safety	Prevention of occupational diseases caused by adverse environmental factors
Quality control department	Ensuring high quality of service while taking into account both technical and environmental aspects.

Conclusions

1. Creating (changing) real estate objects inevitably has a negative impact on the environment.
2. The impact is uneven, there are losses which are specific for each stage.
3. A preliminary analysis allows for obtaining information and assessing risks.
4. Legislative framework contributes to the development of environmental safety of projects of investment and construction processes.
5. In order to improve legislation, it is necessary to perform monitoring and analysis functions.
6. To improve eco-efficiency it is necessary to introduce ecological management in activity of participants of investment-construction cycle.
7. Competent development and distribution of environment protection functions between the divisions of enterprises participating in the investment and construction cycle will help to achieve an increase in environmental safety.
8. This type of activity can provide necessary results only if all the functions of environmental management are constantly performed by participants in investment and construction activities.

9. Creating an environmentally safe construction project is a complex process that covers all stages of its life cycle. At the stage of concept development and design, the theoretical level of environmental safety of the project is laid. During the construction process, the actual environmental safety of the project is ensured. After the project comissioning, it must be maintained at the required level by a proper organization of operation.

References

1. C. Llatas, B. Soust-Verdaguer, A. Passer, Building and Environment **182**, 107164 (2020)
2. S. Seyis, Building and Environment **173**, 106703 (2020)
3. R. Santos, A.A. Costa, J.D. Silvestre, T. Vandenbergh, L. Pyl, Building and Environment **169**, 106568 (2020)
4. M. Najjar, K. Figueiredo, A.W.A. Hammad, A. Haddad, Applied Energy **250**, pp. 1366–1382 (2019)
5. T.D. Johansson, W. Lu, J. Schade, T. Olofsson, Energy and Buildings **128**, pp. 592–604 (2016)
6. C. Cavalliere, G. Habert, G.R. Dell'Osso, A. Hollberg, Journal of Cleaner Production **211**, pp. 941–952 (2019)
7. V.W.Y. Tam, K.N. Le, *Sustainable construction technologies* (Butterworth-Heinemann, 2019)
8. D.T. Doan, A. Ghaffarianhoseini, N. Naismith, T. Zhang, A. Ghaffarianhoseini, J. Tookey, Building and Environment **123**, pp. 243–260 (2017)
9. K. Galal, S. Alkass, Building and Environment **80**, pp. 61–70 (2014)
10. A. Fuertes, M. Casals, M. Gangolells, N. Forcada, M. Macarulla, X. Roca, Journal of Cleaner Production **52**, pp. 425–437 (2013)
11. J. Hong, H. Kang, S. Yung, S. Sung, T. Hong, H.S. Park, D.-E. Lee, Building and Environment **170**, 106636 (2020)
12. H. Bon-Gang, *Performance and improvement of green construction projects* (Butterworth-Heinemann, 2018)
13. A.N. Kochurko, L.G. Sryvkina, Vestnik BSTU **1**, pp. 100–106 (2019)
14. N.V. Mikhailova, E.I. Kisel, Vestnik BSTU **3**, pp. 61–65 (2019)
15. A. Barylka, Inżynieria Bezpieczeństwa Obiektów Antropogenicznych **1–2** (2019)
16. O.A. Vysotski, Vestnik BSTU **3**, pp. 2–5 (2016)
17. O.A. Vysotski, I.M. Garchuk, N.S. Danilova, V.F. Medvedeva, *Basics of sustainable development of a production organization* (Pravo i ekonomika, Minsk, 2015)

 https://doi.org/10.1051/e3sconf/202021202008

Choice of structural and technological solutions for engineering preparation of bases for foundations

Dzmitry Klebanyuk[1,*], *Peter* Poyta[1] and *Peter* Shvedovsky[1]

[1]Brest State Technical University, 224017, Moskovskaya 267, Brest, Belarus

Abstract. The object of the research is a comprehensive assessment of the choice of structural and technological solutions for the engineering preparation of soil foundations using the methods of dynamic compaction with heavy rammers. It is proposed in the studies to choose the optimal solution according to the criteria of both efficiency and preference with sequential or selective implementation of certain target groups: multi-purpose choice from a set of goals; optimization on a set of conditions and in dynamics; multi-vector optimization, etc. Location of the options according to preference, the main directions for optimizing the options for constructive and technological solutions and the best option for the engineering preparation of the base for the foundations were identified on the basis of the proposed method for the experimental site.
Key words: foundations, engineering preparation, dynamic compaction, choice of solutions, optimization.

1 Introduction

Limited territories with favourable geotechnical conditions for construction necessitate the development of complex, in engineering and geological terms, construction sites.

Local [1–4] and foreign [5–8] practice has shown that the most effective preparation of such areas using methods of dynamic compaction with heavy rammers.

A certain problem when choosing the optimal design solution in terms of structural and technological parameters is created by the need for test compaction of soils at the construction site and multiple uncertainties: initial state of soils; influence of technological parameters on the change in soil properties during the compaction process; peculiarity of taking into account soil compaction in time, etc. [9].

It should also be noted that traditional methods for solving single-criterion problems that implement optimizing factors when restrictions are imposed on all other, other alternative constructive and technological solutions only in terms of economic indicators are not always legitimate, since the cost estimate is most often nonlinear in terms of the optimality of the decision taken [10].

*Corresponding author: klebanyuk.dmitri@gmail.com

It should also be noted that traditional methods for solving single-criterion problems that implement optimizing factors when restrictions are imposed on all other, other alternative constructive and technological solutions only in terms of economic indicators are not always legitimate, since the cost estimate is most often nonlinear in terms of the optimality of the decision taken [10].

The authors [4] propose to use multivariate models which allow to take into account to a large extent the direct and feedback links of the main factors, their interrelation, as well as the uncertainty of many subjective and objective conditions and parameters of the compaction of soil massifs.

2 Identification of interdependencies and interrelationships of conditions and factors

Production experience and most studies [11-14] allow us to note that:

– with an increase in the deformation modulus of soils in the natural state, a decrease in the diameter of the rammers is required, but not less than 1 m, since in this case the soil loosens with the formation of soil lifting zones;

– when compacting soil strata of high thickness, simultaneously with an increase in the diameter of the rammer, it is necessary to increase its mass and discharge height;

– the greatest efficiency of compaction is achieved with optimal moisture content and content of clay particles;

– the more homogeneous the soil the more efficient the compaction with the same energy consumption;

– the increase in the rammer weight often does not lead to a high degree of compaction and uniformity of the compacted base;

– with an increase in the impact energy, the density of the soil increases most significantly at the initial stage of compaction;

– the quality of the compaction is largely determined by the layout and the distance between the compaction points;

– the smaller the diameter of the rammer (at constant impact energy) the smaller the size (radius and depth) of the compaction zone;

– the greater the deformation modulus of natural soils the smaller the tamping diameter should be;

– uniformity of the soil structure in the compaction zone can be achieved by increasing the initial density of dry soil;

– to achieve the required density it is necessary to ram additional volumes of soil into the base at relatively low density of natural soil;

– for soft soils the greater their power, the greater should be the mass of the rammer or the height of its dropping and diameter;

– the greatest depth of the compaction zone while minimizing energy costs can be achieved through the use of rammers of large diameters (2 m or more) at the initial stage of compaction with additional compaction with rammers of a smaller diameter upon reaching failure;

– the greater the thickness of the compacted layer, the distance between the points of compaction of the soil should decrease and at certain values of the thickness of the compacted layer compaction must be performed without the presence of bridges between the indentations, i.e. continuous throughout the site;

– in the presence of a solid underlying layer within the compacted strata, an increase in the radius of the compacted zone is characteristic, which makes it possible to increase the distance between the points of soil compaction;

– with an increase in the depth of the imprint of the rammer and its diameter, the distance between the sealing points increases proportionally;

– the formation of the soil compaction zone is greatly influenced by the shape of the rammer sole. The use of a rammer with a stepped bottom allows to significantly increase the depth of the compaction zone in comparison with a rammer with a flat bottom;

– the maximum depth of the compaction zone with a sufficiently uniform degree of compaction can be achieved using a two-stage technological process and rammers with a concave spheroid and a convex spherical bottom.

3 Research method

Hence, selection of the optimal solutions should be carried out on a group of criteria of efficiency $\left(K_{ij}\right)$ and preference $\left(U_i\right)$ (techno-economic, structural,technological and others.), with the need for consistency or selective implementation of certain target groups: multi-purpose selection of a variety of purposes; optimization on multiple conditions; optimization in dynamics, at many stages and many options; multi-vector optimization, etc. [2, 9, 10].

As parameters of preference, it is advisable to operate with a number of preferences $\left(\bar{X}_i\right)$ and vectors of priorities $\left(\lambda_i\right)$ and significance $\left(q_i\right)$ determined by the method of sequential optimization. Note that the preference series determines the ordered set of local criteria $\bar{x} = x_1 > x_2 > ... > x_n$, the priority vector $\lambda = \left(\lambda_1, \lambda_2, ..., \lambda_n\right)$ shows the degree of coincidence of two adjacent performance indicators (PI) in terms of significance and the significance vector - the significance of the *j* PI in comparison with others [15].

At the same time, performance indicators can be set both precisely (the principle of strict priority) and approximately - by a certain area (the principle of flexible priority).

Using the method of expert evaluations to determine the significance of performance indicators and taking the Bernoulli criterion as the most reliable criterion for the optimality of a constructive and technological solution

$$K_i^j = \left\{ a_i \,/\, \max \frac{1}{n} \cdot \sum_{j=1}^{n} \bar{X}_{ij} \right\}, \qquad i = 1, \overline{m};\ o = 1, \overline{n} \tag{1}$$

where value $\bar{X}_{ij}$ of the j indicator for the i option which implements the principle of fair absolute concession and allows one to switch from vector criteria to scalar ones, reducing the multicriteria problem to a single-criterion one.

For significant indicators of the effectiveness of constructive and technological solutions, the following were taken: height of discharge of heavy rammers (H, m); energy consumption $\left[\sqrt{M \cdot g \cdot H},\, (mnf \cdot m)^{1/2}\right]$; parametric characteristic of compaction $\left(h_{opt}^{dep} \,/\, h_{dep}\right)$; index of the degree and uniformity of compaction $\left(\rho_{nat} / \rho_{req} \cdot k_n\right)$; deformability of the base $\left(E_{mp},\, MPa\right)$ and expended work $\left(N, mnf \cdot m\right)$.

The state of the optimal compaction depth is taken as a parametric characteristic of the compaction zone $\left(h_{opt}^{dep}\right)$ and final imprint depth $\left(h_{dep}\right)$, and the index of the degree and uniformity of compaction is natural $\left(\rho_{nat}\right)$ and required $\left(\rho_{reg}\right)$ soil density and uniformity coefficient $\left(k_n\right)$.

The calculations were carried out in the following stages:

• the initial matrix (*P*) of the form was determined –

$$P = \begin{array}{c} \\ a_1 \\ a_2 \\ . \\ . \\ . \\ a_m \end{array} \begin{array}{c} \begin{array}{cccccc} x_1 & x_2 & . & . & . & x_n \end{array} \\ \begin{bmatrix} x_{11} & x_{12} & . & . & . & x_n \\ x_{21} & x_{22} & . & . & . & x_{2n} \\ . & . & . & . & . & . \\ . & . & . & . & . & . \\ . & . & . & . & . & . \\ x_{m1} & x_{m2} & . & . & . & x_{mn} \end{bmatrix} \end{array} \tag{2}$$

where $a_1, ..., a_m$ - compared options for technical solutions $i = \overline{1, m}$; $x_1, ..., x_n$ - polydimensional performance indicators $j = \overline{1, n}$; $x_{11}, ..., x_{mn}$ - values of performance indicators;

• it was normalized $\left(\overline{P}\right)$ by dependence, which made it possible to compose an efficiency matrix $\left(F\right)$ and determine all the sets of agreement $\left(C_{ke}\right)$ and disagreement $\left(H_{ke}\right)$, while the set of agreements consists of indices of those indicators for which the *k* option is not worse than option *e* and the set of disagreements consists of indices of the same indicators, according to which *e* option is superior to *k* option;

• based on the values of the efficiency indicators $\left(q_i\right)$, an efficiency matrix was compiled $\left(F = /\overline{P}/ \cdot /q/\right)$;

• sets of agreement and disagreement were determined for each pair of options;

• indices of agreement $\left(C_{ke}\right)$ and disagreement $\left(d_{ke}\right)$ were determined, reflecting the preference of one option over another, from which the corresponding matrices of agreement (*C*) and disagreement (*D*) are formed;

• the dominant agreement $\left(D_c\right)$ and disagreement $\left(D_H\right)$ matrices were constructed based on the correspondences of the relative weightings of the indicators l_{ke} and d_{ke}:

$$l_{ke} = \begin{cases} 1, \ if \ C_{ke} \geq \overline{C} \\ 0, \ if \ C_{ke} < \overline{C} \end{cases}; \quad d_{ke} = \begin{cases} 1, \ if \ d_{ke} \geq \overline{d} \\ 0, \ if \ d_{ke} < \overline{d} \end{cases}, \tag{3}$$

where $\overline{C}$ is the average index of agreement, and $\overline{d}$ is the average index of disagreement.

As the significance of performance indicators are taken according to [6] -

$$q_1 = 0{,}193;\ q_2 = 0{,}196; q_3 = 0{,}229; q_4 = 0{,}202; q_5 = 0{,}180.$$

4 Results and discussion

The initial matrix was formed for an experimental construction site in the geological structure of which technogenic formations are involved, represented by dumps of soil from sands of various sizes, mixed with sandy loam with inclusions (up to 3%) of gravel. The dumping time is more than 10 years, which determines their sufficiently compacted state. The thickness of the deposits is from 0.5 to 0.7 m. The bedrock deposits are represented by fine and medium sands of low strength and medium strength. The thickness of the sandy strata is from 2.0 to 5.0 m. Loams occur from a depth of 7.0-8.0 m.

Groundwater - free-flow with a steady level in the range of 4.6-5.1 m.Water-bearing soils - fine and medium sands with the following averaged physical and mechanical characteristics:

Based on the production capabilities of the general contractor, the following indicators of the effectiveness of constructive and technological solutions were adopted (Table 1).

Table 1. Significant indicators of the effectiveness of constructive and technological solutions for soil compaction with heavy rammers.

Option No.	Constructive and technological solutions		Efficiency criteria				
			Energy consumption	Parametric characteristic of the compaction zone	Index of the degree and uniformity of compaction	Deformability of the base	Work expended
1	Drop height, m	H_1=8	20,0	4,5	0,82	16,8	12,3
2		H_2=10	23,2	4,7	0,86	19,3	15,6
3	Distance between compaction points, m	l_1=4,0	21,3	4,6	0,84	18,4	10,4
4		l_2=6,5	20,0	4,5	0,82	20,2	11,9
5	Ramming weight, t	M_1=5,0	20,8	4,58	0,79	20,3	13,5
6		M_2=7,0	19,6	4,62	0,82	21,2	15,2

Note: The variant with the index "1" is accepted as the basic one.

According to the average index of agreement $\bar{C} = 0,513$ and disagreement index $\bar{d} = 0,476$ dominant matrices of agreement D_c and disagreement D_n, aggregated matrix the matrix of dominance $D_a = D_c \cdot D_n$ was determined and in accordance with the criteria of agreement and disagreement the non-dominant options are the first, fourth and sixth options.

No less important is the question of ordering options according to their preference. We have used the most acceptable method of ordering which is suitable for a set of cardinal (numerical) and ordinal indicators of performance of the compared options.

As an estimate of the ordering of the variants $\beta_q \left(q = 1, 2, ..., m!\right)$ the following dependence was taken:

$$\beta_q = \sum_{k,e=1}^{m} \cdot \sum_{\ni C_{ke}} q_j - \sum_{k,e=1}^{m} \cdot \sum_{\ni H_{ke}} q_j, \quad q = 1,2,...,m! \tag{4}$$

where

$$\left.\begin{array}{l} C_{ke} = \left\{ j / X_{kj} > X_{ej} \right\},\ k,\ e = 1,\ \overline{m};\ k \neq e; \\ H_{ke} = \left\{ j / X_{kj} > X_{ej} \right\},\ k,\ e = 1,\ \overline{m};\ k \neq e \end{array}\right\} \tag{5}$$

and then the agreed (the best) ordering will be mo, for which the value βq is the largest. It should be borne in mind that the most consistent ordering is identical to the series of preferable options.

Significant cardinal and ordinal indicators are shown in Table 2.

Table 2. Cardinal and ordinal indicators of the effectiveness of structural and technological solutions for compaction of bases with heavy rammers.

Indicator of efficiency	Indicator index	Options				Significance of the indicator
		№1	№2(3)	№3(5)	№4(6)	
1	2	3	4	5	6	7
Work expended	X_1	12,3	10,4	13,5	15,2	0,15
Compaction uniformity index	X_2	0,84	0,85	0,81	0,83	0,11
Deformability of the base	X_3	34,2	39,7	34,7	36,9	0,18
Energy consumption	X_4	20,0	21,3	20,8	19,6	0,16
Parametric characteristic of the compaction zone	X_5	4,5	4,6	4,58	4,62	0,09
Degree of homogeneity of the compacted base	X_6	III	III	II	II	0,06
Relative energy costs	X_7	II	III	II	II	0,07
Operational reliability	X_8	III	III	II	III	0,05
Degree of dynamic impact on existing adjacent objects	X_9	III	III	III	II	0,03
Internal factors beyond the control of the designers	X_{10}	II	III	III	II	0,02
Internal factors beyond the control of the contractor	X_{11}	III	I	II	III	0,02
External factors of the features of the compaction devices	X_{12}	I	III	III	III	0,01
External factors of qualification of work performers	X_{13}	II	II	III	III	0,02
Factors of randomness and uncertainty	X_{14}	I	II	I	II	0,05

Note. Ordinal indicators I, II and III for levels and conditions characterize respectively "high-medium-low", and for needs - "small-medium-large"

The preference of the options was analyzed on the basis of entropy [16, 17] for which transforming the initial matrix (P) into the calculated one $(\overline{P})$ the levels of entropy (E_j), levels of variability (d_j) and the absolute weightings of indicators (q_j) were determined according to the dependences –

$$\left.\begin{array}{l} \bar{P}_{ij} = X_{ij} / \sum_{i=1}^{m} X_{ij}; \qquad E_j = k \cdot \sum_{i=1}^{m} p_{ij} \ln p_{ij},\ j = 1,\bar{n}; \\ d_j = 1 - E_j,\ j = 1,\bar{n}; \quad q_j = d_j : \left(\sum_{j=1}^{n} d_j \right),\ j = 1,\bar{n}. \end{array}\right\} \tag{6}$$

The obtained criteria for the weighted average success of the decision made make it possible to arrange the options according to preference as follows: $a_4 > a_6 > a_1$, i.e. a_4 is "no worse than a_6 ", etc.

No less important is the issue of directed improvement of the selected (out of all competitive) options for constructive and technological solutions, which is based on a common utility function that takes into account both independence in preference and independence in utility [18–21].

The verification of the fulfillment of the conditions of independence in terms of preference and utility, carried out for all pairs of variables, made it possible to obtain equivalent pairs of multidimensional alternatives, and given that the general utility function should have a multiplicative form

$$\underset{i=1,4}{u(x_i)} = \frac{1}{k} \prod_{i=1,4} 1 + k_i \cdot k \cdot u_i(x_i) - 1; \quad \sum_{i=1}^{4} k_i \neq 1, \tag{7}$$

showed that the best constructive and technological option is the a4 option.

5 Conclusions

1. The conducted studies allow us to note that an arbitrary choice of design parameters of heavy rammers and not taking into account the technological features and patterns of soil compaction dynamics leads to an increase in the cost of engineering preparation of a construction site and does not allow achieving the required degree and uniformity of soil compaction at relatively acceptable energy costs.
2. Optimization of the size and shape of the rammers and accordingly the mass, dropping height, the distance between the compaction points, the technology and organization of work requires a complete and reliable account of both engineering and geological conditions as well as design and technological parameters and their attendant factors.
3. Since the indicators of the efficiency of the design option are a large group of possible alternatives, when choosing the optimal resource-saving option, it is advisable to use the scheme we proposed, which allows the main tasks - drawing up a series of preferences and choosing a design option - to be solved at any level using the capabilities of computer-aided design (CAD).

References

1. D.N. Klebanyuk, P. S. Poita A, Vestnik BSTU **1**, pp. 77–81 (2013)
2. P. V. Shvedovsky, D. N. Klebanyuk, P. S. Poyta, Sb. articles II MNTK "Theory and practice of research and design in construction with the use of CAD", pp. 50–58 (2018)

3. P.S. Poita, P.V.Shvedovskiy, D.N. Klebanyuk, Vestnik BSTU **1**, pp. 72–77 (2003)
4. D. N. Klebanyuk, P. S. Poyta, P. V. Shvedovsky, Geotechnics of Belarus: science and practice: materials of the Intern. scientific and technical conferences **2,** pp. 109–120 (2013)
5. L. M′enard, Y. Broise, G′eotechnique **25(1)**, pp. 3–18 (1975)
6. A. Oshima, N. Takada, International Conference on Soil Mechanics & Foundation Engineering **13(4)**, pp 1641–1194 (1994)
7. J. Pan, Z.M. Hwang, Sino Geotech **51**, pp. 35–50 (1995)
8. A. Perucho, C. Olalla, Ground Improvement **10(2)**, pp. 55–68 (2006)
9. P.S. Poyta, Vestnik BSTU **1**, pp. 109–110 (2003)
10. E.K. Zavadskas, *Complex assessment and selection of resource-saving solutions in construction* (Vilnius, Mokslas, 1987)
11. V.B. Shvets, *Compaction of soil bases with heavy rammers* (Rosstroyizdat, 1988)
12. Y.K. Chow, D.M. Yong, K.Y., Yong, S.L. Lee, Journal of Geotechnical Engineering **118(8)**, pp. 1141–1157 (1992)
13. H.Y. Fang, G.W. Ellis, *Frity Engineering Laboratory Report 462-06* (Lehigh University, Bethlehem, 1983)
14. S. Helwany, *Applied Soil Mechanics ABAQUS applications* (New York, 2007)
15. B.G. Mirkin, *Solution of group choice problems* (Moscow, Nauka, 1998)
16. G.K. Bondarik, *Fundamentals of the theory of variability of engineering-geological properties of rocks* (Moscow, Nedra, 1971)
17. V.B. Mikheev, I.V. Shitova, Foundations, foundations and underground structures **71**, pp. 16–21 (1980)
18. M. Smits, J.H. Th, L. De Quelerij, Proceedings, 12th international Conference of Soil Mechanics and Fundations Enginering **2**, pp. 1419–1422 (1989)
19. P.W. Mayne, J.S.Jones, J.C. Dumas, Journal Geotech Eng, ASCE **110(JT6)**, pp. 757–774 (1984)
20. C.J. Poran, K.S. Heh, J.A. Rodriguez, Soils and Foundations **32(4)**, pp 81–92 (1992)
21. A. Scott, W. Pearce, G′eotechnique **25(1)**, pp. 19–30 (1975)

Scheme of soil stability losses in the laboratory tray

Alexander Kremnev[1], *Nikolay* Vishnyakov[1,*] and *Victoria* Ermachenko[1]

[1] Polotsk State University, 211440, Blokhina 29, Novopolotsk, Belarus

Abstract. The object of the research is the diagram of the stability loss of the soil foundation. Currently, there are still discussions about how the stability of the subgrade undergoes the combined action of vertical and horizontal loads. Most often, in the calculation, the sliding surface is taken to be circular. Whether this is true, especially for a heterogeneous or anisotropic base, is an important question today. In the work, chute tests were carried out, with modeling of the scheme of loss of stability of anisotropic and isotropic bases and results were obtained that confirm the formation of a circular cylindrical sliding surface.
Key words: anisotropy, soil bases, circular sliding surface, soil stability losses, laboratory tray.

Introduction

It should be noted that most of the theories currently used for calculating subgrade bases assume that the foundations are homogeneous and isotropic. To analyze the stress-strain state of such bases, the well-known solutions of the linear theory of elasticity, the theory of limiting equilibrium are used. However, it is reliably known that most soils are heterogeneous in nature.

The relevance of the issue of taking into account anisotropic base is due to the fact that at any time due to the action of natural factors (freezing, flooding, flooding, etc.) any homogeneous base can turn, at least, into anisotropic, which, in turn, can cause the need for reinforcement. This procedure is quite lengthy, associated with the cost of a large amount of material and labor resources and the disruption of the normal operation of existing buildings, thus a method that takes into account the factor of anisotropy allows to reduce material and technical resources.

Methods

The tests were carried out in a tray with dimensions of 1.5x1.0x0.35 m, welded from metal corners and pipes, the front side of which is equipped with plexiglass 12 mm thick, which makes it possible to observe the nature of deformation and the formation of a sliding surface in the simulated soil base.

---

*Corresponding author: vnnpsu@mail.ru

A 30x30 cm metal plate was taken as a foundation model. A triangular wedge was used to apply a load at an angle of 20 °. A hydraulic jack, complete with an oil station, was installed on the wedge. The oil station is equipped with a pressure gauge to measure pressure. The fixation of the magnitude of horizontal and vertical displacements was carried out using 6PAO deflection meters. The test setup is shown in Figure 1.

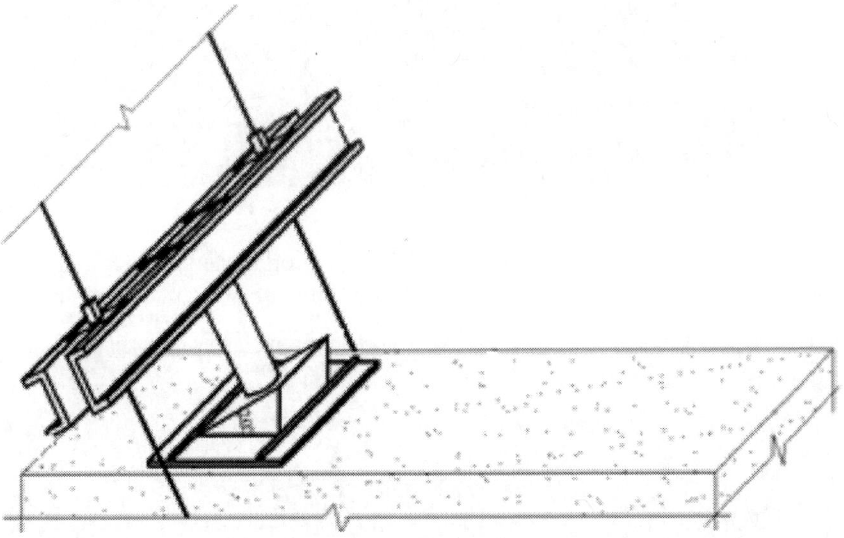

Fig.1. Test setup diagram.

The load on the punch was applied as a resultant from vertical and horizontal forces. At the first stage, tests were carried out for loads applied at the following angles: 30 °, 25 °, 20 °, 15 °. In this case, only the volume of soil involved in the formation of the vortex prism was recorded. It was found that the most optimal loading angle is 20 °. With the accepted dimensions of the tray and the size of the stamp, this angle allows one to observe the most complete picture of the stability loss of the base. In the second stage, at a fixed angle of application of the load of 20°, tests were carried out for homogeneous and anisotropic subsoils.

The die load was transferred in steps with a step of 2.5 kN. Each loading step was maintained until the onset of conditional stabilization. In this case, the vertical and horizontal movements of the stamp were recorded.

The test continued until the complete loss of stability of the foundation model with the formation of a soil shear surface and a surface uplift.

Tests with a homogeneous sand base

The filling of a homogeneous sandy base was carried out with medium sand. To visualize the process of deformation of the subgrade, test strips were made using dolomite flour. The stripes were created only near the plexiglass, the thickness of the filling was 0.5 cm, with a step of 3 cm vertically. A homogeneous sandy base with strips of dolomite flour before testing is shown in Figure 2.

To ensure the uniformity of the base, the soil was compacted manually, using a rammer in layers.

Fig.2. Uniform sandy base before testing.

In the course of the experiment, at 5 and 7.5 kN, a pronounced wedge-shaped figure was observed under the punch bottom, after which significant horizontal displacements were visible and at 12.5 kN the foundation lost its stability (Fig. 3).

Fig.3. Loss of stability of a homogeneous sandy base.

Generalized test results are shown in table 1:

Table 1. Results of tests of the sand base.

Die load, kN	Horizontal movement of the stamp, mm	Vertical movement of the stamp, mm	Notes
2,5	6,58	12,30	Compaction of soil under the stamp sole
5,0	21,55	30,92	Formation of a wedge by compacted soil
7,5	33,18	46,31	Distinct wedge-shaped figure under the stamp sole
10,0	52,72	67,49	Significant die movements
12,5	108,16	141,43	Loss of foundation stability

After the test, graphs of the dependence of horizontal and vertical deformations on the load were plotted (Figure 4.5)

https://doi.org/10.1051/e3sconf/202021202009

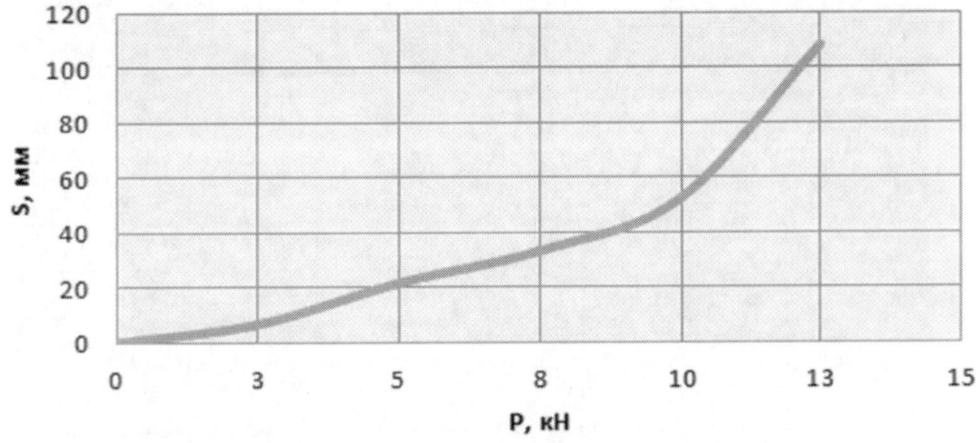

Fig.4. Graph of the deformation of the sand base in the horizontal direction.

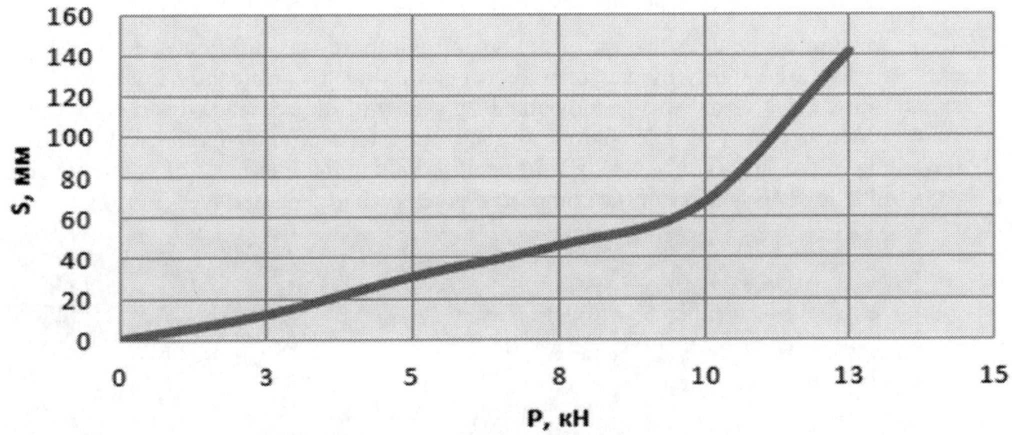

Fig.5. Graph of the deformation of the sand base in the vertical direction.

As a result of tests carried out with a sand base, loss of stability occurred at a load on the punch of 12.5 kN and the failure showed a sliding surface close to circular cylindrical.

Tests with anisotropic sandy-shale base

The anisotropic subgrade model was created from two materials: medium sand and interlayers of bentonite clay.

2 variants of modeling an anisotropic base were carried out:

- In the first case, thick layers of bentonite clay, 2.5 cm thick, settled every 8 cm of sand.
- In the second case, 0.5 cm of clay through 4 cm of sand.

The first variant of modeling an anisotropic sandy-argillaceous base with interlayers of bentonite clay 2.5 cm thick before deformation is shown in Figure 6. The soil, as in the case of a homogeneous base, was compacted manually using a rammer.

Fig.6. Anisotropic sandy-argillaceous base before testing.

The nature of the deformation of the anisotropic base was different from the isotropic one. The loss of stability of the foundation occurred at a significantly lower load on the stamp - only 2.5 kN, and was accompanied by a shift of the compacted soil core under the stamp along a clearly visible sliding surface close to a circular cylindrical and surface upwelling of the soil. In this case, a shift occurred along the interlayer of bentonite clay, as soon as the lower boundary of the compacted core reached the interlayer (Fig. 7).

Fig.7. Loss of stability of anisotropic sandy-argillaceous base with layers of 2.5 cm.

Repeating the experiment with thinner clay interlayers, 0.5 cm thick, it is clearly seen that the nature of the loss of stability of the base is similar to the shear with thicker interlayers, but the shift occurred, but at a greater depth (Fig. 8). The results of the experiment are presented in table 2.

Fig.8. Loss of stability of a sandy-clay base with 0.5 cm interlayers.

Table 2. Test results of sandy-clay base.

Die load, kN	Horizontal movement, mm	Vertical movement, mm	Notes
2,5	4,68	0,54	Compaction of soil under the stamp sole
3,0	54,47	0,92	Significant horizontal movement of the stamp
3,5	79,85	3,58	Loss of stability of the subgrade

Based on the data obtained, graphs of the dependence of horizontal and vertical deformations on the load were constructed (Fig. 9, 10).

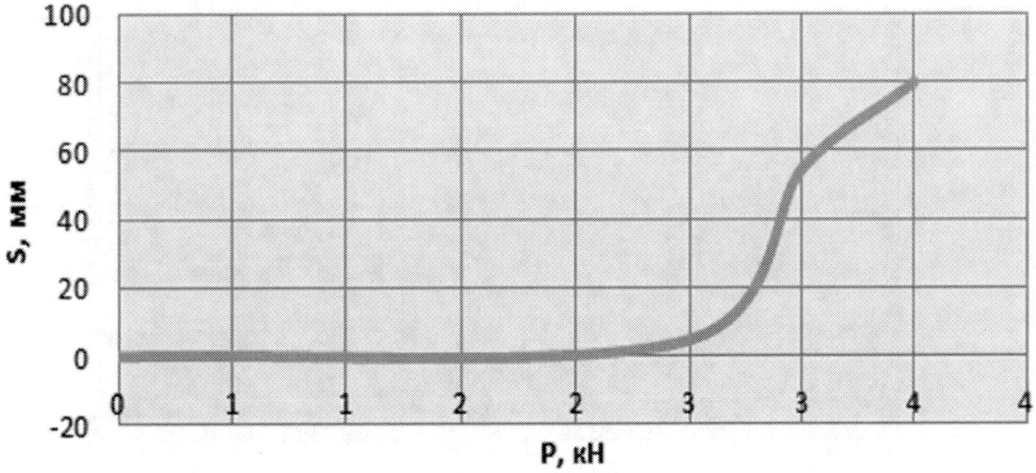

Fig.9. Graph of deformation of anisotropic sandy-argillaceous base in the horizontal direction.

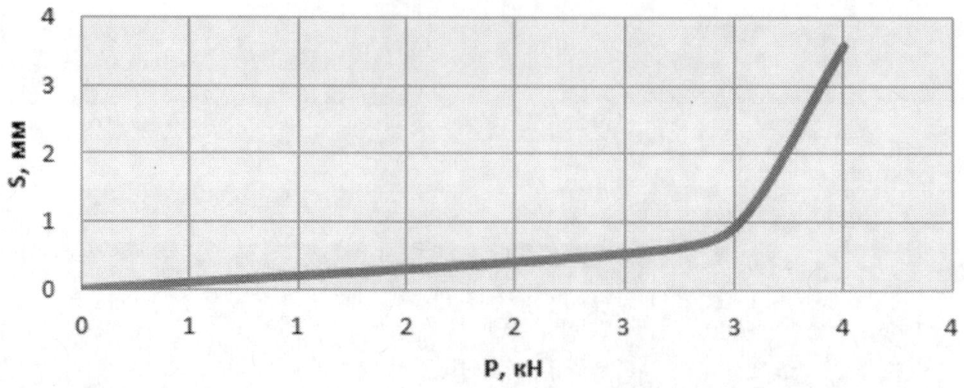

Fig.10. Graph of deformation of anisotropic sandy-argillaceous base in the vertical direction.

During the analysis of the outlines of sliding surfaces, it was found that, in general, the shift of one part of the soil relative to another occurred along a surface close to a circular cylindrical one (Fig. 11).

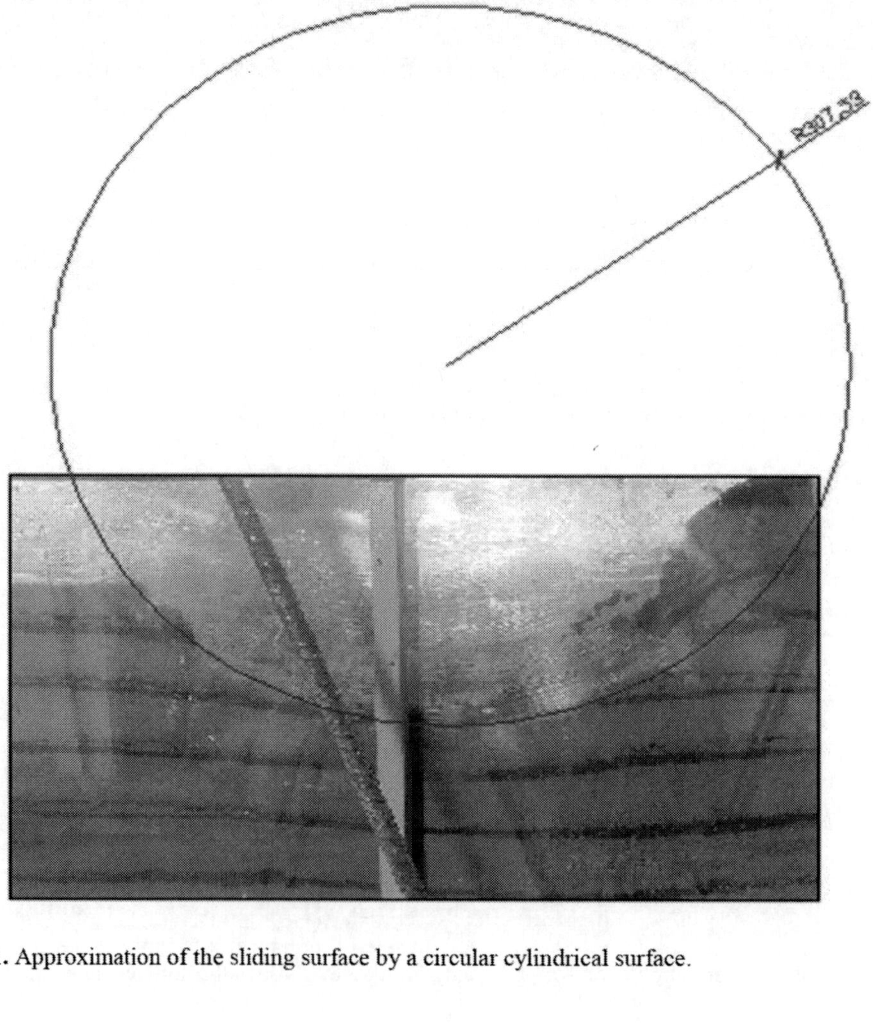

Fig.11. Approximation of the sliding surface by a circular cylindrical surface.

Conclusion

Thus, laboratory studies of soil foundation models in trough conditions have shown that the loss of stability of a soil mass with pronounced horizontal anisotropy under the action of vertical and horizontal loads also occurs on a surface close to a circular cylindrical one. It should be noted that the sliding surface at the base with pronounced anisotropy is initially formed along a weak interlayer. In the first case, the sliding surface was formed in the upper weak layer, in the second case in the second. With respect to weak interlayers, the circular-cylindrical sliding surface in the first and in the second case passed along a tangent.

Reference

1. A. Gadre, R. Dobry, Lateral cyclic loading centrifuge tests on square embedded footing. Journal Geotechnical and Geoenvironmental Engineering. **124**. pp. 1128–1138 (1998)
2. J. D. Murff, T. W. Miller, Stability of offshore gravity structure foundations by the upper-bound method, Proceedings of the Offshore Technology Conference, **3**, pp. 147–154 (1977b)
3. Y. Maeda, Study on application of velocity field method in evaluation of bearing capacity of foundation, Kyushu University doctoral thesis, 2003
4. T. Schanz, Zur Modellierung des mechanischen Verhaltens von Rei bungsmaterialien // Mitt. Inst. fur Geotechnik 45. Universita..t Stutt gart. – Stuttgart. – 152 S. (1998)
5. N.N. Vishnyakov, A.P. Kremnev Bearing capacity of the foundations of spacer structures, taking into account the anisotropy of the strength properties of soils – Materials of the International Conference, Minsk, (2018)
6. A. Kremniou, N. Vishniakov, Anisotropy of the Strength Properties of Clays of the Clays of the North of Belarus//Historical experience and challenges of geotechnical problems in Baltic Sea region//13th Baltic Sea Region Geotechnical Conference, Vilnius, Lithuania, (2016)
7. O.A. Korobova, Analysis of the results of theoretical and experimental studies of the stress-strain state of anisotropic soils and subsoil foundations. – Polzunovsky Almanac, (2001). – [electronic resource]. – access mode: www.vc.secna.ru: 8080/Books/Files/ 2001-02/10/pap_10.html
8. T. Schanz, P.A. Vermeer, P.G. Bonnier, The HardeningSoil Model: Formulation and verification // Beyond 2000 in Computational Ge otechnics. – Balkema, Rotterdam, (1999). – P. 281–290. http://www.unistuttgart.de/igs/content/publications/40.pdf
9. A.P. Kremniov, D.U. Sobolevski, D.V. Skoryk, THE DEFINITION OF FRICTION AND COHESION PARAMETERS OF REINFORCED SOILS. – Доклад на Международной научно-практической конференции, Астана, (2002)
10. I.E. Pavlovskaya, Polotsk glacial-lake basin: structure, relief, history of development / I.E. Pavlovskaya. – Minsk: Navuka i tekhnika. – 128 p., (1994)
11. A.P. Kremnev, Determination of the most dangerous sliding surface when calculating the stability of slopes by the method of circular cylindrical sliding surfaces / A.P. Kremnev, D.O. Glukhov, N.N. Vishnyakov // Bulletin of the Polotsk State University PolotskSU theta. Series F, Applied Sciences. Building. – pp. 37–41 (2011)
12. A.A. Kalyaev, To the calculation of the stability of the foundations of massive structures on sandy soils / A.A. Kalayev // Foundations, foundations and soil mechanics. – No. 5. – pp. 3–6 (1962)
13. G Meyerhof, The bearing capacity of foundation under eccentric and inclined loads, Proc.3rd Int. Conf. Soil Mech. and Found. Eng.: pp. 4–24 (1953)
14. R. Bellotti, M. Jamiolkowski, D.F. Lo Presti, D.A. O'Neill, Anisotropy of small strain stiffness in Ticino sand, Géotechnique, **46(1)**, pp. 115–131 (1996)

15. A. Casagrande, N. Carillo, Shear failure of anisotropic materials, Journal of Boston Society of Civil Engineering, **31**(4), pp. 74–87 (1944)
16. C. Cherubini, Reliability evaluation of shallow foundation bearing capacity on c',φ' soils, Canadian Geotechnical Journal, **37**(1), pp. 264–269 (2000)
17. J.M. Duncan, H.B. Seed, Anisotropy and stress reorientation in clay, Journal of the Soil Mechanics and Foundations Division, ASCE, **92**(SM5), pp. 21–50 (1966)
18. R.E. Gibson, The analytical method in soil mechanics, Géotechnique, **24**(2), pp. 115–140 (1974)
19. M. Jamiolkowski, R. Lancellotta, D.C.F. Lo Presti, Remarks on the stiffness at small strains of six Italian clays, Proceedings of the International Symposium on the Pre-failure Deformation Characteristics of Geomaterials, Hokkaido, Japan, Vol. **2**, pp. 817–836 (1995)
20. V. Jovicic, M.R. Coop, The measurement of stiffness anisotropy in clays with bender element tests in the triaxial apparatus, Geotechnical Testing Journal (GTJODJ), 21(1), pp. 3–10 (1998)

Pile quality assessment by vibroacoustic method

Sergey Leonovich[1], *Nely* Budrevich[2*], *Ping* Lyu[3], *Jijun* Miao[3]

[1]Bialystok University of Technology, 15-351, ul. Wiejska 45 A, Bialystok, Poland
[2]Belarusian National Technical University Minsk, Republic of Belarus
[3]School of Civil Engineering, Qingdao University of Technology (Qingdao, China)

Abstract. In recent years, there has been a steady growth in industrial and civil construction, especially in large cities. Construction is often carried out in increasingly difficult geotechnical and hydrogeological conditions. This contributes to the emergence of new production technologies and diagnostic methods. The use of non-destructive testing methods allows you to quickly obtain information about various characteristics of an object without disrupting the process of its operation. This is especially true for hidden work, since access to the research object is practically closed, and its opening may entail undesirable consequences for the operation of the entire structure.
Keywords: vibroacoustic method, pile field, non-destructive methods, Impact echo method, acoustic contrl.

1 Introduction

When inspecting piles, the most productive and easy-to-use method is the surface acoustic method, known in foreign literature as the Impact echo method (IEM). The IEM method consists in shock excitation of an elastic wave of the acoustic frequency range in the inspected pile and registration of echo signals.

The surface acoustic method has been used for pile diagnostics for over 40 years. During this time, extensive experience in pile inspection has been accumulated. In the simplest case (a pile without defects or discontinuities), the IEM data obtained under the conditions of the long wavelength limit shows only a forward signal and a pulse reflected from the end of the pile. This method is the most productive and easiest to use. It is advisable to use it in cases where there is free access to the pile head and it is necessary to carry out prompt diagnostics of a sufficiently large number of piles [1-8].

Basically, these experimental studies are devoted to the determination of the mechanical characteristics of the material by observing the features of wave propagation on samples in the soil, and do not touch on the range of issues related to the interaction of the rod. Therefore, within the framework of this work, a study was carried out of the phenomena observed during the inspection of piles, on rods in the form of driven and bored piles [9-10].

 https://doi.org/10.1051/e3sconf/202021202010

2 Research program

In the course of the experiment, to determine the lengths, two types of piles were studied - driven and bored (bored). Pile length measuring device - СПЕКТР 3.

3 Experimental results

To ensure the reliability of the results obtained, before driving and shearing off the head after failure when driving driven piles (Fig. 1, 2), their length was measured with a tape measure and amounted to 8 m.

Fig. 1. Measurement process (driven piles): *a* – pile field; *b* – pile surface; *c* – sensor installation; *d* – blow produc

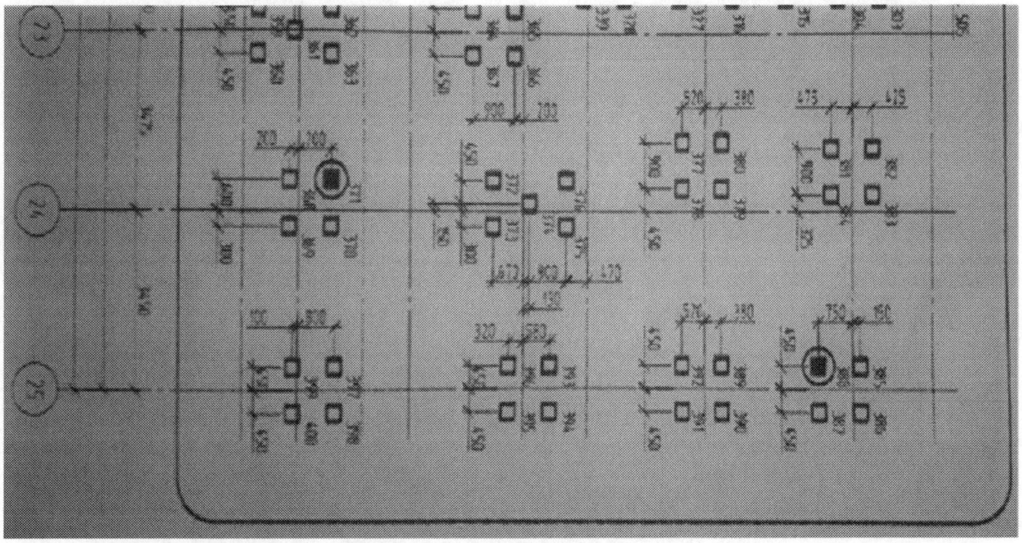

Fig. 2. Layout of driven piles

Before measuring the length of the bored piles, the design and as-built documentation was studied. The design length of the piles should be 9 m.

a *b*

 https://doi.org/10.1051/e3sconf/202021202010

c *d*

Fig. 3. Measurement process (bored piles): *a* – pile field; *b* – pile surface cleaning process; *c* – leveling the surface of bored piles; *d* – measurement process

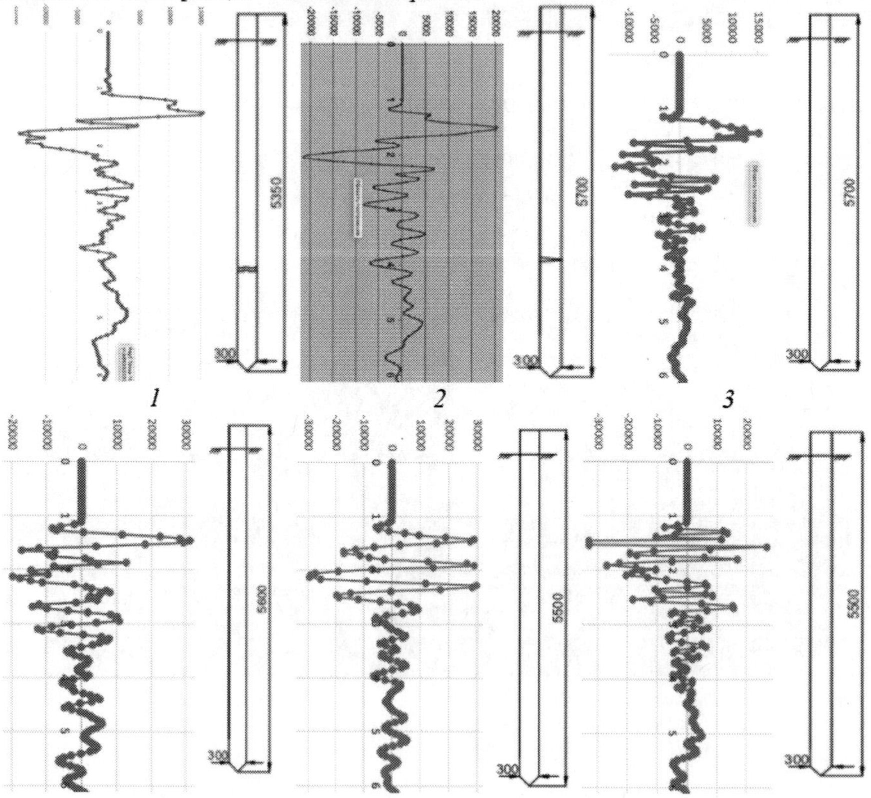

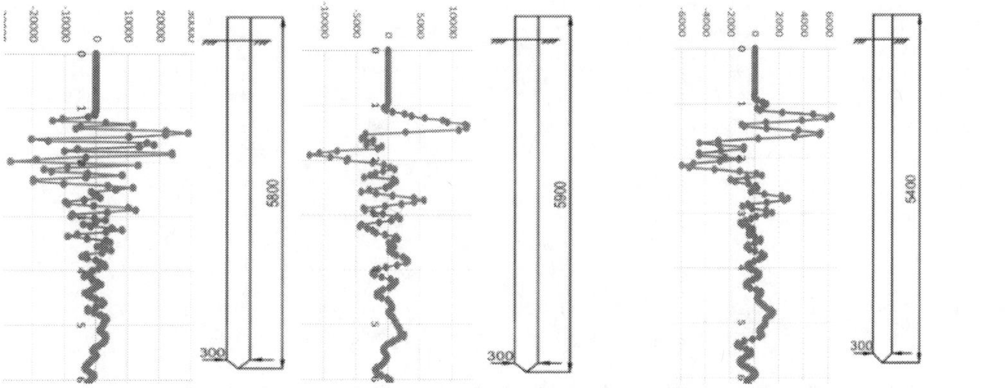

7 *8* *9*

Fig. 4. Results of measuring the length of driven piles by the IEM method

As can be seen in Figure 4, according to balanced reflectograms obtained during the study of driven piles, the average length is in the range of 5.3 ... 5.9 m.

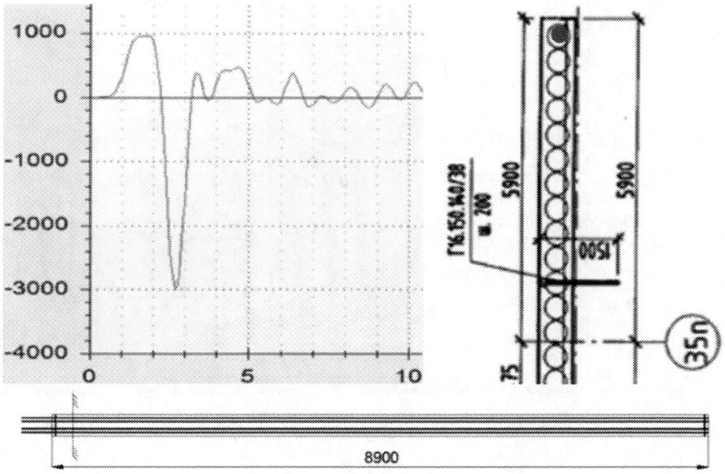

1

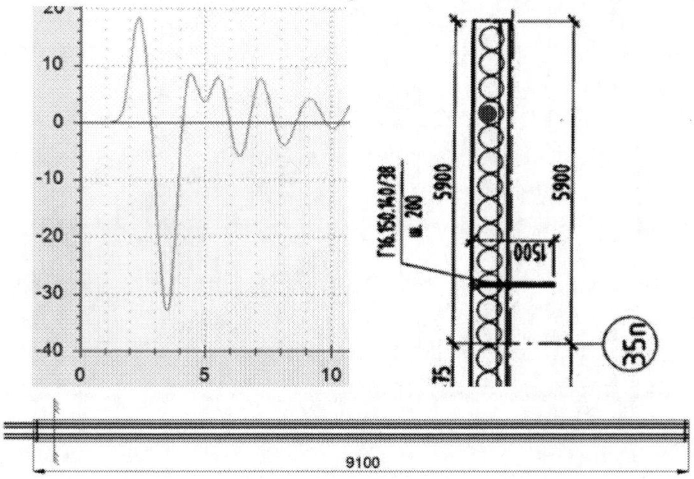

2

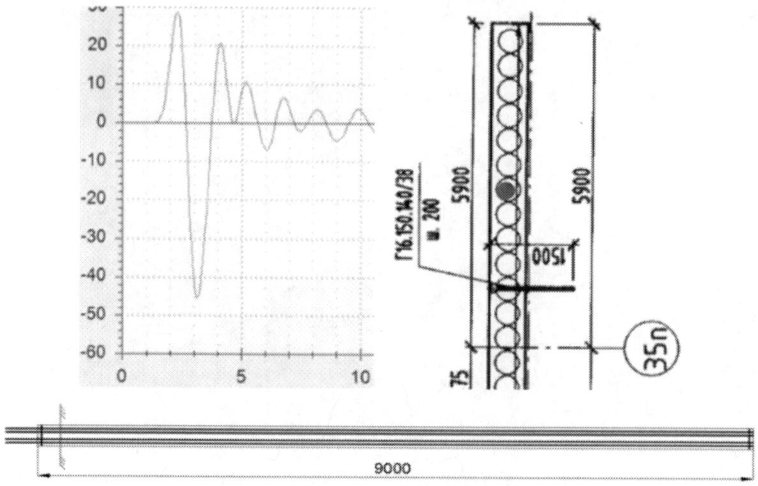
20
10
0
-10
-20
-30
-40
-50
-60
0
5
10
5900
5900
1500
75
35п
9000

3

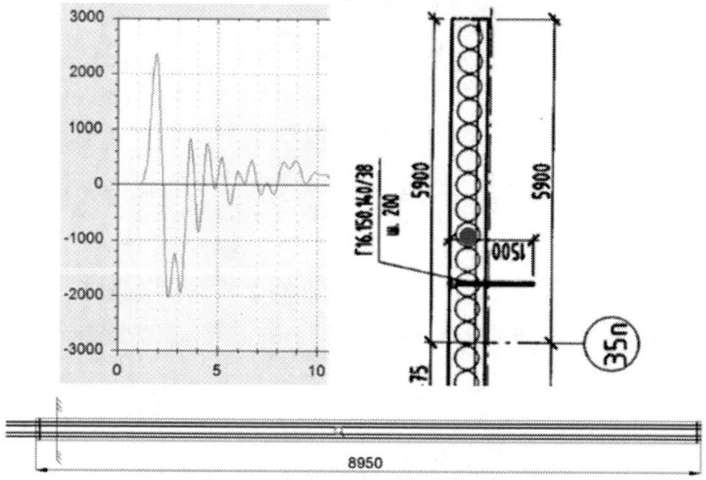
3000
2000
1000
0
-1000
-2000
-3000
0
5
10
5900
5900
1500
75
35п
8950

4

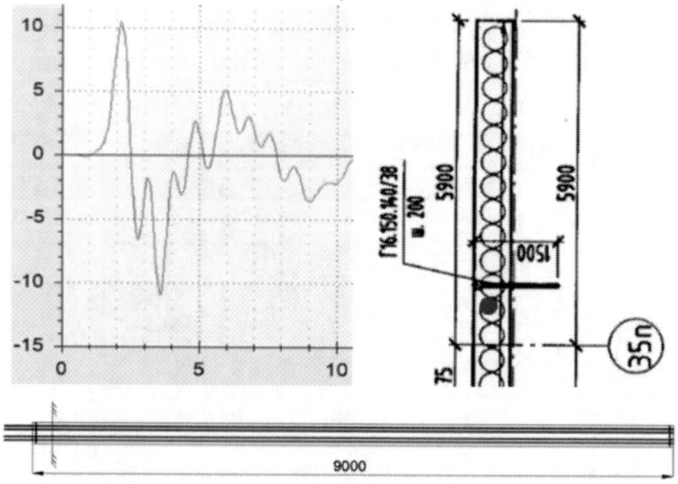
10
5
0
-5
-10
-15
0
5
10
5900
5900
1500
75
35п
9000

5

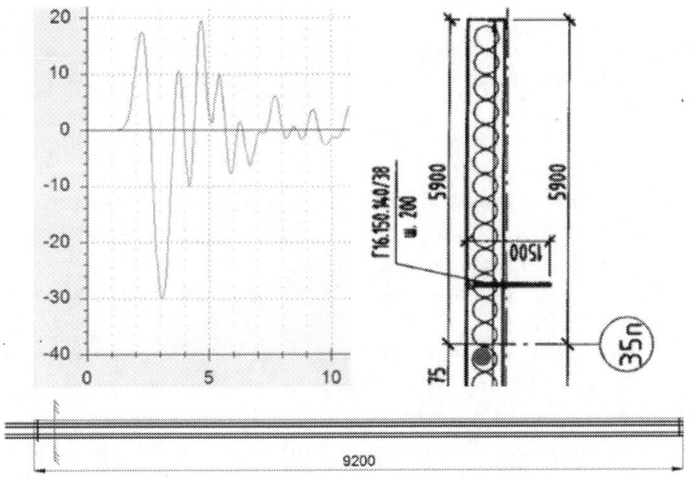

6

Fig. 5. Results of measuring the length of bored piles by the IEM method

As can be seen in Figure 5, according to the balanced reflectograms obtained during the study of bored piles, the average length is in the range of 8.9 ... 9.2 m.

4 Conclusion

Numerous field measurements were carried out on piles [11]. The main purpose of the research was to study the records of acoustic vibrations of piles from the position of field control to assess the possibility of separating the effects associated with the propagation of a plane wave (the pile itself) and the influence of the enclosing soil, and expanding the range of problems solved by the acoustic method.

Based on the results of the conducted field experiments, the following conclusions can be drawn:

1. Controlling the acoustic field allows you to meet or not to fulfill the conditions of the long-wavelength limit. This makes it possible to study only the pile itself or the additional enclosing environment.

2. In order to distinguish between reflections associated with the pile itself and with the enclosing soil, at each measurement it is necessary to control the vibration spectrum, since the fulfillment or non-fulfillment of the conditions of the long-wave limit depends on the frequency composition with which it was possible to excite the fluctuations in this measurement.

3. If the conditions of the long-wavelength limit are not met, the registration of reflections from the boundaries in the host environment allows solving a number of additional tasks, for example, determining the depth of pile burial into the reference horizon, tracking the groundwater level, determining the depth of an artificial embankment, monitoring the quality of filling pipe piles, etc.

4. f there is no data on the core velocity of the longitudinal wave, the technique of different-frequency excitation allows its estimation by observing reflections from boundaries with a known position in depth.

5. In the case of working on real piles, the concept of a "long" rod should have a broader meaning than follows from the theoretical premises. A rod can be considered “long” if the length is more than ten times the radius of the rod and at least 1 or 2 wavelengths fit in the length.

References

1. D.Yu. Snezhkov, S.N. Leonovich. *Monitoring of erected and operated reinforced concrete structures by non-destructive methods* (Minsk, BNTU, 2016)
2. Samokrutov A.A., Kozlov V.N., Shevaldykin V.G., Meleshko I.A. Ultrasonic defectoscopy of concrete by means of pulse- echo technique. 8th European conference for Non-Destructive Tensing (2002)
3. M.Sansalone, J. Carino, "Impact-Echo: A Method for Flaw Detection in Concrete Using Transient Stress Waves", Report NBSIR 86-3452, National Bureau of Standards, U.S. Department of Commerce (1986)
4. K. Nagrodzka-Godycka. Badanie wlasciwosci betonu I zelbetu w warunkach laboratoryjnich. Arcady (1999)
5. S.N. Leonovich, D.Yu. Snezhkov. Non-destructive control of strength of maintained reinforced-concrete structures "Durability design and fracture mechanics of concrete structures". Internatuonal Conference (2003)
6. Christian U. Grosse, Hans W. Reinhart, Ralf Beutel «Impact- Echo measurement on fresh and hardening concrete». Institute of Construction Materials (2003)
7. T.V. Fursa. Non-destructive electromagnetic method of flaw detection of products made of dielectric materials. – Defectoscopy **10**, pp. 27-32 (2003)
8. D.Yu. Snezhkov. Improvement of hardware for acoustic control of elastic-deformative characteristics of concrete. Materials of the XI international scientific - methodological interuniversity seminar "Prospects for the development of new technologies in construction and training of engineering personnel in the Republic of Belarus" (2004)
9. V.S. Mulyarchik, D.Yu. Snezhkov, S.N. Leonovich. Application of a set of non-destructive testing tools for assessing the strength properties of in-situ concrete. In the collection of scientific papers of the international scientific and methodological interuniversity seminar (2005)
10. S.N. Leonovich, D.Yu. Snezhkov, V.S. Mulyarchik, D.M. Markovskiy. Assessment of the stripping strength of concrete when concreting structures under construction conditions using modern non-destructive methods. Bialorusko-Polski Naukowo-Praktyczny Seminarium (2004)
11. S.N. Leonovich, V.S. Mulyarchik, D.Yu. Snezhkov. Applicability of dry contact in ultrasonic testing devices for crack depth control. Materials of the XI international scientific - methodological interuniversity seminar "Prospects for the development of new technologies in construction and training of engineering personnel in the Republic of Belarus (2004)

https://doi.org/10.1051/e3sconf/202021202011

Properties of basalt fiber sulfo-aluminate modified cement binders

Inessa Paulava[1*]

[1]Brest State Technical University, 224017, Moskovskaya 267, Brest, Belarus

Abstract. The present paper considers the influence of basalt fiber method of addicting in to self-stressing concrete, modified by sulfo-aluminate type expansive additive, on main properties of concrete. The use of basalt fiber and expansive additive together allows to obtain a composite with optimal structure and high-quality properties. The combined application allows to achieve several effects such as filling effect, pozzolanic effect and dispersed reinforcement effect (due to chaotic fiber distribution in the concrete composite structure and needle-shaped crystal of ettringite micro reinforcement). Use of even small amount of basalt fiber due to disintegration on a great amount of monofilaments with a great specific surface leads to an increase in concrete composite tensile strength due to spatial dispersed reinforcement effect and changing fracture toughness There is also an effect of simultaneous expansion due to adding an expansive additive and strengthening achieved by binding effect of basalt fiber as a dispersion-reinforcing element. The maximum amount of basalt fiber in the research is limited by 4% to prevent full percolation effect, but at the same time to provide the formation of filament spatial skeleton in the concrete structure. In this study are considered the changes of properties of the concrete composite with different degrees of expansion and self-stressing (varied by energy-activity and amount of sulfo-aluminate expansive additive). In the article the most optimal basalt fiber addicting method in the concrete mix is proposed, preferences and influence degree of this method on strength properties are determined. The integral use of the modifying expansive agent makes it possible to predict an increase in impermeability due to chemical-mechanical reinforcement of the structure, which results in durability of this type of concrete. This predetermines a perspective range of their applications in various kinds of structures including those with high serviceability.
Keywords: Self-stressed concrete, basalt fibre, sulpho-aluminate type additive, ettringite, dispersed reinforcement

Introduction

Nowadays, concrete is known as one of the most widely used composite materials worldwide. Being high performance in different environmental conditions, just economic, components availability, as well as high compressive strength, are among parameters which

* Corresponding author: pavlinna@tut.by

makes it acceptable as a construction material. On the other side, it has tensile and strain weakness that reinforcement or various fibers usage compensate for that [1,2]. Adding fibers to concrete mixture can significantly improve its engineering properties such as tensile strength, flexural strength, impact and fatigue strength, as well as ductility, toughness and post-cracking capacity [3–5]. In recent years, fiber reinforced concrete (FRC) had a wide spread usage in civil engineering including roads pavement, bridges, slabs, explosives storages and so on [6,7]. FRCCs (fiber reinforced cementitious composites) are kind of compounds with appropriate uniformity and cohesion, capable of creating a ductile concrete. FRCCs performance depends on various parameters such as materials or fibers type and properties. Presence of cementitious supplementary materials (e.g. metakaolin, silica fume, fly ash, slag) as well as shape, size, type, content, and fibers distribution affects technical and economic efficiency.

Wide spectrum of fibers with different physical, mechanical and chemical properties are used and proposed for cementitious matrix reinforcement [8]. Among accessible, applicable fibers in cementitious composites steel, polypropylene, glass, carbon and basalt fibers can be mentioned [8,9]. Basalt fiber had recently a considerable development in concrete and cementitious products [8] due to its ecological and technical advantages.

In a literature review it was found many studies on the effect of various fibers including: natural, artificial, hybrid (steel, carbon, polypropylene, glass) on concrete performance. First attempt on concrete dispersed reinforcement was done by Baston and Ramualdi by placing steel fibers in 1960, USA [10]. According to researchers' study results, four main factors are effective in strength and bearing capacity of fibrous cementitious composites; matrix quality, dimensions ratio, volume fraction, and bond strength of fibers [11]. Nevertheless, bond failure related to fiber-matrix transition zone is the initial reason leading to the failure of the whole structure [12]. Many strategies are suggested to improve the bond properties at fiber-matrix transition zone, such as: (a) densification of the cementitious matrix and ITZ; (b) use of deformed fibers; (c) treatment of fibers surface [13–15]. Many densification methods have been proposed which include reduction in water-cement ratio, curing conditions, heat curing, and replacement of pozzolanic materials [16].

Basalt fiber

Basalt fibers, which are considered modern fibers with industrial potentials and less harmful for environment, have become a suitable alternative in the composites industries. Considering their physical and mechanical properties, basalt fibers are considered serious rivals for industrial fibers such as carbon and glass. Considering its properties including: greener, lighter, and reduction of reinforcement bars corrosion, basalt fibers have good potential for development of application in other industries [17,18]. Basalt fiber is a mineral matter with no environmental hazards, plus non-toxic and decomposable [19]. Considering high silica content in basalt stone structure, fibers made from it have become a suitable alternative, compatible with cementitious matrix. The tendency to use mineral basalt fibers in basalt fiber reinforced polymer (BFRP) is increasing and there have been different researches about the BFRPs [17–23].

By evaluating basalt fiber reinforced concrete, Dias et al. [24] reported that adding basalt causes decreased compressive strength and increased flexural or tensile strength. The basalt fibers caused the increased bearing and ductility in geo-polymeric concrete beam and it was also observed that the presence of basalt fiber raised beam's opening displacement which demonstrates less sensitivity at crack presence. The conducted studies imply the good strength of basalt against alkaline attacks [25]. Adding more than 1% basalt decreased the mortar workability. Presence of fibers decreased the autogenous shrinkage of cementitious mortars by 15–20%. The results show that optimal volume of basalt fibers was between 0.5% and 0.8%.

Ayub et al. [26] performed a study on mechanical properties and microstructure of Basalt Fiber Reinforced Concrete (BFRC). The results showed that adding basalt causes decreased concrete workability. The mechanical performance of basalt fibers in BFRC was better than Polypropylene. Compared to control specimen, BFRCs showed higher flexural and tensile strength which intensifies by increased fiber length. After 28 days, the compressive, tensile and flexural strength of BFRC with fiber length of 12 mm increased 0.18% to 4.68%, 14.08% to 24.34% and 6.30% to 9.58% respectively.

But, long term durability of BFRC composites has been a major concern in limiting the widespread application of the material. OPC matrix BFRC demonstrates significant reductions in tensile strength and ductility as time develops, particularly in humid environments [27,28]. The degradation is normally associated with fiber corrosion by the high alkalinity within the matrix [29–31] and/or loss of fiber flexibility by the precipitation of CH within and around basalt fiber strands [28,29,32]. Therefore efforts have been directed towards the study of modified matrix formulations with markedly reduced alkalinity and/or propensity to precipitate calcium hydroxide.

Sulfo-aluminate (SA) modified binder, often considered as a potential low-carbon binder and as a shrinkage-compensating additive, demonstrates potential as an addition to improve the durability properties of BFRC composites. This can be attributed to the intrinsically different hydration process of SA–OPC blended cement, which is CH free and thus provides a less alkaline and less aggressive environment for embedded basalt fibers.

Variation in amount and type of used expansive additive make it possible to obtain self-stressing concretes with different level of linear expansive strains and self-stresses. Expansion mechanism is based on early ettringite formation (see **fig. 1**).

$$3CaO \cdot Al_2O_3 + CaSO_4 \cdot 2H_2O + 26H_2O \rightarrow 3CaO \cdot Al_2O_3 \cdot 3CaSO_4 \cdot 32H_2O$$

Combined introduction of basalt fiber with expansive admixture leads to the advantageous stress-strain relationship due to chemical prestress and spatial reinforcement.

As can be seen from **fig. 2**, the addicting of basalt fiber does not lead to a sharp decline in linear expansion, which allows maintaining the required level of energy activity. At the same time, the binding of a hydrating system to a spatial framework will prevent partial decompaction and strength decreases inherent in energy-active mixes.

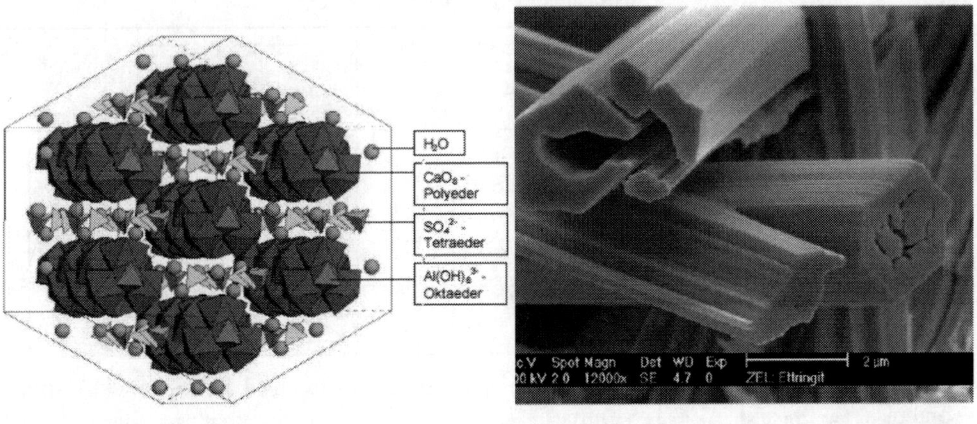

Fig. 1. Ettringite structural model (according to J.Neubauer (left) and crystal microphotography (right).

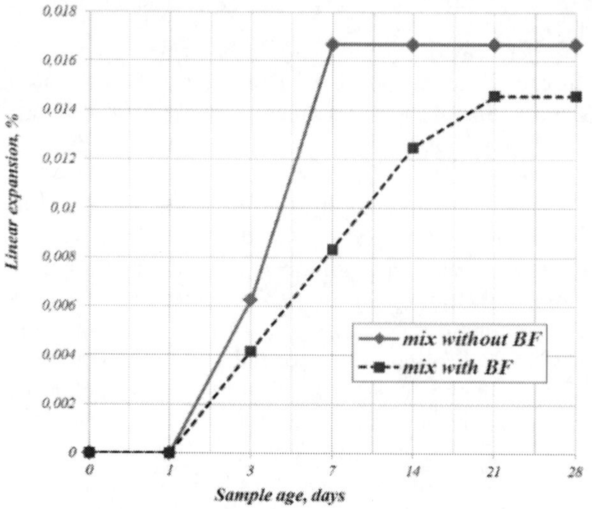

Fig. 2. Dependence of linear expansion development in time (BF – basalt fiber).

This paper compares the microstructure and mechanical properties of BFRC composites made with two different matrix formulations (OPC, and SA–OPC blend). Also, considered method of fiber addicting into the prepared mix.

1 Materials and methods

The type of cement used in all mixtures was CEM I42.5 according to EN 197-1 and its chemical compositions and physical properties presented in **Table 1**. An expansive sulfo-aluminate (SA) additive was used in this study with the chemical composition seen in Table 1. In order to observe the effect of SA additive on mechanical properties and microstructure of composites, it was used as a replacement of cement materials in different value of 10% by weight.

Table 1. Chemical compositions of cement and sulfo-aluminate additive.

	Chemical compositions (% by weight)								Physical properties	
	CaO	SiO_2	Al_2O_3	Fe_2O_3	MgO	K_2O	Na_2O	SO_3	Density (g/cm^3)	Blain surface (m^2/kg)
Ordinary Portland cement (OPC)	62,22	19,73	5,45	2,65	2,09	1,05	0,17	2,99	3,10	320
Sulfo-aluminate addictive (SA)	34,31	8,12	26,55	0,50	0,54	0,09	0,06	30,69	2,92	827

Basalt Fiber with following properties (see table 2) was used throughout the study (at contents 0%, 1% and 1.5% by volume).

In these studies were considered two schemes for loading materials during the preparation of fiber-concrete mix (see **Ошибка! Источник ссылки не найден.**).

Table 2. Technical characteristics of the used basalt fiber.

Name of the indicator	Value
Basalt fiber length, mm	24,5
The diameter of the individual monofilament, μм	16
Modulus of Elasticity, MPa	93200-116000
Rupture strength, MPa	1500-2000
Density, g/cm³	2,7

Table 3. Technical characteristics of the used basalt fiber.

№ of stage	Scheme 1	Scheme 2
	Mix components	Mix components
1	SA-OPC+Sand	SA-OPC +Water+ Plasticizer (0,8%)
2	Basalt fiber (4%)	Basalt fiber (4%)
3	Water+Plasticizer (0,8%)	Sand

2 Results and discussion

2.1 DTA

A prominent composite triple endotherm at 100–180 °C is generally associated with the presence of (in the order of dehydration temperature) weakly-bound pore water, C–S–H gel and/or ettringite. Dehydroxylation of CH occurs between 500 and 550 °C. The low temperature peak for SA additive is significantly larger than that for OPC and the most prominent sub-peak is shifted towards higher temperature, suggesting a relative preponderance of ettringite over C–S–H (**Fig. 3**). The absence of the CH endotherm is clear in SA-OPC compound. Peaks between 600 and 780 °C are indicative of the existence of $CaCO_3$, but very little seems to be apparent in either material.

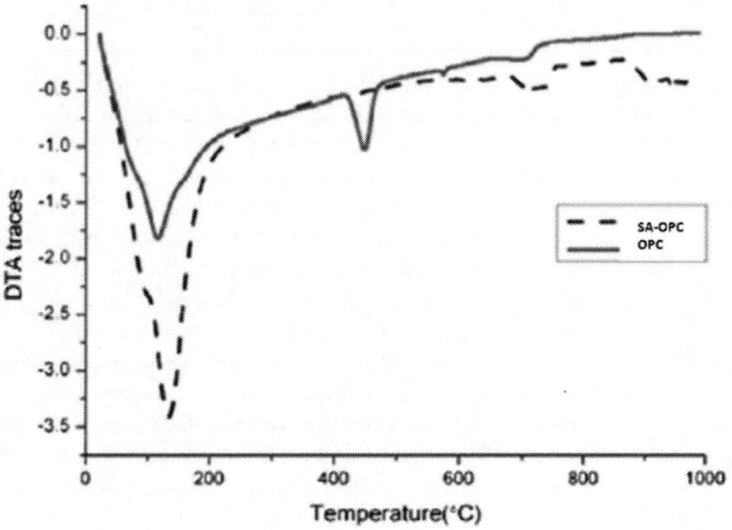

Fig. 3. DTA thermo grams of OPC and SA-OPC matrices.

2.2 X-ray diffraction

XRD results (**Fig. 4**) confirm that SA additive is free of CH and relatively large amounts of ettringite can be observed, confirming the DTA results above. A broad band between 2θ = 33 and 34° reveals significant remnants of unhydrated C_2S in SA-modified cement systems; this may be attributed to rapid water consumption during the hydration of calcium sulfoaluminate within the first 1–2 days, which leaves insufficient water for the hydration of belite at later ages.

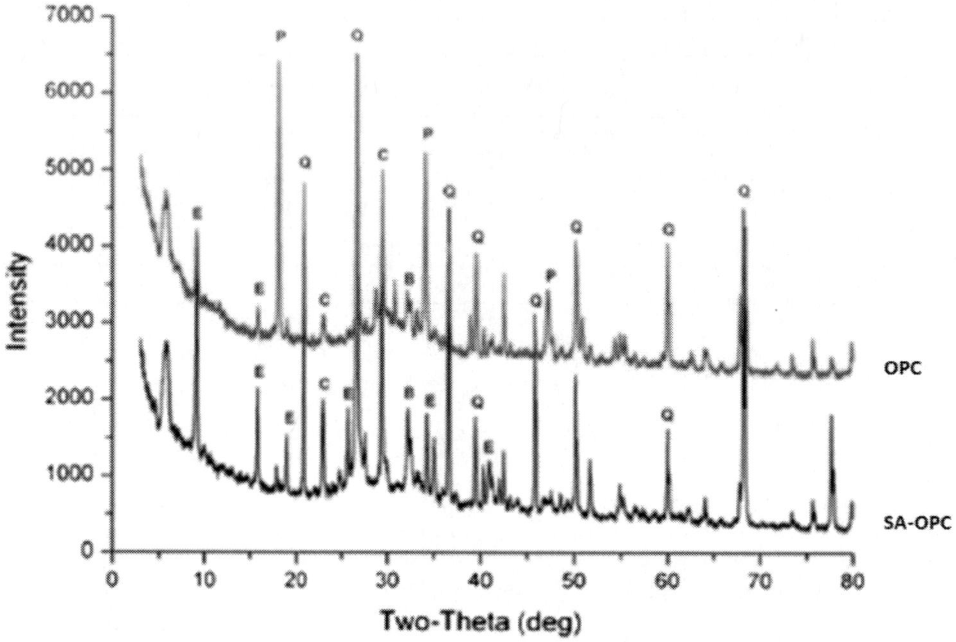

Fig. 4. X-ray diffraction patterns of samples (E–ettringite, P–$Ca(OH)_2$, Q–quartz, C–$CaCO_3$, B–belite).

2.3 SEM

Scanning electron microscope (SEM) is one of the most effective and helpful instruments to understand the morphology of cementitious composites structure. In order to recognize the bond quality between fibers and cementitious matrix, a projective evaluation of cement paste microstructure was made using microscopic images. Composites with SA and basalt fibers were investigated and their SEM images are shown in **Fig. 5. Fig. 5** (a) presents the composites on ordinary PC. It indicates that crystallized flake-shaped structures have a considerable volume including great amount of Portlandite and can be the main reason of relatively low strength of these specimens. It can be concluded that the interfilamentary phases in SA-OPC are mainly C–A–S–H gel finely intermixed with AFt phases at various levels (**Fig. 5** (b)). The observed microstructure of OPC (**Fig. 5** (a)) indicated that the matrix space was significantly filled with hydration products, mainly dominated by Ca and Si according to the elemental mappings. Analysis reveals dominance of CaO (53.0 wt.%) and SiO_2 (42.2 wt.%) at the interfilamentary space. In modified by SA cement paste Al and S could be observed between fibers, thus eliminating the presence of ettringite and/or AFm in this zone (**Fig 5** (c)).

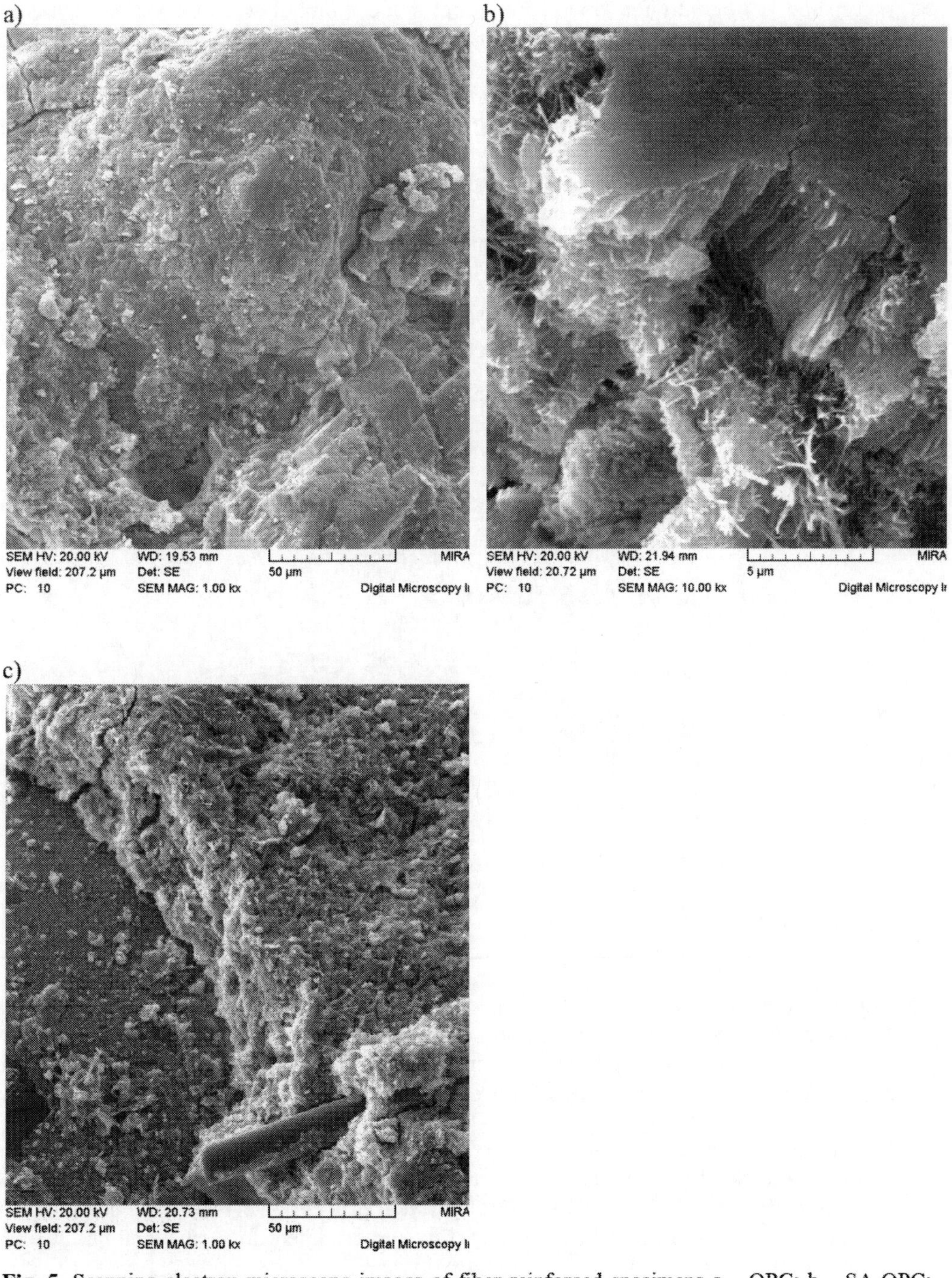

Fig. 5. Scanning electron microscope images of fiber reinforced specimens a – OPC; b – SA-OPC; c – SA-OPC+ basalt fiber.

2.4 Method of fiber addicting

One of the purposes of this study was to determine the most optimal of the methods of basalt fiber addicting into a concrete mix.

There were proposed two schemes of the basalt fiber addicting into the concrete mix:

- scheme 1 – the introduction of basalt fiber in a dry mix of cement and sand, obtaining a homogeneous mixture of dry components and their subsequent mixing with water and a plasticizing additive;
- scheme 2 – the introduction of basalt fiber in prepared suspension consisting of cement, water and plasticizer, obtaining a homogeneous mixture and the subsequent introduction of sand with the joint mixing of all the components of the fiber-concrete mix.

The composition without basalt fiber was taken as the control.

Tensile strength was chosen as control.

Tests of control samples were made at the age of 2, 7 and 28 days. The results are presented in **Fig. 6**.

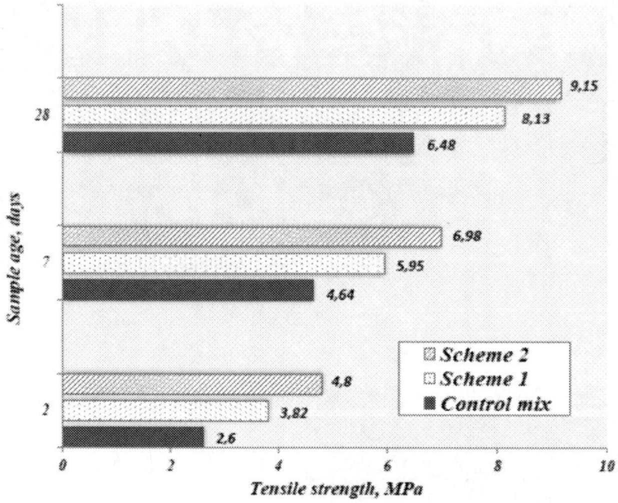

Fig.6. Sample tensile strength.

Conclusions

1. The combined use of basalt fiber and an expansive sulfo-aluminate additive in the hydrating cement system allows to achieve the effect of expansion (or shrinkage compensating) without decompression and decrease in strength, which is usually is shown in energy-active compounds. At the same time, the addicting of basalt fiber slightly reduces the strain of free expansion, which allows that the energy potential of active expansive binders to be fully realized.

2. Dispersed basalt fiber reinforcement of cement systems leads to the creation of a spatial framework, due to the free and chaotic distribution of elementary fibers in the cementitious matrix. Intrinsically different hydration process of SA–OPC blended cement, which provides a less alkaline and less aggressive environment for embedded basalt fibers improves microstructure and leads to durability increase.

3. BFRC on SA-OPC blended binders have an average tensile strength of 25% higher than the strength of control samples on OPC without fiber. The maximum effect in increasing strength was shown at an early age (the increase in strength was from 32% to 45%).

4. Optimal way of basalt fiber addicting into the concrete mix in suspension achieved the uniform distribution of fibers in the cement matrix.

References

1. ACI T. Uygunog˘lu, Mater. Struct. **41**,1441 (2008)
2. E. Güneyisi, M. Gesog˘lu, A.O. Akoi, K. Mermerdas, Compos. Part B-Eng., **56**, 83 (2014)
3. S. Iqbal, A. Ali, K. Holschemacher, T.A. Bier, Constr. Build. Mater., **98**, 325 (2015)
4. K. Hannawi, H. Bian, W. Prince-Agbodjan, B. Raghavan, Compos. Part B-Eng., **86**, 214 (2016)
5. M.L. Santarelli, F. Sbardella, M. Zuena, J. Tirillò, F. Sarasini, Mater. Des. **63**, 398 (2014)
6. J. Kim, D.J. Kim, S.H. Park, G. Zi, Compos. Struct., **122**, 198 (2015)
7. C. Lin, O. Kayali, E.V. Morozov, D.J. Sharp, Cem. Concr. Compos., **51**, 27 (2014)
8. P. Zhang, Q.F. Li, Compos. Part B-Eng., **45**, 1587 (2013)
9. X. Wang, Z. Wang, Z. Wu, F. Cheng, Constr. Build. Mater., **73**, 781 (2014)
10. S.T. Buljan, A.E. Pasto, H.J. Kim, *Advanced Structural Ceramics: Properties, Reliability and Applications* (1986)
11. K. Wille, A.E. Naaman, ACI Mater. J., **109**, 4 (2012).
12. S.F. Lee, S. Jacobsen, Mater. Struct., **44**, 1451 (2011)
13. T. Abu-Lebdeh, S. Hamoush, W. Heard, B. Zornig, Constr. Build. Mater., **25**, 39(2011)
14. Z. Wu, C. Shi, W. He, L. Wu, Constr. Build. Mater., **103**, 8 (2016)
15. V.C. Li, H.C. Wu, Y.W. Chan, J. Am. Ceram. Soc., **79**, 700 (1996)
16. X.H. Wang, S. Jacobsen, S.F. Lee, J.Y. He, Z.L. Zhang, Mater. Struct., **43**, 125 (2010)
17. N. Morova, Constr. Build. Mater., **47**, 175 (2013)
18. V.I. Fiore, G. Di Bella, A. Valenza, Mater. Des., **32**, 2091 (2011)
19. Y. Zhang, C. Yu, P.K. Chu, F. Lv, C. Zhang, J. Ji, H. Wang, Mater. Chem. Phys., **133**, 845 (2012)
20. T. Bhat, V. Chevali, X. Liu, S. Feih, A. Mouritz, Compos. Part A, **71**, 107 (2015)
21. R. Li, Y. Gu, Z. Yang, M. Li, S. Wang, Z. Zhang, J. Nucl. Mater., **466**, 100 (2015)
22. V. Manikandan, J.W. Jappes, S.S. Kumar, P. Amuthakkannan, Compos. Part B-Eng., **43**, 812 (2014)
23. D. Yeboah, S. Taylor, D. McPolin, R. Gilfillan, Constr. Build. Mater., **38**, 962 (2013)
24. D.P. Dias, C. Thaumaturgo, Cem. Concr. Compos., **27**, 49 (2005)
25. B.E. Ramachandran, V. Velpari, N. Balasubramanian, J. Mater. Sci., **16**, 3393 (1981)
26. T. Ayub, N. Shafiq, S.U. Khan, J. Mater. Civ. Eng., **28**, (2015).
27. A. Bentur, S. Mindess, *Fibre Reinforced Cementitious Composites, Elsevier Applied Science, London* (1990)
28. A.J. Majumdar, V. Laws, *Glass Fibre Reinforced Cement, BSP Professional Books, Oxford* (1991)
29. A. Bentur, *Proceedings of Durability of Glass Fibre Reinforced Concrete Symposium* (1985)
30. P. Purnell, et al., Compos. A: Appl. Sci. Manuf., **30**, 1073 (1999)
31. J. Orlowsky, M. Raupach, Mater. Struct., **39**, 635 (2006)
32. R.C. Devekey, A.J. Majumdar, J. Mater. Sci., **5**, 183 (1970)

Experience of practices of urban territorial Planning of Belarus

Tatyana Panchenko[1,*]

[1]Brest State Technical University, 224017, Moskovskaya 267, Brest, Belarus

Abstract. Rapid urbanization is one of the most significant global trends in an increasingly interconnected world. This means a new global approach to assessing the role of cities in the economy, climate change, use and consumption of natural resources as well as their contribution to social and innovative development. Globalization and large-scale urbanization, combined with smart technologies, are rapidly changing the world around us, the structure of economic activity and the way people live. After the publication in 1987 the report "Our Common Future" by the UN World Commission on Environment and Development, the conference in Rio de Janeiro in 1992 and other international initiatives by 2008 declarations, framework directives, strategies and methodologies for sustainable development and the creation of "green" economy were adopted.
Taking into account the historically high level of urbanization in our country (70.08% of the population lives in cities) [1], in recent years, legislative instruments of the local (city) level, aimed at using the tools of integrated territorial planning, have begun to form. The article describes the experience of urban territorial planning practices in Novogrudok, Polotsk, Novopolotsk and Brest for which local strategic documents of a new type have now been developed: "Green Urban Development Plans" and the Concept of Smart Sustainable Development.
Keywords: sustainable development, urban planning, strategic planning, Urban settlements, Republic of Belarus

Introduction

The world experience of recent decades shows the rapid development and diversity of urban development models and the lack of universal tools. At the same time, the fact of the common genesis of the developed largest cities-megalopolises and, accordingly, their experience and practices, as developing cities in the urbanization stage are already in different socio-economic, climatic and environmental conditions.

In 2017, the Republic of Belarus adopted the National Strategy for Sustainable Socio-Economic Development of the Republic of Belarus until 2030, the National Action Plan for the Development of a Green Economy in Belarus. A number of Belarusian cities have joined the European initiative "Covenant of Mayors on Climate and Energy" (now there are

*Corresponding author: tapanchenko46@gmail.com

53 cities). This served as a starting point for finding tools, models, creating framework documents and plans for the sustainable development of urban systems and developing the experience of Belarusian cities. At the same time, the analysis of the practices of strategic planning of urban systems and their development allows the cities of Belarus to choose those methodologies and tools of integrated development which taking into account the historically high level of urbanization will allow integrating spatial, environmental, sociocultural, economic factors, urban systems and institutional aspects in such a way as to achieve a new quality of economic and social development while maintaining and developing the accumulated human capital.

Methods

The study used a systemic and comparative method, natural, graphic-analytical and quantitative research methods.

Results

Sustainable development is based on rational management and intensive land use, an efficient energy and transport system, compactness of living and population density, territorial, functional and social integration, maintaining the integrity and viability of biological and physical natural systems in order to achieve long-term effects. Despite the fact that the tasks of sustainable development are global in nature, the strategies for achieving them at the local level are largely local and, therefore, differ in terms of conditions and content depending on the region and city.

The successful development of cities in the modern period around the world is determined by a long-term, holistic, multidimensional and multifunctional approach to urban development which allows integrating spatial, environmental, sociocultural, economic factors, urban systems and institutional aspects in such a way that they mutually reinforce each other, ensuring synergistic effect of combining energy and financing in one direction. Planning has its general goals, objectives and varieties but the form that they acquire is determined by the social conditions and traditions of specific cities.

The UN-Habitat report, Planning for Sustainable Cities: A Global Report on Human Settlements 2009 [2], provides a rationale for urban planning as a means of addressing the challenges of 21st century cities and improving the quality of sustainable urban development: "Urban planning is a conscious joint activities related to the creation or change of the appearance of a small or large city, urban area or a wider territory as well as with the implementation of the results of such activities, taking into account the priorities regarding investments in the development of the territory, environmental protection measures, new and reconstructed areas of the settlement, financing of development basic infrastructure and principles of regulation of the use of the territory "[2]. The new system of urban planning should: take into account existing problems in the field of ecology, social inequality and contribute to their solution; focus on the fulfillment of national obligations under international treaties (conventions); comply with the implementation of the constitutional rights of citizens, including those with disabilities; be flexible and, if necessary, reorganize to solve urgent problems of the local level; promote the consolidation of the local community through open dialogue with representatives of stakeholders; be aligned with budget, strategic spatial planning plans and land use systems; include monitoring and evaluating city plans.

Thus, the concepts of territorial development and strategies for their implementation reflect the context, preconditions, priorities and needs, in particular, in the social environment (for example, social justice, cultural identity and cultural traditions, cultural heritage, human health, safety and comfort, and social infrastructure). Urban development planning practices are aimed at ensuring openness and security, increasing the attractiveness of cities for life and investment, stimulating development processes and competition. The concepts of urban development can be simultaneously quite diverse and similar to each other, since the biggest problems of cities are simultaneously quite diverse and similar to each other: lack of resources, low rates of economic development, social isolation, climate change and the danger of environmental disasters.

This is an extremely urgent and difficult task that must be solved jointly by residents and city leaders in order to ensure a high quality of life and urban environment in the long term, and a balance of natural and anthropogenic factors. It also addresses the broader question of how to transform a city into a stable political, economic and ecological system, as well as how this city-system will be embedded and can increase equity, inclusiveness, sustainability and environmental friendliness in the country and the world.

Foreign experience in the formation and implementation of strategic documents for territorial planning is very wide. There is currently no generally accepted definition of a "green" city. However, practice shows that in the global experience, more and more attention is paid to the urban environmental agenda and focuses on "green" decentralized integrated solutions and complex tools, creating new values and new opportunities for business and society for citizens, effectively using all tangible and intangible assets, providing productive, inclusive and sustainable economic activities. This scenario forms a system of principles that include an integrated approach, a combination of regional and sectoral aspects of management, overcoming departmental (sectoral) disunity, building up the own potential of cities and moving to indicative planning and assessment methods.

The development in Belarus of the European Initiative "Covenant of Mayors on Climate and Energy" in order to reduce GHG emissions by at least 20% compared to the base year stimulates city administrations to fundamentally change approaches and practices of urban planning, developing innovative integrated approaches and interdepartmental cooperation, taking into account future needs.

In 2016, the European Bank for Reconstruction and Development (EBRD) launched its largest Green Cities initiative to address urban climate change challenges. Acting according to the proposed methodology (based on the "load-state-response" framework concept), city administrations develop "Green Action Plans", the structure and solutions of which cover local environmental priorities, are based on a set of indicators that is unique for each city and contain the implementation timeline and methodology for monitoring their implementation [3].

In 2017–2018 under the expert support of the UNDP Project "Belarus: Support for Green Urban Development in Small and Medium-Sized Cities of Belarus (Green Cities)", "Plans of Green Cities"(PGC) were developed in our country for Novogrudok, Polotsk and Novopolotsk. All of them are participants in the Covenant of Mayors on Climate, their Green Action Plans (GAPs) are strategic documents that define the long-term vision of a green city, strategic goals for each priority area (perspective 10–15 years), and also serve as the basis for decision-making in planning investments and operational work of city administrations in the medium term (3–5 years) [3].

The Green Action Plans for Novogrudok, Polotsk and Novopolotsk were developed based on an analysis of long-term and medium-term documents adopted at the republican, regional and local levels, existing external and internal prerequisites, trends and competitive advantages. They have a general structure, including an analysis of the current state (city passport, analytical overview and profile of the city, analysis of the spatial structure); description of the Mission, Vision and Development Goals; Urban space transformation strategy (spatial development aspects, sectoral strategies, road map and funding sources); a set of actions (localization of green actions) on the territory of the city [4].

Each of the cities formed its own scenarios (for example, for Novogrudok, on the basis of a "base" line, active, advanced and prevailing development for various sectors in terms of their impact on environmental aspects and for Novopolotsk – "mono-industrial development", "multi-industrial development "). Determination of priority directions and areas, strategic goal, Mission and Vision of the city were formulated based on the results of a survey of city residents and consultations with experts, representatives of the administration and determine the Local agenda and purpose of the city.

The spatial development strategies outlined in the PGC are based on the principles of effective land use, form the preconditions for the transition to the concept of a recoverable, smart, convenient city, the implementation of the urban planning policy of transit-oriented development, the integrated improvement of developed territories, the preservation of historical heritage, the formation of a unified water-green system, development infrastructure for a new type of economy - closed type.

For a set of actions for the implementation of urban development priorities for each of the cities a system of complex indicators has been assembled with their subsequent monitoring. The consistent implementation of the Green Urban Planning presupposes targeted interventions in the most problematic or, on the contrary, prepared areas of the city to launch its green development, therefore such points of growth and pilot development projects were identified in cities [4].

Fig. 1. Plans for green urban development in Polotsk and Novogrudok [4]

Brest joined the Covenant of Mayors movement on October 26, 2015. In this regard, one of the activities of the National Action Plan for the Development of the Green Economy in Belarus until 2020 (approved by the Resolution of the Council of Ministers of the Republic of Belarus No. 1061 on December 21, 2016) [5] was the development and implementation of the Concept Brest: Symbio City 2050.

Concept "Brest: Symbio City 2050" was developed to accomplish these tasks - a model of smart sustainable urban development based on the possibilities of synergistic interaction of six urban systems (transport (urban mobility); energy and climate; architecture and "green culture"; waste; water supply and sanitation, biodiversity and landscape) while providing economic, social and environmental benefits and phasing in its implementation in the short, medium and long term. For each of the areas of the Concept, fundamental (general and local) goals and objectives are collected, a set of indicators is determined (taking into account the methods of the EBRD, U4SSC, ISO and TNLA RB) and the localization of pilot projects for spatial development.

The strategic goals of the concept were:

– development of a new planning model for the integrated development of the city of Brest in the short, medium and long term based on sustainable principles, using the possibilities of symbiosis of the main urban systems, introducing the decarbonization of infrastructure systems, ensuring a stable improvement in the quality of life of citizens and future generations, as well as increasing the sustainability of the community and stimulating innovative and technological activities.

– creating a basis for dialogue, partnership and cooperation with all stakeholders, actors and policy-makers aimed at creating sectoral and integrated strategies related to the social and economic systems of the city, as well as for interdisciplinary and sectoral analysis of sustainability at all levels.

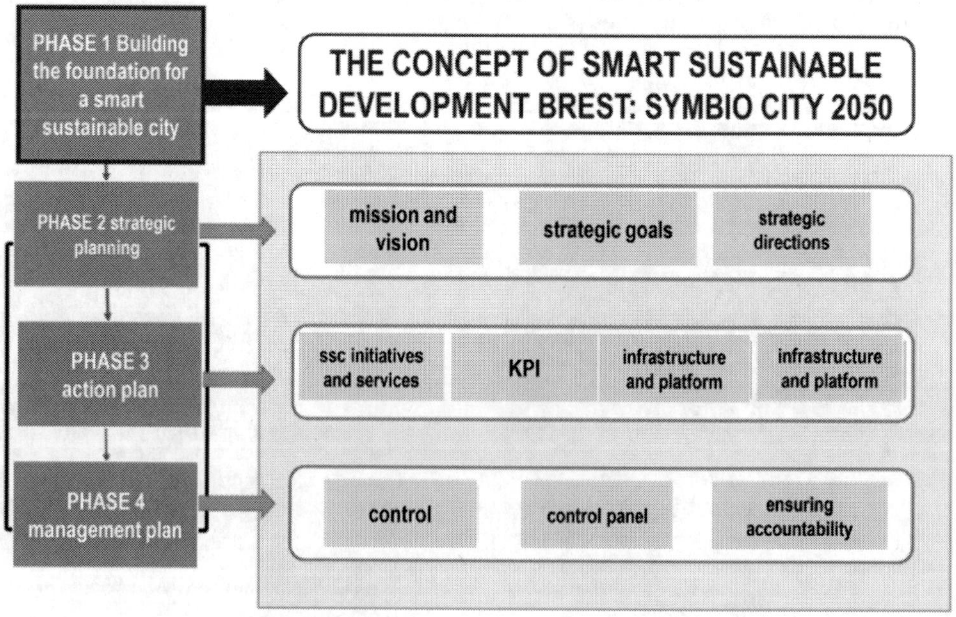

Fig. 2. Concept formation model

The concept was developed in accordance with the provisions of the National Strategy for Sustainable Socio-Economic Development of the Republic of Belarus for the period up to 2030 [6] (in the future until 2035), the Urban Development Policy of the Republic of Belarus for 2016-2020, the legislation of the Republic of Belarus, State, National and regional programs.

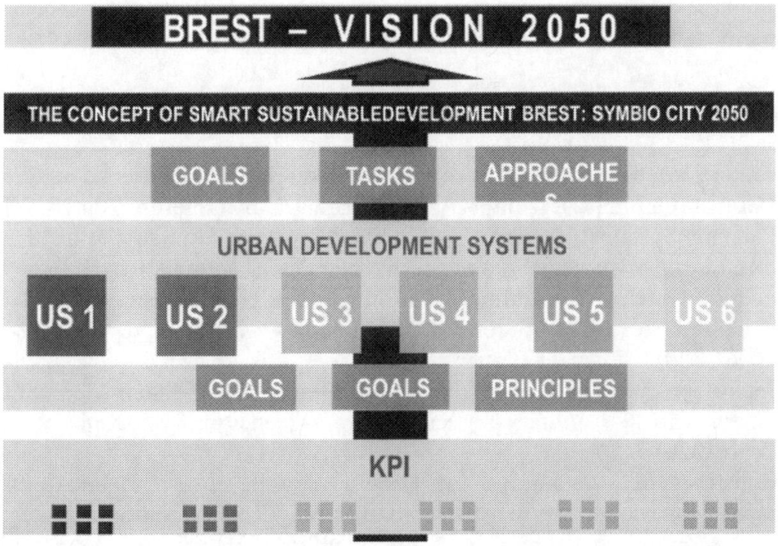

Fig. 3. Concept structure

Its creation and implementation will create conditions for the integrated planning of the development of the city and its infrastructure systems in the short, medium and long term to improve the quality of life of the population, attract investments and innovative development; to connect to this process the current and future state and regional programs, strategies and plans; develop and implement city and sectoral documents based on measurable indicators of both quantitative and qualitative state of development processes; improve existing urban development policies, procedures and practices; develop multi-sectoral and interdisciplinary collaboration between stakeholders, experience in the integrated actions of all urban systems; provide the citizens of the city – a stable improvement in the quality of life, a comfortable living environment, prosperity, security; territories - a high level of comfort and arrangement of the living environment, conditions for optimal spatial development. At the same time, the conditions for the conservation of resources, the development and productive use of innovations / new technologies, a high, socially balanced quality of life are met [5].

Conclusions

The described strategic planning documents highlight the priority areas of urban development policy focused on the development and improvement of all spheres of city life on the basis of existing resources. At the same time, they are documents open to suggestions and adjustments aimed at initiating discussions on promising directions and priorities for urban development.

Determining the long-term innovative development of the city they are a tool for coordinating the actions of urban systems, setting priorities for territorial management of energy and climate issues, urban planning and architecture, mobility and organization of traffic, safety, water supply and sanitation, natural resources and waste management.

The presence of an interconnected set of documents (PGC, city development concept, urban planning documentation, urban strategy, and sectoral strategies) and effective management will create conditions for the transition to a high-quality, measurable assessment of urban planning decisions. Cities of Belarus will receive new opportunities for the development of an integrated sustainable urban system that has synergistic links between sectors based on energy efficiency, smart land use management, reduced consumption and renewable resources, convenient and efficient transport system, increased prosperity and a multi-faceted cultural identity of society.

References

1. National Statistical Committee of the Republic of Belarus: demographic and social statistics [Online] URL: http:// www.belstat.gov.by (reference date: 12.06.2020)
2. Planning Sustainable Cities — Global Report on Human Settlements 2009 [Online] URL: http://www.unhabitat.org (reference date: 11.06.2020)
3. 3.EBRD Green Cities[Online] URL: https://ebrdgreencities.com/assets/Uploads/PDF/e0d6d7606e/Green-City-Action-Plan-Methodology-Russian.pdf. (reference date: 12.06.2020)
4. Belarus: Supporting Green Urban Development in Small and Medium Cities of Belarus (Green Cities) [Online] URL: http://greencities.by/our_project (reference date: 05.06.2020)
5. Ministry of economy of the republic of Belarus [Online] URL: http://www.economy.gov.by/uploads/files/NSUR2030/ National strategy for sustainable socio-economic development of the Republic of Belarus for the period up to 2030.pdf (reference date: 05.06.2020)
6. Ministry of economy of the republic of Belarus [Online] URL: https://www.economy.gov.by/uploads/files/1061r.pdf (reference date: 05.06.2020)

Nanoindentation of cement stone samples

Elena Polonina[1], *Olaf* Lahayne[2], *Josef* Eberhardsteiner[2], *Sergey* Leonovich[1]
[1]Belarusian National Technical University, 220013, Nezavisimosty Ave., 65, Minsk, Belarus
[2]Vienna University of Technology, A-1040, Karlsplatz 13, Vienna, Austria

Abstract. The preliminary results of the studied cement samples were obtained by the nanoindentation method. It was revealed that the elastic modulus M increases in samples that contain a complex additive containing nanosized particles. The effect is also observed with the introduction of an additive containing only one type of nanoparticles (nanosilica sol SiO_2 or carbon nanomaterial MCNT). The selection of the parameters of the nanoindentation method, which ensured the obtaining of the final consistent results, was performed. These results are presented by histograms of the distribution of nanoindentation points in modulus of elasticity M and hardness H and distributions in M and H in the horizontal XY plane perpendicular to the motion of the nanoindentor. The results obtained indicate that there is a change in the nanostructure of the C – S – H gel, which is compared with an increase in strength, Young's moduli and shear, upon the introduction of SiO_2 nanoparticles and MCNT nanoparticles.
Key words: complex nanodispersed additive, nanoindentation method, cement samples.

1 Introduction

Higher requirements for the mechanical properties of materials began to appear. The development of various technologies has become widespread in the modern world, which has contributed to the creation of new building materials. The nanoindentation method has become an alternative approach to studying the mechanical properties of materials at the nanoscale. In this case, it is not the sample size that decreases, but the size of the deformed region. During nanoindentation, most solid and superhard nonmetallic materials are deformed elastically-plastically, which makes it possible to characterize such mechanical properties as hardness and elastic modulus.

Based on the analysis of literature sources, it was revealed that C-S-H exists in at least three structurally different forms: low, high and ultra-high densities, which have different average values of hardness and hardness and different volumetric concentrations. The average values of the stiffness and hardness of various forms of CSH gel turned out to be properties inherent in the C – S – H structure, which do not change in various materials based on Portland cement [1]. Nanoparticles of different chemical composition with high specific surface area and high surface energy are used to target the CSH gel nanostructure. The results of such an influence are studied by a set of methods: calorimetry, thermogravimetry, IR

*Corresponding author:grushevskay_en@tut.by

spectroscopy, X-ray phase analysis, small-angle neutron scattering, scanning and tunneling electron microscopy, atomic force microscopy, nanocondensation, etc. [1-9,13,14-21].

The nanoindentation method makes it possible to assess the effect of nanoparticles directly on the volume fraction of different forms of CSH gel at an early age of 6, 12, 15, 24 h and at an adult age of 4-6 months [2-7]. The aim of the work was to determine the effect of hydrothermal SiO_2 nanoparticles with a high density of surface silanol groups Si-OH and MCNT nanoparticles with an acid-activated surface on the CSH gel nanostructure, separately and in combination, on the CSH gel nanostructure.

2 Methods

The method of nanoindentation was used, when a solid needle of a known shape is pressed into the surface of a cement stone sample at a constant speed. When the specified load or indentation depth is reached, the movement stops for a certain time, after which the needle is retracted in the opposite direction. During loading, the load values and the corresponding indenter displacement are recorded. The resulting relationship is called the loading / unloading curve. From this experimental curve, the hardness and elastic modulus of the material can be determined (a method proposed by Oliver and Far).

In this work, the studies were carried out by an automated Hysitron TI 950 TriboIndenter nanoindenter (a type of nanoindenter with a Berkovich tip with a cone angle of 143 degrees). For mechanical measurements, a diamond probe tip was used.

When constructing the histograms, the M step was 1.5 GPa (m = 100 intervals within 0-150 GPa), the H step was 0.1 GPa (100 intervals within 0-10 GPa).

The results are presented as histograms of the distribution of nanoindentation points in terms of the reduced modulus of elasticity M and hardness H. Deconvolution in three phases J = 1, 2, 3 values x = M, H was performed according to the Gaussian normal distribution for each J-th phase

$$p_j(x) = \frac{1}{\sqrt{2\pi S_J^2}} \exp\left(-\frac{(x-\mu J)^2}{2S_J^2} \right) \quad (1)$$

where μ_J is the mean value of M, H of the J-th phase, s_J is the standard deviation of the distribution in the J-th phase (StdDev). The theoretical probability density function P (x) of the distribution of M, H values was determined taking into account the volume fraction f_J of each phase:

$$P(x) = \sum_{J=1}^{n} f_J p_J(x) \quad (2)$$

where J = 1, 2, 3.

The unknown values of μ_J, s_J, f_J were found from the condition of the minimum of the root-mean-square deviation between the experimental discrete values of the nanoindentation of the probability density Pi and the theoretical values of P (x_j) in each i-th interval under the additional condition that the sum of the volume fractions of the phases f_J is equal to one (m = 100 is the number intervals along the M, H axis) [6]:

$$\min \sum_{i=1}^{m} \frac{(P_i - P(x_i))^2}{m} \quad (3)$$

$$\sum_{J=1}^{n} f_J = 1 \quad (4)$$

where J = 1, 2, 3.

To find the minimum of function (3) under additional condition (4) and unknown values μ_J, s_J, f_J, we used the algorithm of the MathLab program.

During the research, the following materials were used as the main components: binder - Portland cement PC 500 D0 with the following mineral composition, wt%: C_3S – 58.31, C_2S - 18.38, C_3A – 8.01, C_4AF - 10.64. Modifying substances: sol of nanosilica (NC) with the following characteristics: SiO_2 content in ash -120 g / dm^3, density ρ = 1075 g / dm^3, total salt content - 1720 mg / dm^3, pH = 9.2, specific surface area 500 m^2 / g and average diameter of primary SiO_2 nanoparticles 5.5 nm; carbon nanomaterial (CNM): average diameter of tubes and fibers 10-300 nm, average length of tubes and fibers 0.01-20 microns, bulk density 0.15-0.22 g / cm^3, ash content no more than 5%, specific adsorption surface from 60 m^2 / g. Superplasticizer (SP) in the form of an aqueous solution is a polycarboxylate copolymer with a density of 1.1-1.14 g / ml, pH = 6-8, a viscosity of 230-330 cps, a content of non-volatile substances of 39-41%, a water-reducing capacity of over 40%. Water for mixing and subsequent hardening.

Tests by the nanoindentation method and ultrasonic measurements were carried out on cement specimens with dimensions of 10x10x20 mm.

The tests were carried out on 4 compositions: Sample № 1 - contains an additive of superplasticizer, Sample № 2 - containing an additive SP + sol of nanosilica SiO_2, Sample № 3 - containing an additive of superplasticizer SP + carbon nanomaterials, Sample № 4 - containing an additive of superplasticizer SP + SiO_2 + CNM

The additive for samples № 1-4 was introduced in the amount of 0.8% by weight of cement. The amount of mixing water for all samples was selected in such a way as to obtain a dough of normal density in all cases. Samples were made from cement paste of normal density. The water-cement ratio of samples 1-4 was W / C = 0.21, the age at nanoindentation and plotting of histograms was 3 months.

The results of our own studies of samples of heavy concrete [8], a complex nanodispersed system (containing an additive of the superplasticizer SP + SiO_2 + CNM), the compressive strength at the age of 28 days reached 78.7 MPa, which exceeds the strength of the control sample by more than 50% and the sample containing only 37% superplasticizer. The difference in compressive strength (fc) between the control sample №1 and modified samples № 2,3 was also established 4. The difference reached at an early age of 1, 3 days - 13 and 12.7%, at the age of 37 days - 11.9%, and was maximum in in the case of using a combination of nanoparticles (SiO_2 + CNM). The difference in bending strength Rben relative to the control was maximum for the sample with a complex addition of NPs (SiO_2 + CNM) and reached at the age of 1, 3 days -39.6 and 21.6%, at the age of 37 days -23.4%.

Based on the results of ultrasonic measurements, reliable increases in Young's modulus E and shear modulus G were obtained at the age of 1 and 5 months in the samples modified with nanoparticles. The maximum increment of 10% was achieved in the variant of modification with a complex nanoadditive (sample № 4).

Nanoindentation first of all shows differences in structure, and such significant differences between samples with nanoparticles № 2, 3, 4 and sample № 1 without nanoparticles have already been identified, and these differences must be attributed to the action of nanoparticles [9-12].

3 Results and discussion

Time schedule of load: 1. immersion up to 300 nm and maximum force for 10 s; 2. constant load mode for 5 s; 3. raising the nanoindenter to the surface, removing the load to 0 within 10 s.

Immersion speed of the nanoindenter: if the maximum immersion depth of the nanoindenter is 300 nm, the immersion time is 10 s, then the immersion speed is 300 nm / 10 s = 30 nm / s.

For each cycle "loading - unloading", a P - h diagram of the dependence of the load on the penetration depth of the indenter was constructed and stored.

The slope of the hysteresis curves in the upper right corner was used to calculate the values of M at the point of indentation; from the maximum load and the area of the contact zone, H was calculated. The calculated values of M and H were used to construct distributions depending on the contact depth of immersion, from the distributions we went to histograms M and H by intervals (Fig. 1), on the basis of equations 3 and 4, the Gaussian functions were found and deconvolution was carried out in three phases (Fig. 1a, b, c, d).

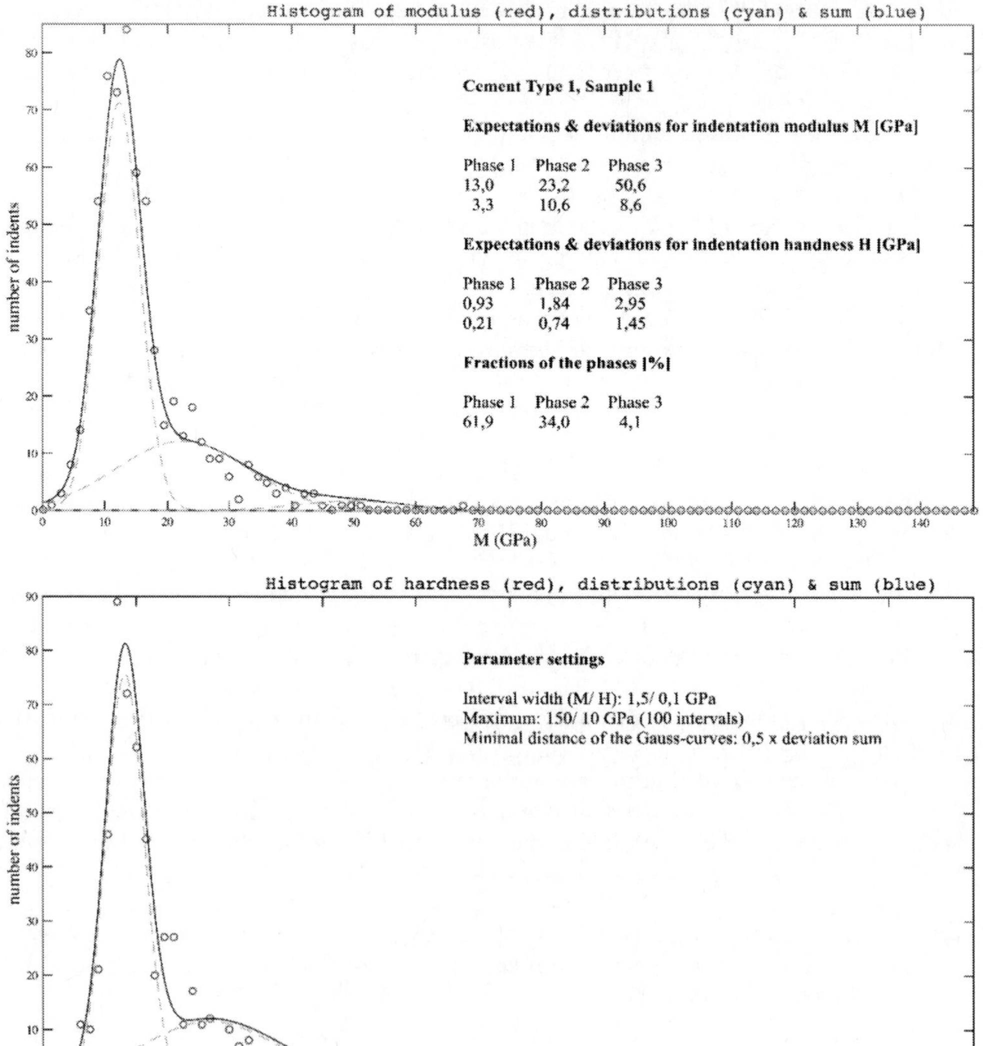

Fig. 1a.Histogram of the distribution of nanoindentation points in modulus of elasticity E and hardness H for sample № 1

https://doi.org/10.1051/e3sconf/202021202013

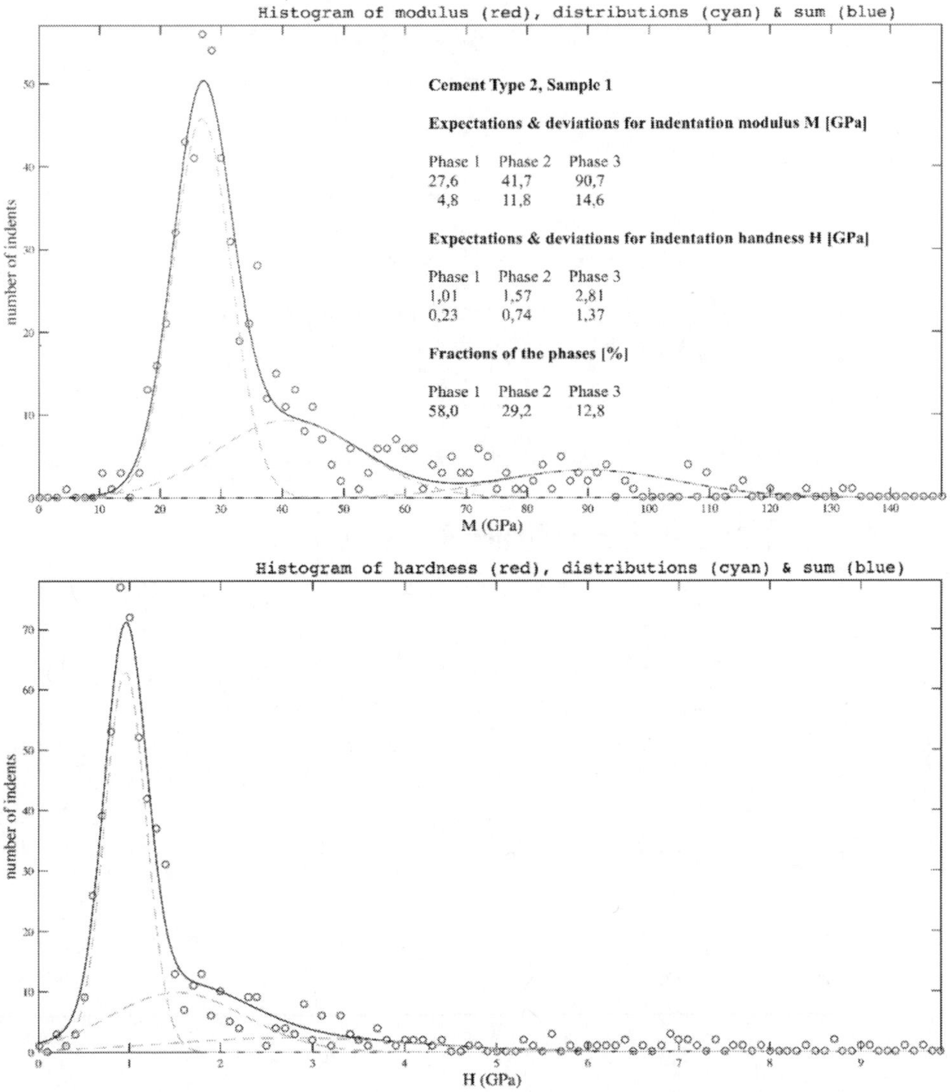

Fig.1b.Histogram of the distribution of nanoindentation points by modulus of elasticity E and hardness H for sample №2

The histograms show phases with their average values of E and H (and moderate statistical scatter within each phase). Then the percentage of each phase was estimated.

The procedure for approximating by Gaussian functions: the starting parameters and the number of phases n are set, and the Matlab program selects the average M and H, the average deviations for each phase, the volume fractions of the phases from the condition of the minimum sum of squares of deviations between the experimental and theoretical points.

The results obtained give a qualitatively correct picture. It is very important that the distribution in the modulus of elasticity M shifted to the right in samples 2, 3, 4 as compared to sample 1. There is also a slight offset to the right in hardness H.

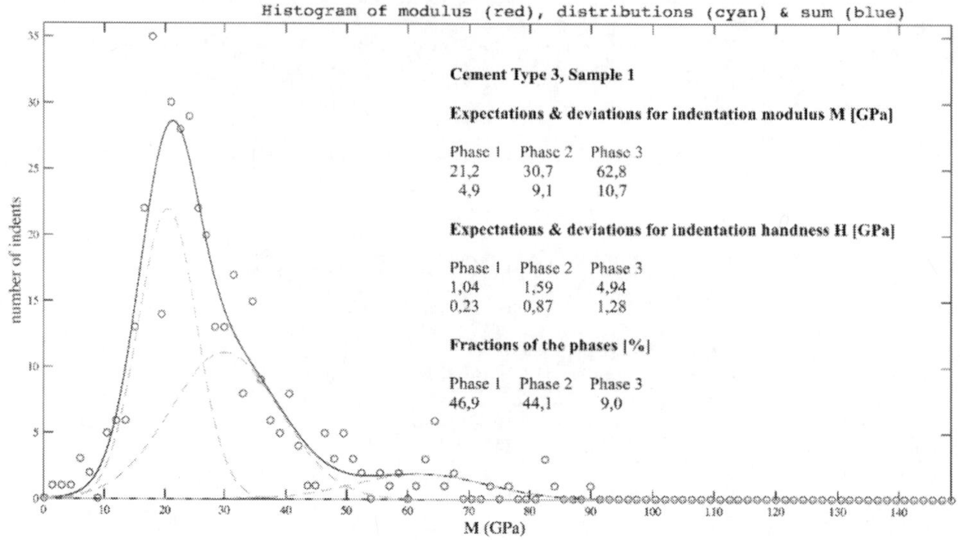

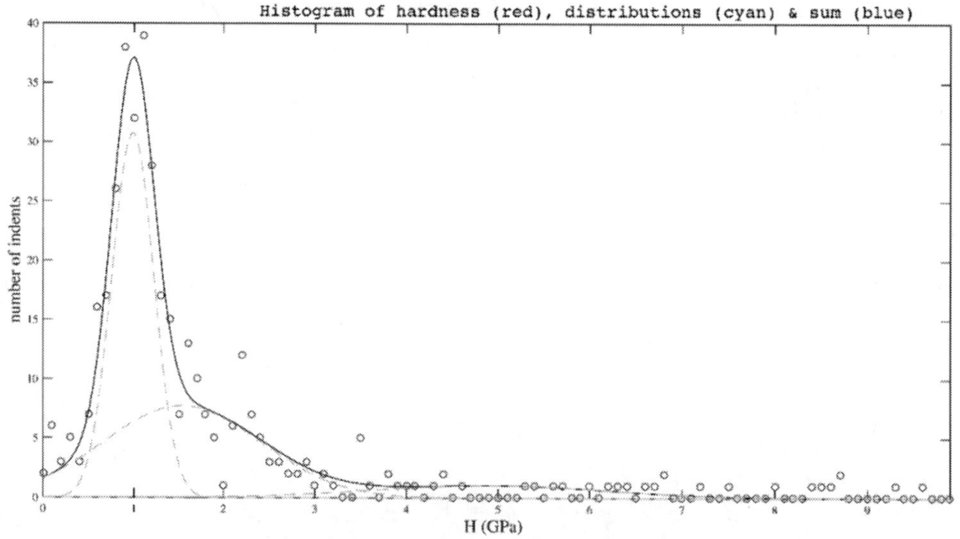

Fig.1c. Histogram of the distribution of nanoindentation points in modulus of elasticity E and hardness H for sample №3

The transition from the distributions M and H to histograms (Fig. 1) showed that the first main maximum for M shifted from the value of 13 GPa for sample № 1 without the introduction of nanoparticles to values in the region of 27.6; 21.2; 22.6 GPa for samples 2, 3 and 4 modified with nanoparticles. The maximum of the second phase in the region of 21.2 GPa for sample № 1 shifted to values of 42.7; 30.7; 36.8 GPa for samples 2, 3, 4. A shift towards higher average values indicates an increase in the volume fraction of the CSH gel phases with a higher packing density of particles and, accordingly, higher mechanical characteristics.

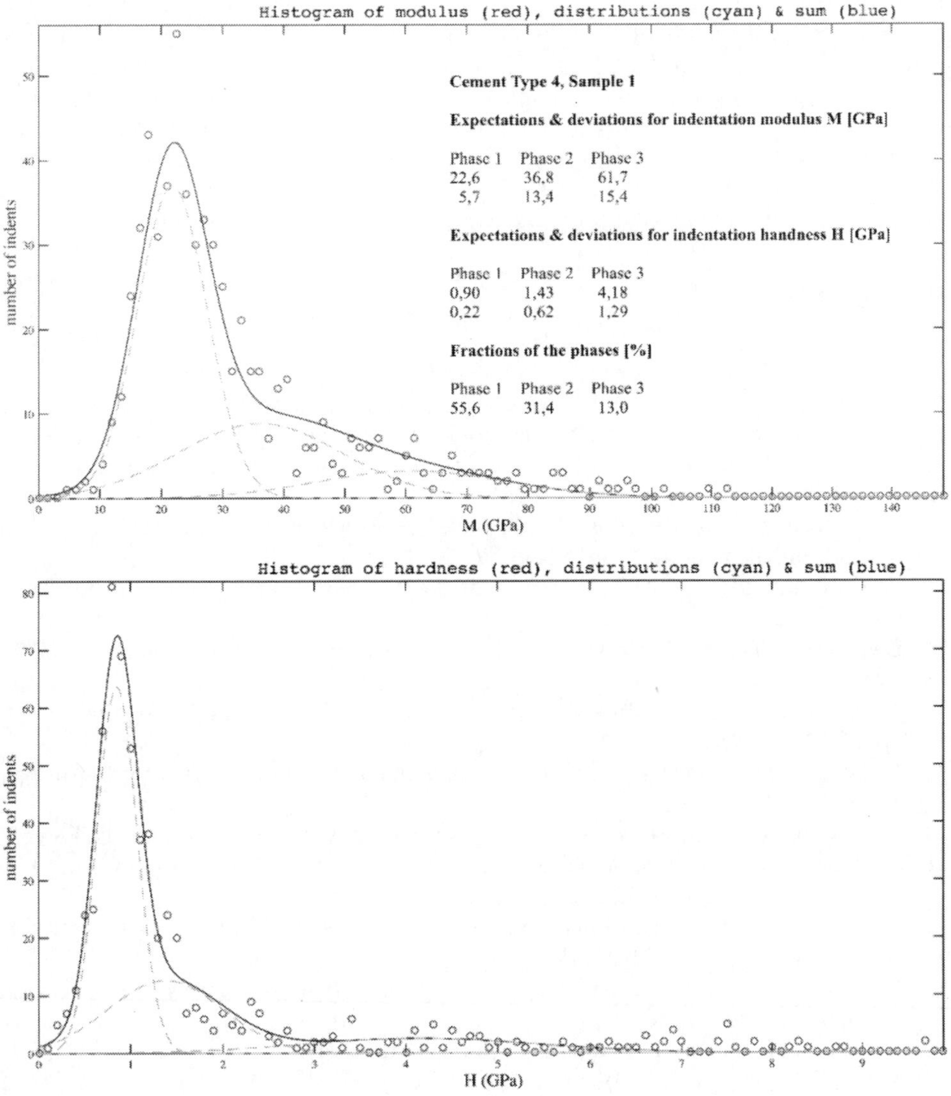

Fig.1d. Histogram of the distribution of nanoindentation points in modulus of elasticity E and hardness H for sample №4

4 Conclusions

1. Obtained hysteresis curves for indentation grid points. For all grid points, the curves were smooth and continuous with a similar shape, which made it possible to calculate the array of M and H values and pass to the distributions of M and H depending on the contact depth of immersion. At the same time, the average immersion depth of about 344 nm provided the following conditions: multiple excess of the minimum size of CSH gel inhomogeneities [6], while having values 10 times smaller than the scale size of the microstructure of the hardening Portland cement sample [6] and 5-fold excess of the size of inhomogeneities of the sample surface (30 nm). The linear dimensions of the indentation area (M) significantly exceeded the scale size of the microstructure of the hardening cement-water sample (1-3 μm). The

number of intervals in the histograms M, H (100) was less than the number of indentation points (595). As a result, these conditions provided a statistically significant set of M, H values, reflecting the inhomogeneous multi-phase structure of the CSH gel - the presence of a number of maxima in the histograms in Fig. 1a-1d with high volume fractions of phases.
2. The shift to the region of higher average values indicates an increase in the volume fraction of CSH gel phases with a higher packing density of particles and, accordingly, higher mechanical characteristics. An increase in the volume of the fraction of the phases of the CSH gel with higher values of M and H corresponds to the results of strength, Young's and shear moduli, and the density of samples of modified SiO_2 nanoparticles and CNM.

References

1. F.Sanchez, K. Sobolev, Vestnik TGASU **3 (40),** pp.262-289 (2013)
2. C. Plassard, E. Lesniewska, I. Pochard, A., Elsevier **100**, pp.331-338 (2004)
3. F.-J. UlmArabian, Journal for Science and Engineering **37(2)**, pp.481–488 (2012)
4. P. Hou, S. Kawashima, ,D. Kong, D. J.Corr, J. Qian, S. P.Shah, Composites Part B: Engineering **45(1)**,pp. 440-448 (2013)
5. J. Eberhardsteiner, S. Zhdanok, B. Khrustalev, E. Batyanovsky, S. Leonovich, P. Samtsov, Science and technology **1**, pp. 52-55 (2012)
6. G. Constantinides, F.J. Ulm, Journal of the Mechanics and Physics of Solids **55(1)**, pp. 64-90 (2007)
7. L.P. Singh, W. Zhu, T. Howind, U. Sharma, Cement & Concrete Composites **(79),** pp. 106-116 (2017)
8. S. Zhdanok, V. Potapov, E. Polonina, S. Leonovich, Engineering Physics Journal **93(3),** pp. 669-673 (2020)
9. V. Potapov, Y. Efmenko, D. Gorev, Nanotechnologies in Construction **11(4)**, pp.308-325 (2019)
10. 10.V. Potapov, E. Grushevskaya, S. Leonovich, Construction Materials, pp.4-9 (2017)
11. S. Zhdanok, E. Polonina, S. Leonovich, B. Khrustalev, E. Koleda, Engineering Physics Journal **92(1)**, pp.14-20 (2019)
12. S. Zhdanok, E. Polonina, S. Leonovich, B. Khrustalev, E. Koleda, Engineering Physics Journal **92(2),** pp.391-396 (2019)
13. L. Krenev, S. Volkov, E. Sadyrin, T. Zubar, S. Chizhik, Engineering Physics Journal **91(3)**, pp.637-644 (2018)
14. U. Sharma, L. Singh., Z. Baojian, C. Poon, Cement and Concrete Composites, pp.312-321 (2019)
15. U. Sharma, L. Singh, D. Ali, C. Poon, Advances in Civil Engineering Materials **8(3)**, pp.346-360 (2019)
16. U. Sharma, D. Ali, L. Singh, European Journal of Environmental and Civil Engineering, pp 1-13 (2019)
17. L. Singh, W. Zhu, T. Howind, U. Sharma, Cement and Concrete Composites **79**, pp.106-116 (2017)
18. I. Flores-Vivian, R. Pradoto, M. Moini, Front. Struct. Civ. Eng. **11**, pp. 436–445 (2017)
19. J. Björnström, A. Martinelli, A. Matic, L. Börjesson, I. Panas, Chemical Physics Letters **392(1–3)**, pp. 242-248 (2004)
20. T. Shi, Y. Gao, P. Surendra, Advances in Cement Research, pp.353-361 (2019)
21. I. Chulkova, I. Selivanov, V. Galdina, Proceedings of universities. Building **10**, pp. 15-27 (2019)

Physical parameters of insulation with a structure-forming material from flax noils

Sergey Romanovskiy[1*], *Aleksandr* Bakatovich[1]

[1]Polotsk State University, 211440, Blokhina 29, Novopolotsk, Belarus

Abstract. The results of studies to evaluate the effective operation of the obtained thermal insulation slabs made of flax fibers or noils were carried out. The sorption moisture content of insulants based on flax fibers or noils with modified liquid glass was determined at a relative humidity of 40–97%. The influence of humidity on the thermal conductivity of the test materials was studied. The coefficient of vapor permeability of insulation made of flax fibers or noils was established. The results of field tests of experimental thermal insulation materials on the attic flooring of a residential building during the cold season were considered. The graphs of the temperature distribution over the structure of the attic flooring and heat flow indicators were obtained. The values of the heat transfer resistance of the attic flooring were calculated at an air temperature of -20 °C to -2 °C. The moisture content of insulation at the end of the cold period was determined and the dependence of the distribution of moisture over the thickness of the thermal insulation layer was plotted. The results of the tests carried out indicate the highest efficiency of thermal insulation slabs made of flax fiber noils in comparison with insulants based on flax fibers.
Key words: flax fiber, flax noils, physical parameters.

Introduction

Plant waste is a continually renewable, large-tonnage product produced all over the world. Technologies for the use of plant raw materials are being implemented in many countries for the production of fibrous thermal insulation materials. This is due to the high physical and mechanical properties and environmental cleanliness of the resulting insulation. The research results, as well as the characteristics of the produced thermal insulation materials, confirm that plant-based compositions compete with traditional insulation materials, including mineral wool and expanded polystyrene.

In countries with large timber reserves, including Russia, research is being conducted to develop insulants based on wood fibers and various binders, such as bio-based adhesive [1], polyvinyl acetate glue [2], paraffin emulsion [3], as well as without binder with the use of antiseptic and fire retardant [4]. At a density of 30-250 kg/m^3, thermal insulation materials have a thermal conductivity of 0.041 W/(m·K) to 0.06 W/(m·K).

* Corresponding author: s.romanovskiy@psu.by

Compositions of insulation slabs based on recycled fiber mass from containerboard or fibrillated waste paper have been developed in Krasnoyarsk [5, 6]. Thermal conductivity of the materials ranges from 0.038–0.051 W/(m·K) with a density of 30 to 70 kg/m^3.

The most famous solution for the use of recycled waste paper for the production of insulation is ecowool, which consists of 80% of fibrous filler and 20% of fire retardants [7; 8]. The thermal insulation material has a density of 30-75 kg/m^3, providing a thermal conductivity of 0.032-0.041 W/(m·K) and a vapor permeability of 0.3 mg/(m·h·Pa).

The possibility of using eucalyptus fibers as a filler for insulation was investigated in Chile [9]. The binder was phenolic resin. Thermal insulation material with an average density of 80 kg/m^3 to 250 kg/m^3 provides a thermal conductivity of 0.05–0.07 W/(m·K).

An innovative development is an insulant based on sphagnum moss and liquid glass [10]. The thermal conductivity of the material is 0.034-0.04 W/(m·K) at a density of 155–170 kg/m^3. However, thermal insulation slabs have a drawback, which is the material shrinkage during the drying process. To eliminate this problem, it was proposed to add crushed straw in an amount of 20–30% of the moss mass [11]. The new thermal insulating material has a thermal conductivity of 0.044-0.046 W/(m·K) and provides a compressive strength of 0.2-0.21 MPa with an average density of 156-190 kg/m^3.

Fibrous aggregates of agricultural origin are a significant reserve in the production of local thermal insulation building materials. For example, hemp-based slabs "Thermo-Hanf" are produced in Germany [12; 13]. Insulation includes the following components: 83-87% of fibrous filler, 10-12% of polyester, and 3-5% of soda, used as a fire retardant. With an average density of 35-40 kg/m^3, the thermal insulation material has a thermal conductivity coefficient of 0.038-0.04 W/(m·K).

Insulation made from cotton waste and sodium water glass is of considerable interest for the countries of Central Asia. When varying density in the range of 40-100 kg/m^3, the thermal conductivity of the slabs changes from 0.037 to 0.041 W/(m·K) [14]. Under conditions of a relative humidity of 60–70%, the moisture content of the insulating material is 11–12% [15].

Petronas University of Technology (Malaysia) studied the possibility of using oil palm bark fibers as a structure-forming insulation material [16]. According to the test results, it was found that when the density varies from 66 kg/m^3 to 110 kg/m^3, the thermal conductivity coefficient of the material changes in the range of 0.03–0.09 W/(m·K).

Research on the production of thermal insulation slabs made of oil palm bark fibers up to 4 cm long and sodium water glass were carried out at Polotsk State University [17]. Lime and gypsum additives were used to increase the water resistance of the binder. The insulation has a density of 135-168 kg/m^3, providing thermal conductivity of 0.046-0.047 W/(m·K) and compressive strength of 0.2-0.24 MPa at 10% linear deformation. High resistance of fibers to decay is noted at a relative humidity of 97%.

Ecological insulation from flax fibers "Ecoteplin" is produced in Russia. Flax fibers are the filler in thermal insulation slabs. Starch is used as a binder. Boron salts are added to increase fire protection and prevent the formation of mold and fungi [18]. The material has the following characteristics: average density of 32-34 kg/m^3, thermal conductivity of 0.038- 0.04 W/(m·K), sound absorption coefficient of 0.74-0.98, vapor permeability of 0.403 mg/(m·h·Pa), flammability class G1 (slightly combustible) [19].

In Belarus, thermal insulation slabs of the "Akoterm Flax" trademark are produced [19]. The insulation contains 85% of flax fibers and 15% of polyester fibers. The thermal conductivity of the material corresponds to 0.038–0.04 W/(m·K), vapor permeability is 0.4 mg /(m·h·Pa), sound absorption is 0.98 with an average density of 30 kg/m³.

The physical and mechanical characteristics of the described thermal insulation materials are shown in Table 1.

Table 1. Physical and mechanical characteristics of the thermal insulation materials

№	Fillers	Binders	Additives	Density, kg/m³	Thermal ductivity, W/(m
1	wood fibers	bio-based adhesive	–	50–100	–
2	wood fibers	polyvinyl acetate glue	foaming agent	81	0,042–0,05
3	wood fibers	paraffin emulsion	–	150–250	0,06
4	wood fibers	–	antiseptic and fire retardant	35–40	0,041
5	recycled fiber mass	polyvinyl acetate glue	–	69	0,049–0,051
6	recycled fiber mass	–	boric acid and borax	30–50	0,038–0,041
7	recycled waste paper	–	fire retardant	30–75	0,032-0,041
8	eucalyptus fibers	phenolic resin	–	80–250	0,05–0,07
9	sphagnum moss	liquid glass	–	155–170	0,034–0,04
10	sphagnum moss and crushed straw	liquid glass	–	156–190	0,044–0,046
11	hemp	polyester	soda	35–40	0,038–0,04
12	cotton waste	liquid glass	–	40–100	0,037–0,041
13	oil palm bark fibers	–	–	66–110	0,03–0,09
14	oil palm bark fibers	liquid glass	lime and gypsum	135–168	0,046–0,047
15	flax fibers	starch	boron salts	32–34	0,038–0,04
16	flax fibers	polyester fibers	–	30	0,038–0,04

The purpose of the complex studies carried out is to determine the physical parameters of thermal insulation slabs with a structure-forming flax noils material. For comparison, the physical and mechanical characteristics of insulants based on flax fibers are investigated. The use of flax noils for the slabs manufacture solves the problem of utilizing plant waste and expands the range of effective heat-insulation materials that are safe for the environment and human health.

Method

Samples of thermal insulation slabs were made from flax fibers up to 25 cm long or flax noils 5 to 10 cm long, obtained at a flax factory in Postavy (Belarus). Flax fiber consists of bundles of elementary fibers interconnected. The bundle diameter is 50-70 μm. Each bundle contains from 10 to 20 connected along the length and tightly adjacent filaments. The length of an elementary fiber varies from 10 to 40 mm, with a diameter of 8-12 μm. Flax fiber noils are loose bundles of elementary fibers. The structure of the bundle in the noils is formed from disheveled elementary fibers with chaotically located contact connections. As a binder in the production of thermal insulation materials, sodium liquid glass was used, produced at JSC "Domanovsky Production and Trade Plant". To ensure the insolubility of liquid glass, hydrated lime and gypsum were used as modifying additives.

The main physical and mechanical indicators of the materials under study, namely the average density and compressive strength at 10% linear deformation, were determined in accordance with GOST 17177 [20].

The thermal conductivity coefficient of insulants was measured with an ITP-MG4 "250" device in accordance with the requirements of STB 1618 [21].

The values of the sorption moisture content of thermal insulation materials were determined according to STB EN 12088 [22].

The study of the moisture index influence on the coefficient of thermal conductivity of samples of thermal insulation materials was carried out according to the method described in [23]. For the experiment, samples of insulation were made from flax noils and fibers with modified liquid glass in the form of slabs with dimensions of 250×250×50 mm. Additionally, for comparative analysis, we used samples of thermal insulation slabs "Akoterm Flax" based on a mixture of flax and polyester fibers. After determining the density and thermal conductivity of the slabs in a dry state, each sample was placed in an individual sealed chamber with pre-filled water below the material installation level by 40 mm. The air temperature in the chamber corresponded to 20±2 °C, and the relative air humidity was 97%. The test conditions reflect the most unfavorable humidity conditions for thermal insulation slabs - the period during and after rainfall. The thermal insulation slabs were kept in a chamber above water for 2, 5, 10 and 20 days, after which they were removed and the thermal conductivity coefficient was determined. Complete stabilization of the moisture index occurred after 20 days, which determines the maximum period of keeping the samples in a sealed chamber. The frequency of indicators determination is due to the most characteristic changes in the dynamics of humidity and thermal conductivity of the samples.

The vapor permeability of thermal insulation materials was determined in accordance with STB EN 12086 [24].

The thermophysical properties of thermal insulation slabs under operating conditions were determined using the data measurement complex RTP-1-16T. The values of changes in the ambient air temperatures and the materials under study were recorded with thermocouples, and the values of the heat flux were determined using heat flux converters. Samples from flax noils or fibers were made in the form of slabs with dimensions 1000×600×100 mm. Insulation materials were placed between the wooden beams of the attic flooring of a one-storey residential building (Fig. 1). Inter-girder construction of the attic flooring included an edged ceiling covering board, a vapor barrier layer, thermal insulation slabs based on flax fibers (flooring 1), flax noils (flooring 2) and a mixture of flax and polyester fibers (flooring 3), as well as a hydro- and windproof membrane. The cross-sectional diagram of the attic flooring is shown in Figure 2.

Fig. 1. Thermal insulation slabs made of flax fibers, flax noils,mixture of flax and polyester fibers (from left to right) on the attic flooring

The moisture content of the insulants after testing under operating conditions was determined in accordance with GOST 17177. Prism samples 50×50×300 mm in size were preliminarily cut through the thickness of the thermal insulation materials. The prism sample was cut in the transverse direction into individual fragments 50 mm long, obtaining samples in the form of cubes. Then each cube sample was weighed and placed in an drying cabinet. Upon reaching a constant mass, the sample cubes were weighed again. The moisture content was determined by the amount of change in the sample cubes mass before and after drying. The sampling scheme is shown in Figure 3.

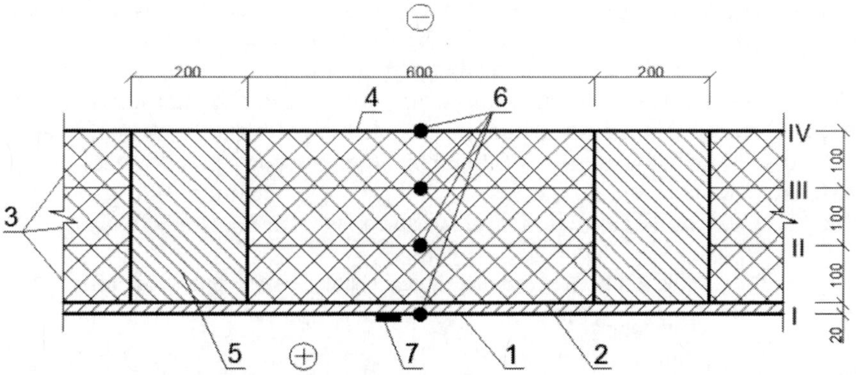

1 – edged ceiling covering board; 2 – vapor barrier layer; 3 – thermal insulation slabs; 4 – hydro- and windproof membrane; 5 - wooden beam of the flooring; 6 – thermocouples; 7 – heat flux converter; I, II, III, IV – thermocouple installation site boundaries
Fig. 2. Cross-sectional diagram of the attic flooring

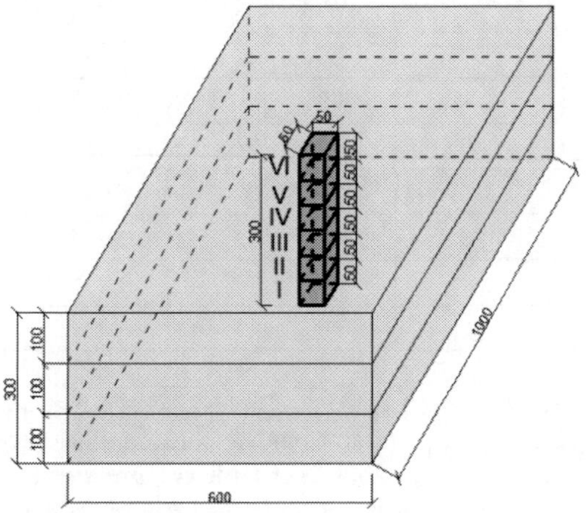

I–VI – sample numbers
Fig. 3. Sampling scheme from thermal insulation boards

Results and discussion.

To determine the quantitative composition of the experimental samples, structure-forming material consumption was changed at a constant consumption of a binder equal to 10 kg and consisting of liquid sodium glass - 9 kg, lime - 0.5 kg, and gypsum - 0.5 kg. The accepted consumption of the binder had a minimal effect on the thermal conductivity coefficient, while providing a coherent structure of the insulation. Research and results on the selection of binder consumption are discussed in a previously published work. The physical and mechanical characteristics of the samples obtained are shown in Table 2.

According to the research results, it has been found that an increase in the consumption of the structure-forming material leads to an increase in the density of the insulation up to 70 kg/m³ (compositions 5 and 6) and, regardless of the type of fibers, provides a decrease in the thermal conductivity coefficient. A further increase in the density of samples (compositions 9 and 10) causes an increase in the thermal conductivity of insulants. At a density of 50 kg/m³ for slabs based on flax (composition 2), the thermal conductivity is 0.041 W/(m·K). A decrease in the thermal conductivity coefficient by 15% occurs with an increase in density by 1.4 times to 70 kg/m³ (composition 6). With an increase in the average density to 90 kg/m³ (composition 10), an increase in thermal conductivity up to 0.037 W/(m·K) is noted. For slabs made of flax fibers, a similar dependence has been established, so the minimum value of the thermal conductivity coefficient of 0.042 W/(m·K) corresponds to a density of 70 kg/m³ (composition 5). A decrease in density to 50 kg/m³ (composition 1) and an increase to 90 kg/m³ (composition 9) leads to an increase in the thermal conductivity coefficient by 14% and 7% W/(m·K), respectively.

Table 2. Physical and mechanical characteristics of the samples

Composition No.	Consumption of structure-forming material per 1 m³, kg		Density, kg/m³	Coefficient of thermal conductivity, W/(m·K)	Compressive strength at 10% deformation, kPa
	flax fiber	flax noils			
1	40	–	50	0,048	1,8
2	–	40	50	0,041	1,1
3	50	–	60	0,045	2,4
4	–	50	60	0,039	1,7
5	60	–	70	0,042	3
6	–	60	70	0,035	2,1
7	70	–	80	0,042	3,4
8	–	70	80	0,035	2,4
9	80	–	90	0,045	3,7
10	–	80	90	0,037	2,6

The analysis of the results presented in Table 2 shows that an increase in the density of thermal insulation slabs at a constant binder consumption provides an increase in the compressive strength at 10% deformation. Thus, the compressive strength of flax noils insulants (composition 2) is 1.1 kPa at a density of 50 kg/m³. An increase in the average density by 1.8 times (composition 10) leads to an increase in strength by 2.4 times to 2.6 kPa. Samples of materials based on flax fibers at a maximum density of 90 kg/m³ (composition 9) reach a strength of 3.7 kPa. A decrease in density to 50 kg/m³ (composition 1) causes a 2-fold decrease in the compressive strength at 10% deformation. It should be noted that in the investigated range of densities, the compressive strength is higher for thermal insulation materials made of flax fibers by an average of 33% compared to slabs based on flax fiber noils.

Due to the low strength characteristics of insulants based on flax fibers and noils, when choosing the compositions of thermal insulation slabs for further research, the minimum thermal conductivity coefficient was taken as the determining indicator. Thus, further studies of insulants were carried out on samples with an average density of 70 kg/m^3, which provide the best thermal conductivity. Additionally, samples of "Akoterm Flax" slabs with a thermal conductivity coefficient of 0.04 W/(m·K) at a density of 34 kg/m^3 were used in the studies. Insulation molding technology does not allow the production of higher density material. The quantitative composition, average density and thermal conductivity of the dry samples for further research are shown in Table 3.

Table 3. Compositions, average density and thermal conductivity of samples

Composition No.	Component consumption per 1 m^3, kg						Average density, kg/m^3	Coefficient of thermal conductivity, W/(m·K)
	flax fiber	flax noils	liquid glass	polyester fiber	lime	gypsum		
1	60	-	9	-	0,5	0,5	70	0,042
2	-	60	9	-	0,5	0,5	70	0,035
3	29	-	-	5	-	-	34	0,04

Sorption moisture.

To study the sorption moisture content of thermal insulation materials based on plant fibers, the compositions given in Table 3 were taken. Experimental data were used to plot the dependences of the material sorption moisture content on the relative air humidity in the form of sorption isotherms (Fig. 4).

From the analysis of the obtained dependences (Fig. 4), it follows that at an air humidity of 40%, the sorption humidity of the thermal insulation material from the noils is 10.2%, which is 10% lower than the values of the samples based on flax fibers and higher than the indicator of the "Akoterm Flax" insulation material by 59%. The sorption humidity of composition 1 at a relative air humidity of 60% is 15.4%. For composition 2, the sorption moisture is less than the material index (composition 1) by 9%, and more samples (composition 3) by 24%. An increase in the sorption moisture content of samples from flax noils to 18.3% occurs at an air humidity of 80%, which practically coincides with the value of samples based on flax fibers and exceeds the indicator of a material made from a mixture of flax and polyester fibers by 17%. The values of the sorption moisture content of the samples (compositions 1, 3) are less by 9% and 15% in comparison with the samples based on flax noils, equal to 26.8% at a relative air humidity of 90%. At an air humidity of 97% in a desiccator, the sorption of the material from flax noils increases to 37.6%, which is 13% and 19% more than for samples based on flax fibers and "Akoterm Flax" thermal insulation material.

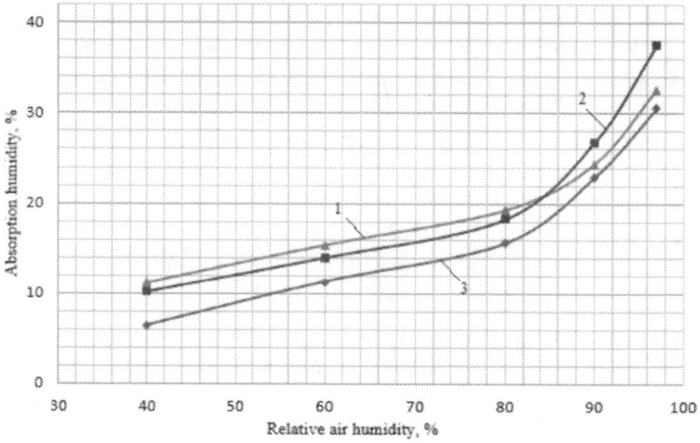

1 - samples with flax fibers; 2 - samples with flax noils; 3 - "Akoterm Flax" samples
Fig. 4. Isotherms of water vapor sorption

The excess of the sorption moisture content of the samples based on flax noils relative to the indicators of compositions 1 and 3 at a relative air humidity of 90-97% is due to the significant value of the sorption moisture content of flax noils in comparison with flax and polyester fibers. Separated elementary fibers in noils have a large geometrical area of surface contact with the external environment relative to the indicators of flax fiber, consisting of dense bundles of elementary fibers. The presence of polyester fibers in "Akoterm Flax" slabs leads to an additional decrease in the moisture absorbed by the material from the air and, accordingly, reduces the rate of water vapor sorption by insulation samples.

Influence of humidity on the coefficient of thermal conductivity

The quantitative composition, average density and thermal conductivity of the samples for the study are shown in Table 3. The dry weight of the samples from flax noils or fibers is 219 g, and "Akoterm Flax" samples are 106 g. The laboratory test results are shown in Table 4.

Table 4. Indicators of mass, moisture and thermal conductivity of wet samples

Composition No.	Holding time of samples in the chamber, days (hours)	Wet sample indicators		
		weight, g	humidity, %	thermal conductivity coefficient, W/(m·K)
Flax fiber insulation				
1	2 (48)	238,9	9,1	0,049
2	5 (120)	264,3	20,7	0,06
3	10 (240)	278,6	27,2	0,072
4	20 (480)	290,3	32,1	0,08
Flax noils insulation				
5	2 (48)	239,4	9,3	0,043
6	5 (120)	269,8	23,2	0,058
7	10 (240)	290	32,4	0,067
8	20 (480)	299,4	36,9	0,075
"Akoterm Flax" insulation				

 https://doi.org/10.1051/e3sconf/202021202014

9	2 (48)	114,4	7,9	0,048
10	5 (120)	124,9	17,8	0,063
11	10 (240)	131,9	24,4	0,074
12	20 (480)	137,8	30	0,081

From the experimental data (table 4) it follows that in 48 hours the moisture index of samples 5 based on flax noils reaches 9.3% and is almost identical to the value of samples 1, while the moisture content of the "Akoterm Flax" insulation is 18% higher (samples 9). For samples 6 based on noils, after 5 days from the start of testing, the humidity increases to 23.2%, which is 12% and 30% more than the values of slabs 2 and 10. During the next 5 days, a gradual increase in the moisture index for all materials occurs. The moisture value of slabs 7 based on flax strips after 10 days is 32.4%, which exceeds the values of samples 3 and 11 by 19% and 33%, respectively. Relative to the dry materials indicators, the moisture content of samples 8 has increased to 36.9% after 20 days of testing, and the increase in the moisture content of flax noils relative to samples 4 and 12 is 15% and 23%.

The experimental data obtained indicate that after 48 hours of testing, the value of thermal conductivity of thermal insulation slabs (samples 5) based on flax noils increased by 23%. Over the same period of time, the thermal conductivity of flax fibers samples 1 and 9 increased by 20%. After 5 days of keeping the slabs in a sealed chamber, the thermal conductivity of samples 6, 2, and 9 based on noils and flax fibers increased by 46%, 66%, and 58%, respectively. The change in thermal conductivity for the second 120 hours is 16% for samples 7 from flax fiber noils, which is 61% less than the change in the index of the same material for the first 120 hours. For slabs 3 and 10 based on flax fibers, the change in the value of thermal conductivity over the same period of time is 20% and 17%, respectively. The thermal conductivity of samples 4, 8 and 12 after 20 days of testing increases by 2-2.1 times relative to the indicators of dry materials.

Lower coefficients of thermal conductivity of flax fiber noils samples are achieved due to the multidirectional arrangement of elementary fibers in the structure of the insulation. This prevents convective air transport as a result of a decrease in the size of thin air layers of irregular shape and their partial localization in the form of separate closed microvoids. Thus, in spite of the higher moisture content of the noils slabs, the obtained structure of elementary fibers allows to provide identical thermal conductivity values with flax fiber slabs and "Akoterm Flax" insulation.

Water vapor permeability

For testing, samples of thermal insulation slabs based on flax noils and fibers with dimensions 110×110×50 mm were made. The composition, density and vapor permeability of the samples under study are presented in Table 5.

Table 5. Composition, density and vapor permeability of samples

Composition No.	Component consumption per 1 m^3, kg			Density, kg/m^3	Water vapor permeability, mg/(m·h·Pa)
	flax fiber	flax noils	binder		
1	30	–	10	40	0,43
2	–	30	10	40	0,41
3	70	–	10	80	0,41
4	–	70	10	80	0,37

5	110	–	10	120	0,38
6	–	110	10	120	0,34

According to the research results (Table 5), it has been found that with a minimum average density of 40 kg/m^3, the vapor permeability index of samples of flax fiber noils is practically identical to that of flax fiber insulation. For samples based on flax noils or fibers with an average density of 80 kg/m^3, a decrease in vapor permeability is observed relative to the indicators of samples 1 and 2. The vapor permeability coefficient for slabs made of flax noils is 10% less than that for materials based on flax fibers at the same density value. With an increase in the average density of samples to 120 kg/m^3, the vapor permeability coefficient of thermal insulation slabs made of flax fibers is 12% higher than that of samples made of flax noils.

Thus, with an increase in the average density of thermal insulation materials, the vapor permeability index decreases. The investigated parameter of the noils insulation at a density of 40 kg/m^3 decreases by 17% with an increase in the average density to 120 kg/m^3. For samples based on flax fibers with a minimum density of 40 kg/m^3, the vapor permeability is 12% higher than that at a maximum density of 120 kg/m^3.

Thermal performance of insulants under operating conditions

During the autumn and spring period 2018-2019, observations were made of the thermophysical processes occurring in the construction of the attic flooring of an operated one-story residential building with the use of experimental thermal insulation materials. As an example, the period from January 20 to January 29, 2019, equal to 10 days, with the lowest night outdoor temperatures in the Polotsk district of the Vitebsk region (Belarus), is considered. The compositions of the investigated thermal insulation slabs are shown in Table 3. According to the results of measurements, it has been found that the temperature distribution for thermal insulation slabs made of a mixture of flax and polyester fibers ("Akoterm Flax" insulation) is identical with insulation made of flax fiber noils. Figure 5 shows the distribution of the outside air temperatures in the attic and the space between the beams of the attic flooring (including insulation) at night. Temperature values are taken as average values of temperature readings from 22:00 to 6:00.

a

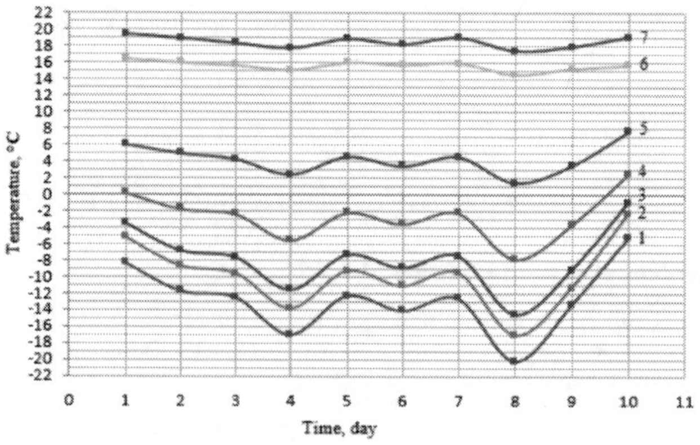

b

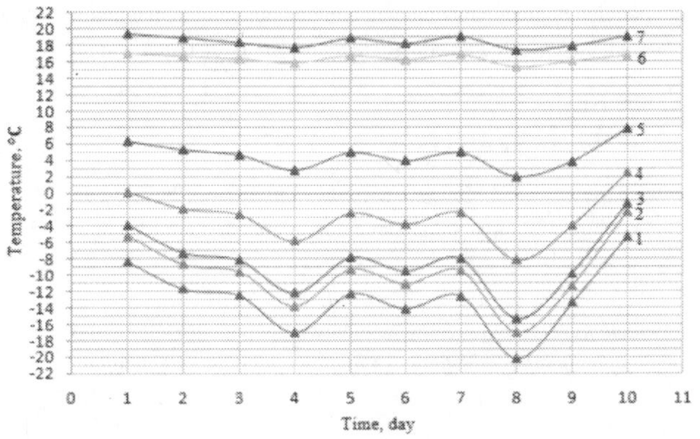

a – flax fiber insulation; b –flax noils insulation
1 – outside air temperature; 2 - temperature in the attic, °C; 3 – temperature of 3 layers at border-line IV; temperature of 3 layers at border-line III; 5 – temperature of 3 layers at border-line II; 6 – temperature of layer at border-line I; 7 – air temperature in the dwelling, °C
Fig. 5. Temperature distribution over the thickness of the attic flooring for a period of 10 days

From the temperature distribution graphs (Fig. 5) it has been found that the coldest hours are the night hours of the eighth day. At night on the eighth day, the temperature amplitude on the surfaces of the attic flooring with thermal insulation slabs made of noils is 30.7 °C, which is identical with the attic flooring 3 and practically coincides with the values of flooring 1, equal to 29.1 °C.

Based on the temperature distribution graphs in table 6, the average values of the temperature distribution in the living room, attic flooring, as well as outside air and attic temperatures for the time period under consideration are given.

Table 6. Average temperatures

Indicator	Flooring 1	Flooring 2	Flooring 3
Outside air temperature, °C	–12,7	–12,7	–12,7
Attic air temperature, °C	–9,8	–9,8	–9,8
Temperature of the third layer at border-line IV, °C	–7,8	–8,6	–8,5
Temperature of the third layer at border-line III, °C	–1,8	–2,8	–2,6
Temperature of the third layer at border-line II, °C	3,7	4,6	4,2
Temperature of 1 layer at border-line I, °C	15,6	16,4	16,4
Indoor air temperature, °C	18,5	18,5	18,5

From the data presented in Table 6, it follows that for the attic flooring with flax noils insulation, the average temperature of the inner surface at the border-line I for 10 days is 0.8 °C higher than for the flooring with thermal insulation material based on flax fibers and is identical to the attic flooring with "Akoterm Flax" slabs. When comparing the average temperatures for 10 days at border-line II, it has been found that for the flooring with thermal insulation slabs made of flax fiber noils, the average temperature value is 0.9 °C higher than for the attic flooring with insulation based on flax fibers and practically coincides with the flooring 3 indicator. For a similar period of time, the average temperature at the border-line III of flooring 2 is 1 °C less than that of attic flooring 1 and is identical to the value of flooring 3. At border-line IV for 10 days, the average temperature of the outer surface of the attic flooring with the material based on noils does not differ from indicators of flooring with insulation (composition 3, table. 3) and 0.8 °C lower than that of flooring with slabs of flax fibers.

Based on the obtained values of heat flux densities, the thermal resistances of the attic flooring structures with experimental insulants were determined. Figure 6 shows the dependence of thermal resistance on air temperature in an unheated attic.

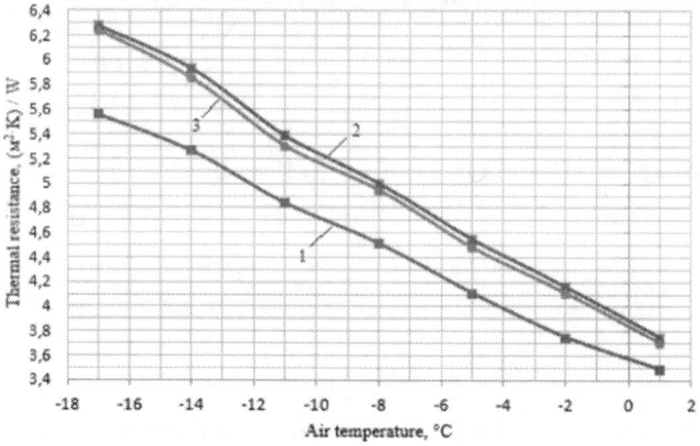

1 – flax fiber insulation; 2 – flax noils insulation; 3 – insulation made from a mixture of polyester and flax fibers

Fig. 6. Dependence of the thermal resistance of the attic flooring structure on the air temperature in an unheated attic

From the dependences shown in Fig. 6 it follows that the thermal resistance value of the attic flooring with flax fiber noils insulation is 3.53 (m^2·K)/W at an attic air temperature of +1 °C, which practically coincides with the value of the flooring with thermal insulation materials based on flax and polyester fibers and 0.27 (m^2·K)/W more than the flooring with flax fiber slabs. The thermal resistance value of flooring 2 at an air temperature in the attic of -8 °C is 5 (m^2·K)/W, which is 11% more than the thermal resistance of flooring 1 and does not differ from the flooring 3 indicator. At an air temperature in the attic of -17 °C thermal resistance of an attic flooring with thermal insulation materials (composition 2) is identical to the value of an attic flooring with "Akoterm Flax" insulants and is 13% higher than the value for an attic flooring with a flax fiber material equal to 5.56 (m^2·K)/W.

In the spring, after the end of the monitoring, the values of the moisture content of thermal insulation materials were determined. The change in moisture content across the thickness of the insulation is shown in Figure 7.

On the basis of the obtained data on the distribution of moisture content over the thickness of thermal insulation slabs (Fig. 7), the average moisture content of the material from flax noils was established to be equal to 12.1%. The obtained value for composition 2 is lower than the moisture index of the fiber-based slab (composition 1) by 14% and higher by 14% than the value of the insulation based on a mixture of flax and polyester fibers (composition 3). At the same time, in layer I, the moisture content of the thermal insulation material made of flax noils is 17% less than the moisture content of the insulation (composition 1), equal to 11.3% and 1.25 times more than that of "Akoterm Flax" slab. In layer VI, the moisture content of the material based on flax noils is 13.8%, which is 13% less than the moisture content of the flax fiber insulation and 10% higher than that of the slab based on flax and polyester fibers.

According to the results of previous studies of the effective insulants operation on the attic flooring without a hydro-wind protection device, it has been found that the moisture content of insulants containing flax noils is 17.6%, flax fibers 19.3%, a mixture of flax and polyester fibers 14.1%. The hydro-wind protection device reduces the moisture sorption of the insulation from the air of the attic space, which makes it possible to reduce the moisture content of the investigated thermal insulation materials by an average of 31% and increase the thermal resistance to 6.27 (m^2·K)/W with a thermal insulation layer thickness of 300 mm.

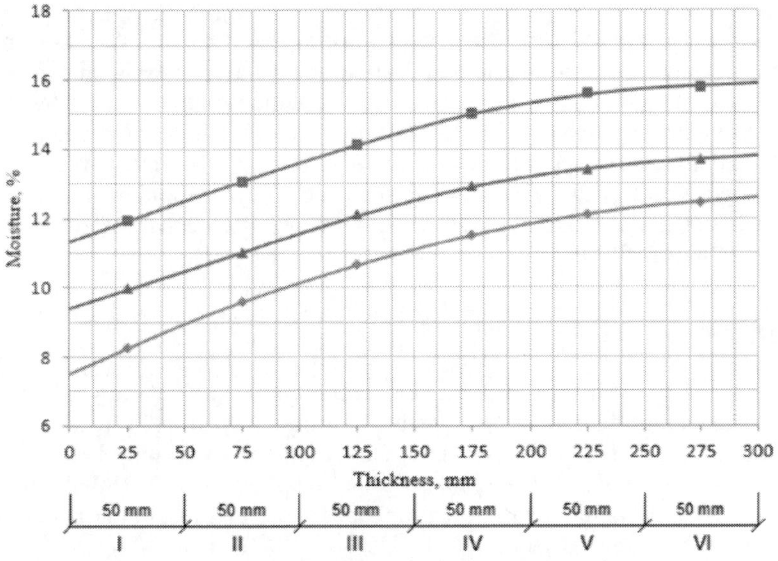

1 – flooring with flax fiber insulation; 2 – flooring with flax noils insulation;
3 – flooring with insulation made of a mixture of polyester and flax fibers
Fig. 7. Distribution of moisture in the thickness of the insulation

Conclusions

According to the research results, effective thermal insulation slabs from flax fiber noils with a density of 70 kg/m^3, characterized by a low thermal conductivity coefficient of 0.035 W/(m^2·K), were obtained.

According to the results of tests of sorption humidity by the desiccator method, it was found that at a relative air humidity of 97%, the indicators of the sorption humidity of samples based on flax fibers and noils are 32.7% and 37.6%, which exceeds the value of "Akoterm Flax" slabs equal to 30.6 % by 7% and 23%, respectively. The obtained data on the kinetics of water vapor sorption by thermal insulation slabs will allow calculating the humidity regime of the structure. Also, the experimental data is supposed to be used in the future in the development of a predictive model of changes in the thermal characteristics of insulants during the operation of buildings and structures.

As a result of the conducted studies on the effect of moisture on the thermal conductivity of fiber-structure insulants, it has been determined that a change in the moisture index has a significant effect on the thermal conductivity coefficient of slabs based on noils, flax fibers, or mixture of flax and polyester fibers. An increase in the moisture content of thermal insulation materials made of flax noils provides an increase in thermal conductivity by a factor of 2.1 to 0.075 W/(m·K) relative to dry samples. For insulants made of flax fibers and "Akoterm Flax" slabs, moisture absorption leads to an increase in thermal conductivity to values of 0.08 W/(m·K) and 0.081 W/(m·K), respectively, which exceeds the indicators of dry materials by 2 times. Thus, with a moisture content that is 19% higher than the value of "Akoterm Flax" slabs equal to 30%, thermal insulation material based on noils provides an identical coefficient of thermal conductivity in a wet state.

To perform calculations of the enclosing structures moisture regime, the vapor permeability of thermal insulation materials made of flax fibers and noils was determined. With an average density of 40-120 kg/m^3, the vapor permeability of noils-based insulants is

0.34-0.41 mg/(m·h·Pa), and for flax fiber samples it ranges from 0.38 mg/(m·h·Pa) to 0.43 mg/(m·h·Pa).

The use of hydro-wind protection in the construction of the attic flooring allows to reduce the moisture content of thermal insulation materials during the operation of the building by an average of 31%. This constructive solution reduces the moisture sorption by insulation from the air in the attic. As a result, in spite of the higher moisture content of flax noils slabs compared to materials made from a mixture of flax and polyester fibers, when using gyro-wind protection, the resulting structure of an insulation made of elementary fibers and liquid glass makes it possible to provide high heat transfer resistance values comparable to those of "Akoterm Flax" slabs.

References

1. V. Strikun, M. Strikun, A. Bayandin et al. Obtaining heat-insulating material from wood fiber based on bio-glue. *Actual problems of forestry*, **48,** 86–87 (2017)
2. P. Ermolina, A. Mironov, A. Byvshev, Obtaining heat-insulating board material based on wood fiber. *Actual problem of the forestry complex*, **25,** 186–189 (2010)
3. L. Zhuravleva, A. Zhuravleva, Soft fibreboard – thermal insulation material. *KrasGAU Bulletin,* **11,** 181–184 (2010)
4. V. Beryukov, V. Danilov, N. Pashkov, Thermal insulation material *RU 2149148* (2000)
5. S. Luchinin, V. Kozhukhov, Y. Alashkevich, Production of heat-insulating materials based on recycled cellulose fiber. *IVUZ. Forest Journal,* **6,** 151–155 (2017)
6. A. Ermolina, P. Mironov, Obtaining and properties of thermal insulation material based on recycled fiber mass. *IVUZ. Forest Journal*, **4**, 109–114 (2011)
7. S. Goreglyad, Environmentally friendly building materials. *Construction Materials* **4**, 5–6 (1996)
8. G. Ivanov, New environmentally friendly thermal insulation material – ecowool. *Construction Materials,* **1,** 21 (1995)
9. A. Bialosau, A. Bakatovich, F. Gaspar, *Materiais compositos para isolamento termico de materias- primas naturais e aglutinantes minerais. Livro de Resumos 3 0 Congresso Luso – Brasileiro de Materiais de construcao sustentaveis. Coimbra, Portugal*. 16–27 (2018)
10. A. Bakatovich, Microstructure as the main criterion for the use of sphagnum moss as a filler for an effective plate heat-insulating material. *Polotsk State University Bulletin. Series F*, **8**, 42–46 (2017)
11. C. Becerra, J. Montory, A new biobased composite material using bark fibres eucalyptus. *The 13th pacific rim bio-based composites symposium «Bio-based composites for a sustainable future*, 46–50 (2016)
12. E. YAkunina, Modern heat-insulating materials, as one of the trends in environmental construction. *Synergy of Science*, **24,** 625–634 (2018)
13. T. Bogatova, A. Dvojcina, Advantages and features of safe natural heaters. *Engineering systems and facilities*, **3-4 (24-25)**, 14-19 (2016)
14. M. Rozyev, A. Bakatovich, Thermal insulation material, using waste cotton production as a placeholder. *XI Junior Researchers, Conference. European & national dimension in research. Architecture and civil engineering,* **11,** 64–66 (2019)
15. M. Rozyev, A. Bakatovich, Heat-insulating material based on cotton fiber waste. *Polotsk State University Bulletin. Series F*, **8,** 29–33 (2019)
16. S. Hassan, A. Tesfamichael, M. Mohd, Nor Comparison study of thermal insulation characteristics from oil palm fibre. *ICPER 2014 – 4 th International Conference on Production, Energy and Reliability,* **13,** 5 (2014)

17. S. Romanovskiy, A. Bakatovich, Insulating material on the basis of bark fibre of the olive palm tree. *IX junior researchers' conference «European and national dimension in research»*, 104–17 (2017)
18. D. Sovetnikov, D. Semashkina, D. Baranova, The optimal thickness of the outer wall insulation to create an energy efficient and environmentally friendly building in St. Petersburg. *Construction of unique buildings and structures*, **12 (51)**, 7–19 (2016).
19. Sound-absorbing heat-insulating plates. Technical conditions *TU BY 391129716.001-2015*, 10 (2015)
20. Materials and products for construction thermal insulation. Test methods. *GOST 17177-94*, 56 (1996)
21. Building materials and products. Methods for determining thermal conductivity in a stationary thermal regime. *STB 1618-2006*, 9 (2006)
22. Building thermal insulation materials. Method for determination of sorption moisture. *STB EN 12088-2008*, 9 (2009)
23. N. Davydenko, A. Bakatovich, Influence of the humidity index on the thermal conductivity coefficient of straw and wood-straw heat-insulating materials. *Polotsk State University Bulletin, Series F*, **8,** 73–78 (2013)
24. Building thermal insulation materials. Method for determining vapor permeability. *STB EN 12086-2007*, 16 (2007)

Experience in conducting field tests of metal structural constructions of coatings of the "BrSTU"

Mikalai Shalabyta[1,*], *Andrei* Shuryn[1], *Tatsiana* Shalabyta[1] and *Viacheslav* Dragan[1]

[1]Brest State Technical University, 224017, Moskovskaya 267, Brest, Belarus

Abstract. There are quite often statically complex structures, the reliability of theoretical calculations of which needs experimental verification in the world practice of design and construction. Studies of building structures functioning can be full-scale or carried out on their models in laboratory conditions depending on the goals set. Full-scale tests of building structures are more expensive, therefore, their implementation requires a special approach and justification.

However, to check the operation of complex, critical structures which primarily include metal spatial large-span structures of coatings of buildings and structures especially those used for the first time to assess their bearing capacity, to establish the degree of participation in the construction of each of the elements, to study the operation of individual new nodal elements etc., only full-scale tests are required.

Key words: metal structural constructions, nodal connection, numerical model, stress-strain field tests.

Introduction

In accordance with the requirements of EN 1990 [1], one of the methods in the design of complex building structures is design based on the test results while depending on the goals set they can be carried out either on their models or full-scale tests of built objects. Full-scale tests of building structures are more expensive than tests on models, however, to check the operation of unique structures used for the first time, to assess their bearing capacity, to establish the degree of participation in the work of the structure of other elements (girders, etc.), to study the operation of separate elements is possible only with this type of test.

Considerable experience has been accumulated in the design and construction of buildings and structures that are unique in the architectural and structural terms in world practice. In the overwhelming majority of cases, metal spatial structures are used when designing such objects. The development of systems of spatial structures is based on the fundamental architectural and structural prerequisites: a single unified range of elements with maximum compositional capabilities; use of rolled profiles with optimal shape; use of high-

* Corresponding author: nnshalobyta@mail.ru

strength materials; complete industrialization of production based on the use of high-performance automatic equipment, casting, stamping, etc .; compactness of system elements and the possibility of their transportation by any means of transport; high reliability, quick assembly and installation of elements, including large-block assembly; wide possibilities of volumetric-spatial composition in the form of an almost unlimited variety of constructive forms based on a limited set of initial types of elements. Possessing the above-mentioned features, spatial metal rod structures with rational manufacturability and low material consumption, provide the possibility of wide unification of structural elements, organize in-line production of a limited nomenclature of the same type of elements, from which a wide variety of spatial structures with high architectural and aesthetic properties can be created.

A number of architecturally unique and structurally complex constructions were designed and built using a new metal structural structure developed at the Brest State Technical University, called the "BrSTU" system in the new century in the Republic of Belarus, they are the covering of the summer amphitheater in Vitebsk, covering of the amphitheater in the park Victories in Molodechno, the dome of the National Olympic Committee of the Republic of Belarus, etc. Most of the designed and built objects have statically complex structures of spatial coverings, the reliability of the calculation of which needed experimental verification not only of separate elements (primarily of the nodal connection), but also full-scale tests of the entire structure [2 ... 8].

The structural design of the "BrSTU" system has a structurally unique nodal element which is made in the form of a hollow sphere with holes in the wall into which through special rigid washers with spherical surfaces facing the sphere, bolts are installed to connect with rod elements. High-strength bolts designed to connect the rod elements of the structural plate they are passed from the side of the ball cavity through the holes and have the ability to freely rotate around their own axis which ensures free assembly of the structure lattice elements. The sphere is made of two hemispheres, obtained by hot pressing from sheet metal and joined by a butt seam with cutting edges. Hemispheres are made by hot pressing from sheet metal with milling of diametrical edges. In more detail, the design solution of the key elements of the "BrSTU" system is described in [3, 4, 5, 8].

Brest State Technical University has accumulated vast experience in carrying out field tests of structural constructions. This article discusses only a few (significant), of more than three dozen structural coatings designed and tested under the guidance of Professor of the Department of Building Dragan V.I. with nodal connections of the "BrSTU" system, as well as the goals and objectives of carrying out field tests [3 ... 8].

Methods

The analytical determination of forces and displacements in the elements of the structural construction of the "BrSTU" system for most of the designed objects was carried out using the "LIRA-Windows" program [3, 5, 6, 8]. The unit of the structural construction of the "BrSTU" system refers to complex combined joints in which each component makes a certain contribution to the overall stress-strain state of not only the nodal connection but also the structure as a whole. In this regard, structural calculations were performed in several stages. The first and second stages included calculations of the spatial structure of the structural slab taking into account the effect on it of static design loads and loads corresponding to full-scale tests, respectively. The calculation was carried out for a numerical model in which all the bar elements are hinged at the nodes taking into account the elastic stage of the construction material, i.e. as a hinge-rod system without taking into account the elastic compliance of the nodal elements. As a result of the static calculation, the maximum design forces that the elements of the structural slab are perceived were determined, their sections were selected with the appropriate geometric and stiffness characteristics and the expected values of the forces in the stubble elements of the structure and the movement of nodes at the points designated for experimental research were estimated.

At the third stage, the actual work of the nodal elements of the structural construction was simulated under the action of a static test load in the elastic stage of the work of the structure material. The flexibility of the structural slab nodes was taken into account by a decrease in the stiffness characteristics of the structure rods.

The last stage included the evaluation of the obtained numerical data on the stress-strain state of the structural slab by means of field tests under the action of a static test load.

Results and Discussion

Roof construction of the Summer Theater in the Park of Culture and Leisure in Brest. The structural construction of the roof was designed and built in 2006 and is made in the form of a cantilever roof slab with plan dimensions of 22.5x18 m with an orthogonal grid of chords, the nodes of the lower grid of which are shifted relative to the nodes of the upper one by half the size of the cell. The mesh cell for the upper and lower chords is 1.5x1.5 m and the height along the axes of the chords is 1.06 m. This object is the first in the practice of using a fundamentally new nodal element of the "BrSTU" system. The structural slab was used as a cantilever roof cover at the "Summer Theater in the Park of Culture and Leisure in Brest" (Fig. 1) and later found application as a typical project at other objects in the parks of culture of cities of the Republic of Belarus [2, 3 , 8].

The supporting nodes of the structural slab are rigidly coupled with the column heads of two types: solid - from pipes with a section of 273x7 mm and cross-cutting 4-branchs made of pipes with a section of 102x6 mm.

The nodes of the upper and lower chords of the structure are connected by braces. All rods of the supporting space system have the same nominal length and consist of electrically welded pipes, into the ends of which round nuts with M24 thread are welded. The rods are taken in two standard sizes: structural elements (chord and spacing) –76x4 mm, support braces on through columns - 102x4 mm.

The forces in the bars of the structural slab were determined using mechanical strain gauges. The indicators were installed on the most loaded elements of chords and braces of the structural slab, located symmetrically to the axis of the slab. To eliminate the influence of random factors and increase the reliability of the results obtained, duplication of the placement of indicators was carried out.

To observe the deformations of the structural plate (vertical displacements of nodes), two independent measurement methods were used together: linear and angular, which made it possible to obtain results with an accuracy of ± 1.0 mm (to eliminate errors in the form of inclination and displacement of the rail heel from the spherical node marker, changing instrument height, distance measurement errors, etc.).

The test was carried out according to two loading schemes: an asymmetric load acting on the cantilever part of the structural slab, and a full load evenly distributed over the slab surface. During loading with a full load, the structural plate was held for a day to take into account the creep of the bolted joints after which the load was gradually removed.

Fig. 1. General view of the structural slab in Brest.

The test load was created using concrete foundation blocks with a mass of 0.485 t, laid on wooden panels along the girders (Fig. 2) while the created level of the test nodal load was 2.495 kH, the achieved overload factor of the structure during testing was $Kq = 1.32$.

Fig.2. Testing the structural plate by static load.

The analysis of the structural slab operation was carried out both by the criterion of using the bearing capacity of all structural slab elements which was assessed using the expected coefficient of additional safety, and by the criterion of displacement deformations of the nodes of the upper belt lattice (Fig. 3).

Comparison of the results of theoretical calculations and a full-scale experiment in all the considered elements of the lattice of the structural plate shows that when the actual compliance of nodal joints is taken into account in the numerical model, there is a decrease in forces by 10–20% and deflections by 6.7%. This effect is explained not only by the redistribution of forces in a statically indeterminate system but also by the real value of the compliance of nodal joints in a structural slab under conditions of a complex stress state [4].

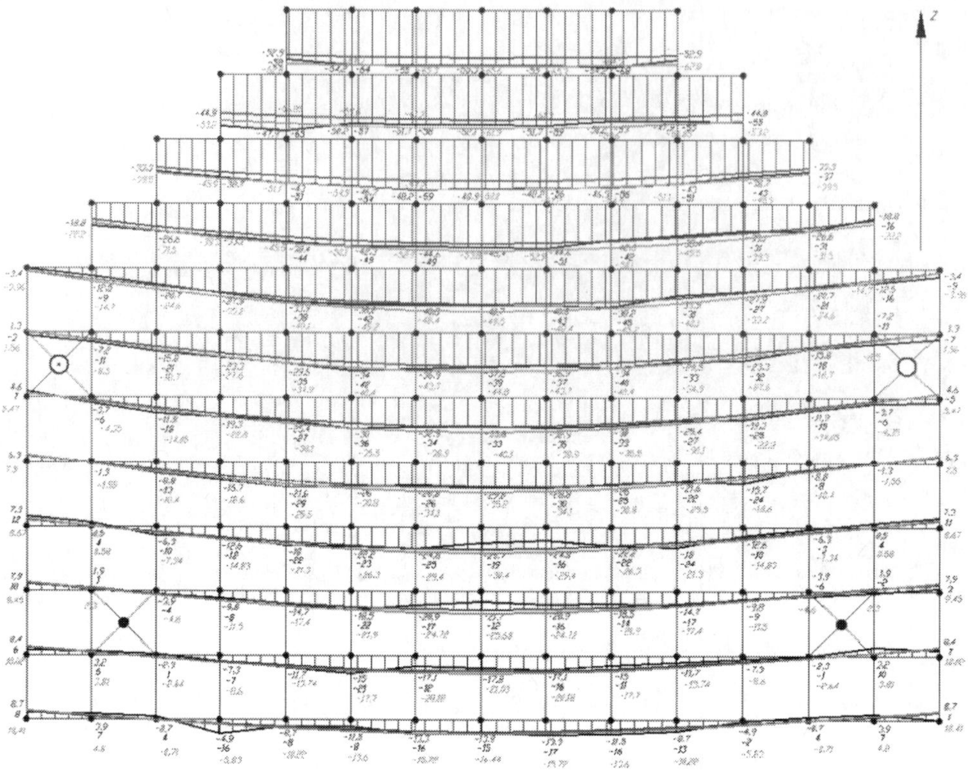

- theoretical deflection without considering units compliance;
- experimental deflection;
- theoretical deflection taking into account units compliance

Fig. 3. Experimental and calculated displacements of the nodes of the lower chord of the slab.

Ice arena in Pruzhany. Large supporting construction of the structural slab in the town of Pruzhany was designed in 2005 and has an orthogonal mesh of the upper and lower chords with cells of 3.0 x 3.0 m and a height of 3.0 m along the axes of the chords. The tilt angle of the structural board is 8 °. The supporting nodes of the structure are hingedly coupled with the column heads. The nodes of the upper and lower chords of the structure are connected by braces. The rods are connected into a spatial structure using a nodal element of the "BrSTU" system. The structural slab includes an adjustable tightening force (Fig. 4). This roof is typical for more than five objects designed and built in the Republic of Belarus [5].

To determine the vertical displacements of the nodes of the upper chord mesh of the structural slab after laying the profiled roofing the surface was leveled along a mesh consisting of 77 control points. As a result of leveling, the geometry of roof coating was obtained which was formed after the installation of the structure and the application of its own weight of the structure, purlins and profiled roofing. Re-leveling on the same grid was performed after applying the full test load.

When modeling the stress-strain state of the roofing, taking into account the operation of the purlins and the profiled roofing, the nature of the stress state changed somewhat. The most loaded were the rods of the middle panels of the upper chord mesh. Initially, maximum compression forces were set for the third panels from the ridge. This change is quite natural, since the roofing located in the ridge above the central panels was not included in the work.

The discrepancies in the values of forces obtained by numerical modeling and as a result of field tests did not differ significantly - the experimental forces amounted to 86.1% - 98.6% of the calculated forces. Analysis of the data showed that there was a redistribution of forces in the elements of the structural construction which is explained primarily by the high internal static uncertainty of the system as well as by some simplifications adopted in the compilation of the numerical model.

The theoretical tightening forces determined in both design models are higher than the measured forces. The inclusion of girders and profiled decking in the work allowed reducing the discrepancy in the values of the compared forces by 3.9% and 4.4%.

Fig.4. Ice arena in Pruzhany, general view and design.

The analysis of the deformed state of the roofing was carried out according to the diagrams of vertical displacements obtained for the nodal elements of the upper chord mesh while the theoretical displacements were determined based on the calculation in the numerical modeling of the roofing without including the girders and profiled roofing into the work of the decking. Actual displacements exceed theoretical ones by 1.3-1.6 times. There is a difference in the outlines of the diagrams in addition to differences in the magnitude of deflections. In this case, the maximum experimental deflections are not on the axis of symmetry but are shifted from it to the ends of the slap by two cells. The inclusion of purlins and gratings in the design model led to a decrease in the displacement values. The main reason for the excess of the actual deflections of constructions over the theoretical is the actual compliance of the nodal connections.

The difference between the actual stress-strain state of the roofing and the stress-strain states obtained in numerical models does not affect the bearing capacity of the roofing. High reliability of the roofing is provided by safety factors of at least 4, calculated for most of the most critical elements [5].

Roof construction of the Summer Amphitheater in Vitebsk. The roofing of the summer amphitheater was designed at the Department of Building Structures of the BrSTU in 2007. The combined system consists of a spatial rod cylindrical shell of a regular structure truncated by two inclined planes and supporting it with nine vertical arches (five lower and four upper), two inclined arches as well as two lattice columns of round tubes located behind the stands (Fig. 5).

Spatial rigidity of the roofing is provided by pinching the arches in the foundations, installing inclined arches and columns, joint work of the rod shell and all arches of the roofing. The span of the lower vertical arches is 120.014 m, the roofing width is variable and varies from 24.0 m on supports to 71.0 m in the middle part of the roofing.

The main role in creating architectural attractiveness is assigned to the structural shell which is based on the key elements of the "BrSTU" system. The height of the shell section is 2.51 m, the size of the cells of the upper chord mesh is 3.0 x 3.0 m, the lower chord mesh is 3.0 x 2.931 m. The construction solution of the structural shell of the roofing is given in more detail in [5, 6].

Testing of the structural shell of the roofing of the summer amphitheater building in Vitebsk was carried out in order to determine the actual operation and the possibility of normal operation of the roofing structures; revealing the actual stress-strain state of the roofing elements under load and determining the bearing capacity reserves under full standard load and checking the quality of manufacturing and installation of the roofing. The test load exceeded the standard value of the design load by 1.18 times (Fig. 5).

Fig.5. General view of the roof under test load.

From a comparison of the results of geodetic survey and the results of calculating a numerical model for a roofing loaded with a test load, an almost complete correspondence of the outlines of experimental and calculated (numerical) diagrams of vertical displacements of the nodes of the upper chord lattice was established (Fig. 6).

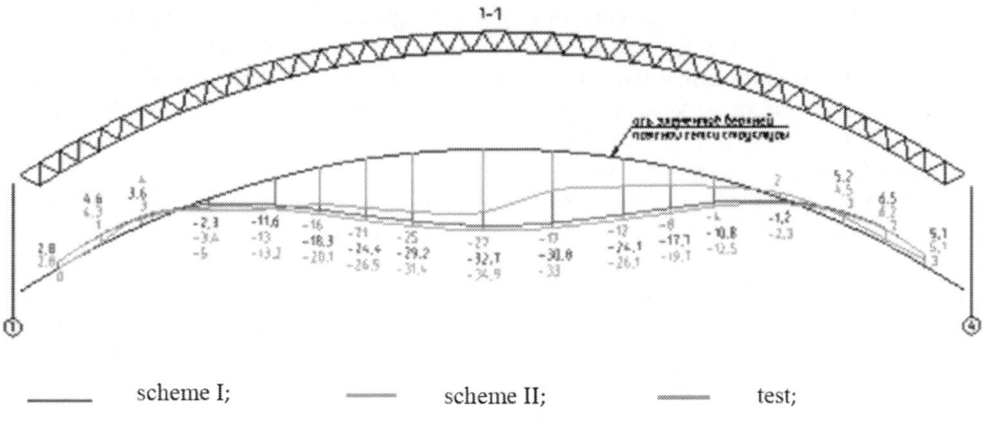

Fig. 6. Diagrams of vertical displacements of the nodes of the upper chord grid.

It should be noted that during the full-scale test, a significant effect of temperature on the deformation of the roofing was noted. A temperature change of 10 ° C caused additional displacements of the structural shell nodes by up to 20 mm.

Structural shell of roofing of the ice rink in Gomel. Roofing was designed at the Department of Building Structures of the UO BrSTU in 2007. The roofing has dimensions in plan of 49.092x57.82 m and is a combined spatial structure consisting of a structural shell and support frames with curved beams, in a longitudinal section close in outline to a sinusoid. The joining of the rods into a single structure is performed using the nodes of the "BrSTU" system (Fig. 7).

When designing the roofing some questions arose that went beyond the regulatory documents in force on the territory of the Republic of Belarus. The complexity of the geometrical surface of the shell did not allow to accept the normalized values and diagrams of wind and snow loads. Full-scale tests pursued the following goals: experimental assessment of the structural bearing capacity reserves; confirmation of the safety, functional suitability and durability of the structure; based on the data of the results of experimental studies, verification of the calculation model, to assess the accuracy of the calculation methodology and the accepted calculation assumptions; assessment of the effect of joint work of girders and structural construction on the stress-strain state of the roofing.

Fig. 7. General view of the ice rink roof in Gomel.

The structural shell rests with the lower nodes on the support platforms of the composite beams to reduce the construction height of the roof and the consumption of metal.

Numerical analysis of the stress-strain state of the structural slab was carried out taking into account the experimental load acting on it. The obtained numerical values of the forces with and without taking into account the malleability of the nodal connections have a sufficiently high convergence. The largest deviation in the results is no more than 5% which indicates the high accuracy of the numerical model adopted in the calculations. The experimental forces in the most loaded rods of the lower chord of the structural slab are less than the calculated values by 11.8%, and in the least loaded ones - by 9.2% (Fig. 8).

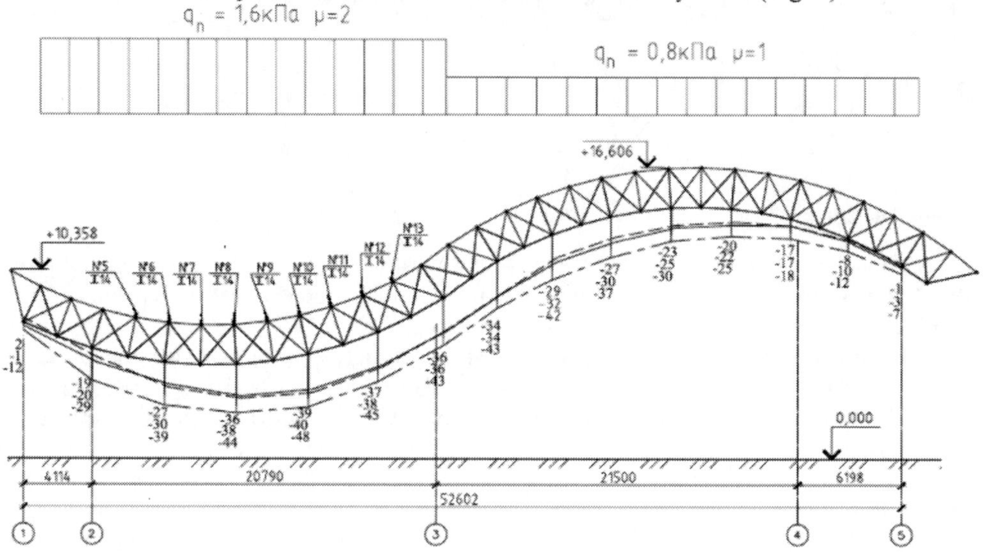

Fig. 8. Scheme of loading the roof with a test load and diagrams of vertical displacements of the nodes of the lower chord mesh.

As a result of full-scale tests, it was found that the diagrams of vertical displacements of the lower nodes of the structural shell under load, constructed from the results of measurements, repeat the outline of the diagrams obtained as a result of static calculation. The experimental displacement values are more consistent with the finite element numerical model that takes into account the work of the runs as part of the roof (Fig. 8).

Experimental studies have also confirmed that an increase in the rigidity of the upper chord of the structural shell by including the girders in the work of roofing leads to a redistribution of internal forces in the structure. Location of girders along the diagonal of the shell cell across the span made it possible to unload the central part of the upper belt of the roof on average up to 20%.

Conclusions

The specialists of the Department of Building Structures have accumulated significant experience in field testing of large-span combined structural constructions. As a result of full-scale tests of objects with key elements of the "BrSTU" system, an assessment of the bearing capacity and reliability of structural constructions was made on the basis of an experimental determination of their stress-strain state. The adopted computational models are verified against the background of the results of experimental studies to correct their computational schemes in numerical simulation.

Analysis of the stress-strain state of structural elements showed that under conditions of complex loading, the picture of the stress-strain state of the nodal joint changes sharply, namely, its work becomes tougher. Taking into account the fact that according to the design features of the assembly the structural plate and bolted joints are initially included in the work (fixed value of the interference) in the end a favorable situation develops for reducing the overall value of the nodal joint compliance. As a result, a redistribution of forces occurs in the structure which we observed during tests and which, ultimately, has a beneficial effect on the operation of the structure as a whole.

In the course of field tests of the structural construction with new nodes, it was found that the bearing capacity of the rod elements of the metal structural plate of the roof is ensured. The average value of the additional safety factor was higher than that established in the studies of the overload factor and amounted to at least 1.50, which allows us to assert a high degree of reliability of the operation of the nodal and rod elements as well as the structure as a whole.

References

1. EN 1990–2011. Eurokod 0. Basis of Structural Design (2011)
2. V.I. Dragan, A.B. Shuryn, M.M. Shalabyta, Promyshlennoe i grazhdanskoe stroitel'stvo [Industrial and Civil Engineering], **6**. Stress-Strain State of a Nodal Element of the «BrSTU» Structural Construction System. pp. 39–44 (2017)
3. M.M. Shalabyta, Napryazhenno-deformirovannoe sostoyanie uzla iz pologo shara novoj metalliche-skoj strukturnoj konstrukcii. Avtoref. dis. ... kand. tekhn. nauk: 05.23.01; BrGTU. 24 (2009)
4. V.I. Dragan, A.B. Shuryn, M.M. Shalabyta, Promyshlennoe i grazhdanskoe stroitel'stvo. **6**. Napryazhenno-deformirovannoe sostoyanie uzlovogo elementa strukturnoj konstrukcii sistemy «BrGTU», pp. 39–44 (2017)
5. A.B. Shuryn, M.M. Shalabyta, Budownictwo o zoptymalizowanym potencjale energetycznym: praca zbiorova. Issledovanie dejstvitel'noj raboty bol'sheproletnogo kombinirovan-nogo strukturnogo pokrytiya s uchetom podatlivosti uzlovyh soedinenij. pp. 290–296 (2010)
6. V.I. Dragan, M.M. Shalabyta, A.V. Muhin, A.B. Shuryn. I.V. Zinkevich. Novaya metallicheskaya strukturnaya konstrukciya sistemy «BrGTU». Opyt proektirovaniya, issledovanij i stroitel'stva v Respublike Belarus'. Promyshlennoe i grazhdanskoe stroitel'stvo v sovremennyh usloviyah : sb. nauch. tr. Mezhdunar. nauchn.-tekhn. konferencii, Moskva, 19–21 aprelya 2011. pp. 34–37 (2011)
7. V.I. Dragan, A.B. Shuryn, A.V. Muhin, M.M. Shalabyta, Kombinirovannaya strukturnaya obolochka pokrytiya ledovogo katka po ul. Golovackogo v g. Gomele. Perspektivy razvitiya novyh tekhnologij v stroitel'stve i pod-gotovke kadrov Respubliki Belarus' : sb. tr. XVIII mezhdunar. nauchn.-metod. seminara : v 2-h t. pod obshch. T.I, pp. 53–59 (2012)
8. M.M. Shalabyta, T.P., Shalobyta. Budownictwo o zoptymalizo-wanym potencjale energetycznym: praca zbiorova. Napryazhenno-deformirovannoe sostoyanie i metodika rascheta na proch-nost' uzla soedineniya sterzhnevyh elementov metallicheskih strukturnyh konstrukcij tipa «BrGTU», pp. 279–290 (2010)

Defects of steel crane beams and methods of their strengthening

Andrei Shuryn[1,*], *Anatoli* Mukhin[1] and *Alexandr* Bryantsev[2]

[1]Brest State Technical University, 224017, Moskovskaya 267, Brest, Belarus
[2] International Educational Corporation (KazGASA), 050043, Ryskulbekova 28, Almaty, Kazakhstan

Abstract. Crane structures are the most vulnerable in buildings and structures because of their early wear in comparison with other building structures. The wear of crane structures is because of the appearance and the development of fatigue damage nature.
The article presents the constructive solutions developed by the authors for strengthening the structures of steel crane beams with conventional and corrugated walls which allow to reduce the influence of the negative influence of installation, operation and technological processes on their operational durability.
The most dangerous installation defect is the displacement of the crane rail axis relative to the axis of the crane beams wall. Value of this shift is exceeded several times in the cases considered below. It is proposed to use crane beams with corrugated walls to solve this problem.
The authors have developed and implemented a project to restore the performance of steel crane beams by strengthening the walls of damaged beams without stopping production. Strengthening was performed on the outside of the eccentricities of the application of loads from crane rails by creating a truss structure parallel to the I-beam with chords of I-beams, which made it possible to significantly reduce the magnitude of local stresses from moments, the magnitude of transverse forces and consequently shear stresses, which are the main factors in the appearance and development of fatigue cracks in their walls.
Key words: crane structures, brake structures, steel crane beams, shear stresses.

Introduction

Crane structures are the most vulnerable in buildings and structures because of their early wear in comparison with other building structures. The term of their normal operation with direct participation in the production cycle is usually several times less than that of other structures. At the end of the service life, if reinforcement is impossible, replacement of damaged crane structures is required which is also not always possible because of the peculiarities of production processes.

The durability of building structures is significantly influenced by the features of crane effects, the presence of structural imperfections in the crane beams, as well as defects in

* Corresponding author: shuryn@mail.ru

their manufacture and installation. The wear of the crane structures is also because of the appearance and development of fatigue damage. Multiple changes in the stress-strain state of crane structures lead to damage which significantly reduces their durability.

The article presents the constructive solutions developed by the authors to strengthen the structures of steel crane beams with conventional and corrugated walls, which make it possible to reduce the influence of the negative influence of installation, operation and technological processes on their service life.

Methods

During the production of certain types of products because of imperfection of technological processes or their violations there are deposits of industrial dust on brake structures which negatively affect the durability of not only crane structures but also crane wheels (Fig. 1).

Fig. 1. Deposits of clinker on brake structures.

To prevent clinker accumulation on horizontal brake beams the authors recommend unloading them by installing holes in them. Work on the study of steel beams with walls with holes is currently underway in Kazakhstan. A round hole is possible at $\tau \leq 0.5R_s$, but with a diameter of not more than half the height of the wall while the hole should be located in the center relative to the height of the wall which are bordered by angle or sheet steel reinforced by paired vertical stiffeners. The following design schemes can be used in the case of high lateral loads (Fig. 2).

A numerical parametric study of steel beams with a triangular corrugation shape [1] included the analysis of 55 beam models of which 1 beam model without a hole, 27 models with three holes in the wall and 27 models with two holes with a diameter of $0.25h_w$, $0.5h_w$ and $0.75h_w$, spaced between the centers of holes 2d, 3d and 4d. Analysis of the operation of models of beams with three holes under a concentrated load and reinforced with edging and parallel stiffeners showed the efficiency of placing holes in the corrugated wall with a hole pitch of 2d and with a hole diameter of $0.25h_w$ and $0.5h_w$ and for two holes reinforced with edging and parallel stiffeners, hole placement efficiency with hole spacing 2d and 3d and hole diameters $0.25h_w$, $0.5h_w$ and $0.75h_w$.

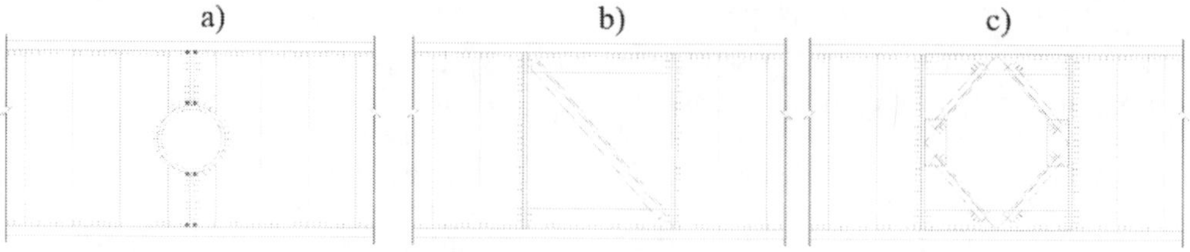

a) a hole reinforced with edging and stiffeners;

b) a hole reinforced with sheets and one brace from the corner;

c) a hole reinforced with a rhombic paired corner lattice.

Fig. 2. Constructive solutions for holes in corrugated walls.

In 2019, in the laboratory of KazGASA LLP (Almaty), on the load-bearing floor, 5 models of beams with corrugated walls (BGS-1,2,3,4,5) were tested, weakened by round technological holes with various reinforcement options holes [2]. Tests [3] were carried out according to a single-span scheme for lateral bending by a concentrated force applied in the middle of the span (Fig. 3) while the beams are based on the supports.

Fig. 3. General view of the installation for testing the BGS-4 beam.

Based on the research program and in accordance with the theory of optimal planning of experiments, 5 models were made, including 4 models with two round holes 0.5hw and a step between the centers of the holes 2d [1] and 1 model without holes. The length of all beams is assumed to be 4.2 m, the height of the walls is 0.6 m, the wall thickness is 2 mm, the width of the shelves is 0.15 m, the thickness is 6 mm. The effective step and high corrugation in accordance with the results of the third chapter of this dissertation work for a wall thickness of 2 mm were taken as 280x45 mm. In the course of the experiment on models with round bordered holes which are used in design practice for the passage of technological equipment within the height of beams the influence of various methods of reinforcing the holes on the bearing capacity and deformability of the models was investigated.

The reinforcement method described in SNiP RK 5.04-23-2020 "Steel structures" was chosen as the basis for reinforcing round holes, consisting of bordering the hole with sheet steel and installing paired vertical stiffeners on both sides of the wall (Fig. 4).

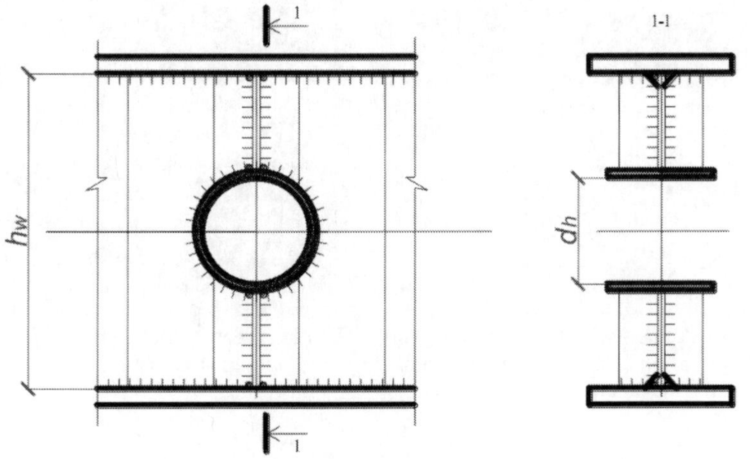

Fig. 4. Corrugated I-beam with a round hole bordered by sheet steel and paired vertical stiffeners.

This method, described in the regulatory documents of the Republic of Kazakhstan, does not contain specific data on the thickness of the metal for edging and stiffening ribs. Therefore, in the course of the study, it was determined that the most effective thickness of the edging and stiffeners for a hole with a diameter of 0.25hw and a step of 2d is a thickness of 2 to 4 mm; for a hole with a diameter 0.5hw and a step 2d, the thickness is from 4 to 6 mm; for a hole with a diameter 0.75hw and a step 2d, the thickness is from 6 to 8 mm.

Improved methods of reinforcing the holes were proposed during the study. Tests of beams together with finite element modeling allowed to draw conclusions about the effectiveness of improved methods of reinforcing circular holes [4, 5]. For the BGS-3 beam [4] - reinforcement of the hole by bordering together with the thickening of the wall in the area of the hole by 2tw and for the BGS-5 beam [5] - the inclusion of an external bend of the border into the existing method of strengthening holes with stiffeners in accordance with the requirements SNiP II-23. Recommendations have been developed for introducing the two proposed options for strengthening the holes of the BGS-3 and BGS-5 models into the corresponding codes of rules (SP) or building codes (SN) and, thus, allowing to expand the range of issues of practical application.

The most dangerous installation defect is the displacement of the crane rail axis relative to the axis of the crane beam wall. In the cases considered below, the value of this displacement is exceeded several times which led to the need to develop special design solutions of an unconventional nature which made it possible to reduce the effect of this installation defect.

It can be used crane beams with corrugated walls to solve this problem. Kazakhstan has accumulated a lot of experience in the use of such beams (Fig. 5, 6).

In 2007, Kazakhstan put into operation a range of welded I-beams of the usual type and with corrugated walls RDS RK 6.04-24-2006 [6], developed by Institute Proektstalkonstruktsiya LLP which corresponds in strength characteristics to rolling I-beams according to STO ASChM 20-93, manufactured in Russia.

To assess the actual bearing capacity of welded I-sections with cross-corrugated walls of the new assortment in the laboratory of Institute Proektstalkonstruktsiya LLP, strength and stiffness tests were carried out for the types of welded I-beams given in the assortment (beam - BGS, wide-flange - ShGS and column - KGS I-beams) on full-scale samples. The experiments performed on large-scale models revealed the efficiency of using corrugated walls and made it possible to increase the cyclic durability by changing the corrugation parameters [6].

Fig. 5. One-story steel frame of the industrial building, in which the roof girders, columns and crane beams are made of welded I-beams with corrugated walls. Seismicity 10 points. 2007.

Fig. 6. Crane beams in the main building of the plant in Almaty region. The height of the beams is 640 mm, the thickness of the corrugated walls is 6 mm. 2001.

Results and Discussion

Constructive solution for warehouse. The clinker warehouse of JSC Krasnoselskstroymaterialy is a one-storey one-span frame building with dimensions in the plan of 24x317 m. The frame is designed according to the frame-link scheme. The connection between the crossbar and the column and the columns with the foundations is rigid. The height to the lower chords of the trusses is 25.95 m. The elevation of the crane rail head is 22.42 m. The frame is mated at different elevations with other workshops associated with it by a single round-the-clock technological cycle. The building is equipped with three overhead cranes 16 ton lifting capacity with grabs.

The frame of the warehouse is formed of steel flat frames, including stepped columns, on which the steel crane beams with a span of 6 and 12 meters with brake beams are supported, steel trusses with a span of 24 m from rolled corners or bent-welded profiles. The trusses of the frame frames are rigidly connected to the columns, which are fixed in reinforced concrete foundations. Steel structures of the roof are united into a spatial system by a hard disk made of prefabricated ribbed reinforced concrete slabs, ties along the roof and along the columns.

Welded crane beams of the 1.426.2-3.1 series are used in the warehouse building, they were made of steel 09G2S-12-1 GOST 19281, with the exception of stiffening ribs, for which VSt3ps-7 is used in accordance with GOST 380. Crane beams with a span of 12 m have a cross section the upper belt is –400 × 14 mm, the lower belt –360 × 10 mm, the walls are –1390 × 12 mm (steel 09G2S-12-1 GOST 19281). Vertical stiffeners are installed along the length with a step of 1.5 m. Beams with a span of 6 m have a cross-section of the upper belt –400 × 14 mm, the lower belt –250 × 10 mm, walls –990 × 10 mm.

Technical expertise of steel crane beams. Examination and observation of the technical condition of the steel crane beams of the clinker warehouse of JSC "Krasnoselskstroimaterialy" have been carried out by the authors of the article since 2006 to the present. Violations of the requirements for the installation of crane structures were found. The survey results showed the presence of large eccentricities between the vertical axes of the rails and the axes of the walls of the crane beams, which should be attributed to defects in the installation of structures. The value of eccentricity for individual beams reached 70 mm, which significantly exceeds the limit value of deviations for the rail axes of 15 mm, provided for in TCP 45-5.04-49 [8]. The curvature of the crane girders in the horizontal plane reaches 30 mm, which also exceeds the limiting values f = 20 mm [9]. These defects, as well as the constant presence of clinker on the crane rails, cause wear of the crane wheels, damage to the crane rails, vertical and horizontal cracks in the walls of the crane beams which is accompanied in some cases by their destruction.

The features of the facility operation include the constant presence of a dust suspension of clinker ingredients in the air, due to the peculiarities of the technological process. These operations lead to dust deposits on all structural elements of the building. The height of dust deposits in some areas of the brake beams after the introduction of the dry cement production line in 2012 reaches 1.5 m (Fig. 1). This leads to the movement of the wheels of overhead cranes over a thin layer of clinker and distortions of the bridge of the crane during its movement along the building.

Defects and damages were found both in the frame structure and in the crane beams and their brake structures; significant wear of the crane wheel flanges and crane rails was noted. The most common defects in crane beams were fatigue cracks in their walls, which were found in 40% of crane beams. Cracks were found only at the ribs of the walls of the crane beams (Fig. 7).

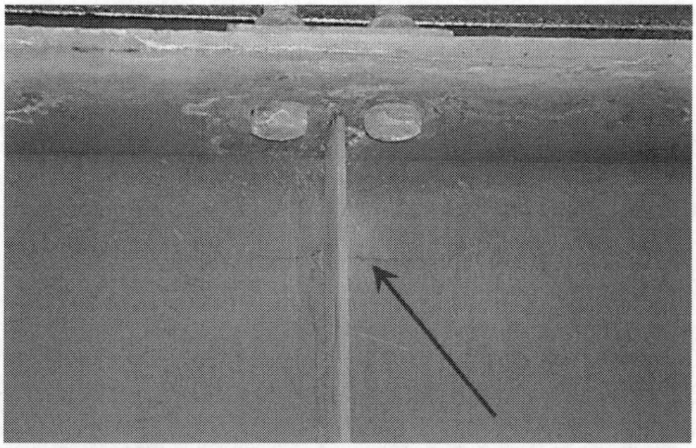

Fig. 7. Crack in the wall of the crane beam.

Defects in the form of cracks in the welded seams of brake structures, destruction of the beams-to-column attachment points were found in some beams. In accordance with the requirements [8], defects and damage of a significant and critical nature [10] in crane structures were eliminated by means of reinforcement.

During the technical examination of the crane beams carried out by the authors in 2009 and 2013 an increase in the number of damages was found some of which were of a critical nature. Cracks in the walls of the crane beams were found in 52% of the beams. New defects appeared in the form of local fractures of welded joints between brake structures and the upper flanges of crane beams (this defect was noted in 25% of crane beams).

It should be noted that when designing the clinker warehouse in accordance with GOST 25546-82 the operating mode of the cranes was taken as particularly difficult. In accordance with GOST 34017-2016, these overhead cranes have a utilization class of at least U7, the Q2 load class is light and, therefore, the A6 operating mode.

Verification calculations of crane structures carried out in accordance with the requirements of SNiP II-23 made it possible to establish that the main reason for the appearance of fatigue cracks in the walls was the presence of unacceptable eccentricities between the rail axes and the wall axes. The massive appearance of cracks in the walls of the crane beams was facilitated by the increased deformability of the frames caused by the lack of an assembly connection of the I-beams of the upper and lower parts of the individual columns of the frame along the A axis which also refers to installation defects of the frame structures.

Strengthening of the crane beams along the A axis. After 18 years of operation, fatigue cracks in the walls no more than 200 mm long and symmetrical with respect to the stiffeners were observed in 40% of the crane beams. It should be noted that the replacement of defective crane beams was immediately ruled out due to the peculiarities of the technological cycle of cement production. And the reinforcement of the walls by inclined lamellas led to the appearance of cracks in the welded seams of the lamellas in a period of 1–2 years after reinforcement.

To prevent clinker accumulation on horizontal brake beams, the authors recommended their unloading by arranging rectangular holes in them with edging and filling from meshes.

The authors developed and implemented a project to restore their operability by strengthening the walls of damaged steel beams without stopping production which made it possible to significantly reduce the stresses in walls of beams from the moment caused by the eccentricity of load application from the crane wheels. Reinforcement of the crane beams is made on the outside of the eccentricities of the application of loads from the crane rails by creating a truss structure parallel to the I-beam (Fig. 8). The truss belts are made of I-beams, the diagonal lattice is made of steel bent closed welded square profiles (GOST 30245).

The constructive solution of reinforcement made it possible to significantly reduce the magnitude of local stresses from moments, the magnitude of transverse forces and, consequently, shear stresses, which are the main factors in the appearance and development of fatigue cracks in their walls [7]. When the reinforced beams were re-examined along the A axis in 2017 after 8 years of operation, no new fatigue cracks were found in their walls.

Unloading of brake structures along the B axis. 25 years after commissioning, in some crane beams of the clinker workshop along the B axis, frequent occurrence and rapid development of fatigue cracks began to be observed, followed by merging in adjacent compartments. Strengthening the walls of the crane beams with inclined and vertical overlays did not give a significant effect. An increase in the number of defects and damages in the brake structure of crane beams and their attachment points to the frame columns was also found.

Fig. 8. General view of reinforced steel crane beams along the A axis with spans of 12 and 6 m.

Based on the research carried out by the authors and to prevent clinker accumulation on horizontal brake beams with spans of 6 and 12 m the authors have developed a project that provides for making rectangular holes in them covered with a mesh of expanded metal sheets of PV 510 grade according to GOST 8706 or a steel mesh round hot-rolled d = 10 mm in accordance with GOST 2590 with a cell of 90x80 mm which is welded to frames made of rolled corners L50x5 in accordance with GOST 8509 (Fig. 9).

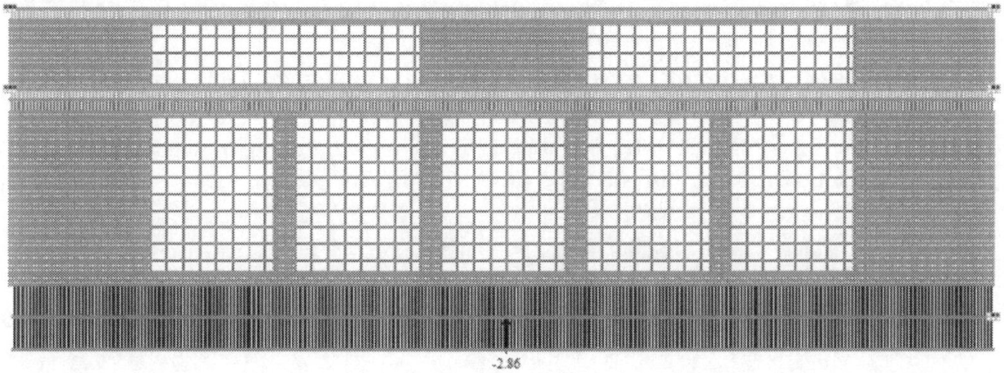

Fig. 9. Finite-element model of a brake beam with a span of 6 m with holes bordered by corners L50x5 in accordance with GOST 8509 and a mesh.

It has been established that the edging of holes filled from meshes or expanded steel in brake structures makes it possible to increase their endurance due to a decrease in horizontal displacements.

Conclusions

1. The strengthening of the steel crane beams developed by the authors by creating a truss structure parallel to the I-beam made it possible to significantly reduce the magnitude of local stresses from moments, the magnitude of transverse forces and, consequently, shear stresses which are the main factors in the appearance and development of fatigue cracks in their walls. When the reinforced beams were re-examined along the A axis in 2017 after 8 years of operation, no new fatigue cracks were found in their walls.
2. The authors have developed a project for unloading the brake beams by making bordered rectangular holes in them filled from meshes. The performed analysis of the stress-strain state of such a structure made it possible to establish:
   - making holes in brake structures without edging increases the probability of fatigue cracks appearing in them;
   - edging of holes in brake structures allows to increase their endurance due to a decrease in horizontal displacements by 2.5 times for BK3 beams 6 m long and 5 times for BK1 beams 12 m long;
   - a grid of cross bars made of round hot-rolled steel d = 10 mm in accordance with GOST 2590 with a cell of 90x80 mm, welded to frames made of rolled corners L50x5 in accordance with GOST 8509, allowed to reduce horizontal deformations of beams by 1.25 times compared with a grid of expanded metal sheets grade PV 510 in accordance with GOST 8706.

References

1. A.A. Bryantsev, V.E. Absimetov, V.V. Lalin, Magazine of Civil Engineering. **87(3)**. The effect of perforations on the deformability of welded beam with corrugated webs. pp. 18–34 (2019)
2. A.A. Bryantsev, Vestnik Kazntu. Almaty. **5.** Full-scale tests of welded corrugated beams with round holes. Vestnik Kazntu. pp. 261–268 (2019)

3. A.A. Bryantsev, V.E. Absimetov, Environmental and Construction Engineering. **70**. Laboratory tests of welded corrugated beams with perforations. Lecture Notes in Civil Engineering. Proceedings of EECE 2019. Energy, pp. 55–67 (2020)
4. Patent 4318 Republic of Kazakhstan, I-beam with reinforced holes / Bryancev A.A. – № 2019/0422.2, publ. 18.09.2019
5. Patent 4540 Republic of Kazakhstan, I-beam with reinforced holes / Bryancev A.A. – № 2019/0774.2, publ. 11.12.2019
6. RDS RK 5.04-24-2006. Mix of fabricated i-beam sections of regular type and with corrugated walls meeting strength properties of rolling sections. – TOO «Institut Proektstal'konstrukciya». **20** (2007)
7. K.K. Nezhdanov, A.S. Lashtankin, I.N. Gar'kin, Structural Mechanics and Analysis of Constructions. Novosibirsk. **3**. Teams crane beams of rolled profiles, pp. 69–74 (2013)
8. TKP 45-5.04-49-2007. Steel designs. Inspection and diagnostics of a technical condition. Ministry of Architecture and Construction of the Republic of Belarus. **127**. (2008)
9. Guidelines for refurbishment of crane structures. USSR Ministry of Metallurgy. **118** (1991)
10. TKP 45-1.04-305-2016. Technical condition and maintenance of buildings and facilities Basic requirements. Ministry of Architecture and Construction of the Republic of Belarus. **107** (2017)

Shear resistance of expensive concrete beams reinforced with FRP longitudinal bars

Viktar Tur[1*], *Aliaksandr* Varabei[1]

[1] Brest State Technical University, 224017, Moskovskaya 267, Brest, Belarus

Abstract. This paper presents the results of testing beams reinforced with FRP in combination with expansive concrete. Failure modes under uniformly distributed loading and ultimate forces recorded in tests are described. The results of comparison of shear resistance models against the background of experimental data are presented.
Keywords: Expensive concrete, FRP reinforcement, Shear resistance, Uniformly distributed load, Beams,

Introduction

Recently, concrete elements reinforced with fiber reinforced polymers (FRP) bars widespread in the building industry. This type of FRP-reinforcement has become alternatives to conventional steel reinforcement for concrete structures owing to their non-corrosive and non-magnetic properties [1]. Classification, main physical and mechanical properties of the FRP reinforcement are presented in [2]. It should be pointed, that despite some advantages over steel bars utilization of the FRP bars have some limitations, mainly in case of the bending elements. Concrete members reinforced longitudinally with FRP bars develop wider and deeper cracks than those reinforced with steel due mainly to the relative low elastic modulus of FRPs [3]. As shown in numerous publications [4, 5] wider cracks decrease the shear resistance contributions from aggregate interlock and residual tensile stresses, where as deeper cracks reduce the shear resistance contribution from the uncracked concrete in compression. Additionally, owing to the relatively wider cracks and small transverse strength of FRP bars, dowel action contribution to shear resistance can be very small compared with that of steel reinforcement [6]. Hence, the overall shear resistance of concrete members (elements) reinforced with longitudinal FRP bars is lower than that of concrete members reinforced with steel reinforcement.

The main practical difficult that is faced during such construction is the stirrups. It is seen that most of the FRP reinforced beams make use of steel stirrups and minimum cover thickness is provided, which nullifies the possibility of overall cross section reduction. Due to the difference in mechanical properties in between FRP bars and steel bars, the failure mode of FRP-reinforced concrete beams is different from that of RC-beams [6]. For FRP RC beams without stirrups, shear failure generally occurs in association with the formation of one or more diagonal cracks, which for simplicity are assumed to form linearly as suggested by Hoang and Jensen [7], Hong and Nelsen [8], Jensen et al [9]. The applied

* Corresponding author: profturvic@gmail.com

shear stresses in a cracked reinforced concrete member without transverse reinforcement are resisted by various shear mechanisms. Most of the shear predictions models and design procedures and standards assume that the shear resistance mechanisms for FRP reinforced member will contribute in a similar manner to the nominal shear capacity of concrete members reinforced with steel [3]. The ACI Committee-445 [2] report identified that shear transfer mechanisms mainly depend on the concrete strength (f_c) and the depth of uncracked zone which is a function of the longitudinal reinforcement properties. Also, arch action occurs in deep beams or members in which shear to depth ratio (*a/d*) is less than 2,5.

The excessive amount of FRP needed to resist shear could be both costly and likely to create reinforcement congestion problems. The following main parameters were considered to affect the shear strength of FRP-reinforced concrete members such as, the concrete compressive strength (f_c), beam width (b_w), effective depth (d), shear span-to-depth ratio (a/d), reinforcement ratio (ρ_f), axial stiffness of reinforcing bars ($\rho_f \cdot E_f$).

One of the possible solutions to avoid shear failure of such type beam is prestressing of the FRP bars. The physico-chemical method of FRP bars pretensioning is considered as an alternative to traditional mechanical method that allows to achieve an effective volumetric stress-strain state in the prestressed member, doesn't need for special devices and anchorage systems as well as being a "gradual and soft" pretensioning process, allows to avoid excessive relaxation strains development in reinforcement. It is theoretically assumed that achieved in the self-stressed FRP reinforced beams initial volumetric stress-strain state has a positive influence on the member crack resistance moment increasing, limitation of exceeding deformation (deflections) and crack opening development [10].

This paper reports the experimental study carried out to determine the behaviour of FRP reinforced beams under uniformly distributed load, verify and compare various proposed shear resistance models of FRP reinforced members based on obtained test results.

1 Experimental investigation

Present study investigates the behaviour of members reinforced with FRP (two types GFRP and CFRP) in combination with expansive (self-stressing) concrete. The main aim for incorporating expansive concrete into FRP reinforced beams is to explore the possibility of increasing of shear resistance of members.

Program of experimental investigations listed in Table 1. Experimental studies were carried out on two series of rectangular cross section self-stressing concrete beams reinforced with GFRP and CFRP bars. All investigated beams had rectangular cross-section b x h=120 x 300 mm and length l=3000 mm. Beams were reinforced with GFRP (Maperod G) and CFRP (Maperod C) bars. Geometry, reinforcement arrangement and basic parameters of beams shown in Figure 1.

Program of experimental investigations composed in such way that adopted amount of FRP-reinforcement (reinforcement ratio) provided near the equal value of ultimate load under shear failure.

Main mechanical parameters of two types of utilized of FRP bars listed in Table 2.

Expansive concrete mix nominal composition per 1 m^3 are listed in Table 3. Average values of the self-stressing concrete characteristics are listed in Table 1.

Demolding of the specimens was carried out the next day after casting, when self-stressing concrete achieved compressive strength 11 MPa at average. Immediately after demolding, specimens were immersed in water and cured there until stabilization of self-stressing concrete expansion was reached. And then self-stressed beams were tested under static loading with monotonically increasing uniformly distributed load (see Figures 2–3).

Table 1. Program of investigations

Series	Unit code	$f_{c,cube}$, MPa	$f_{cm,\ prism}$, MPa	$f_{cm,\ cyl}$, MPa	E_{cm}, GPa	Self-stresses, $f_{CE,k}$, MPa	Reinforcement		
							Quanti-ty	A_{frp}, mm²	ρ_l, %
I	БІ-1СД18	7,6	51,1	7,2	31,7	2,55	2Ø10 GFRP	142,5	0,44
	БІ-2СД18						2Ø10 GFRP	142,5	0,44
II	БІІ-3СД12	77,1	76,5	56,1	43,3	0,45	2Ø10 GFRP	142,5	0,44
	БІІ-4УД12						Ø10 CFRP	73,9	0,23

Notes:
$f_{c,cube}$, – cube compressive strength at 28 days;
$f_{cm,\ prizma}$, – mean concrete compressive strength based on prisms (100x100x400 mm);
$f_{cm,\ cyl}$, – mean concrete compressive strength based on cylinders (Ø150, h=300 mm);
$f_{CE,k}$ – self-stresses grade based on prisms (100x100x400 mm);
A_{frp} – area of FRP;
ρ_l – longitudinal reinforcement ratio.

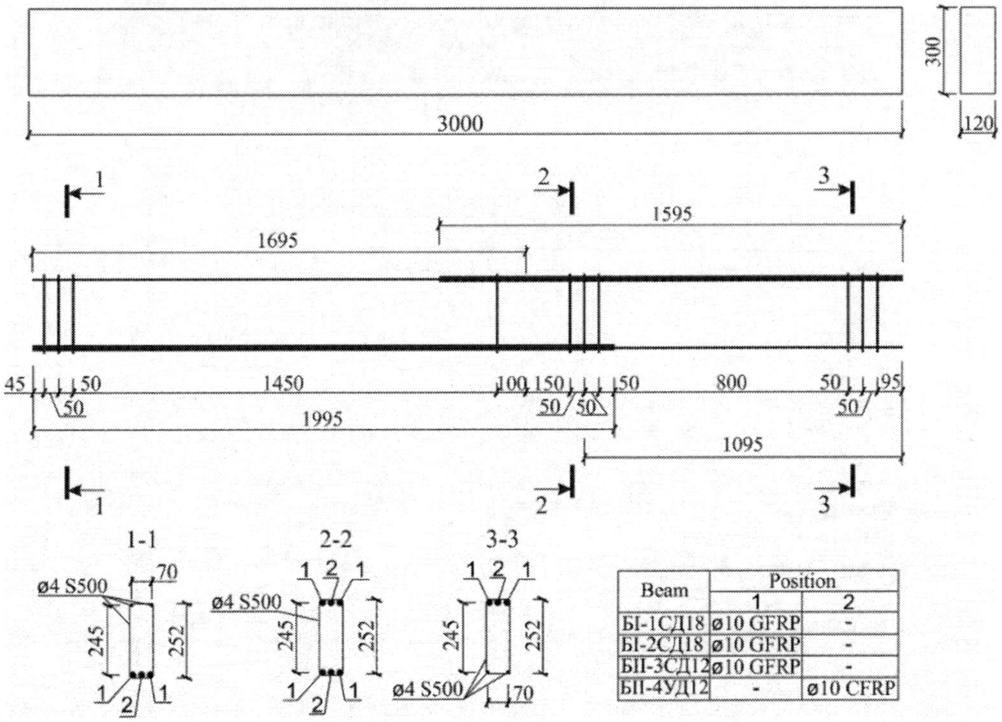

Beam	Position	
	1	2
БІ-1СД18	ø10 GFRP	-
БІ-2СД18	ø10 GFRP	-
БІІ-3СД12	ø10 GFRP	-
БІІ-4УД12	-	ø10 CFRP

Fig.1. Geometry, reinforcement arrangement and parameters of tested beams

Table 2. FRP reinforcement mechanical parameters

Class	Diameter Ø, mm	Strength, MPa	Modulus of elasticity $E_f x10^3$, MPa	Ultimate strain, %
Maperod G	10	760	40,8	2
Maperod C	10	2000	155	1,5

Table 3. Expansive concrete composition

Series	1 $м^3$ in dry conditions, kg					
	Cement	Denka CSA	Fine aggregate	Coarse aggregate	Water, l	Stachement 2010
I	510	90	600	960	240	-
II	410	50	805	990	175	7,0

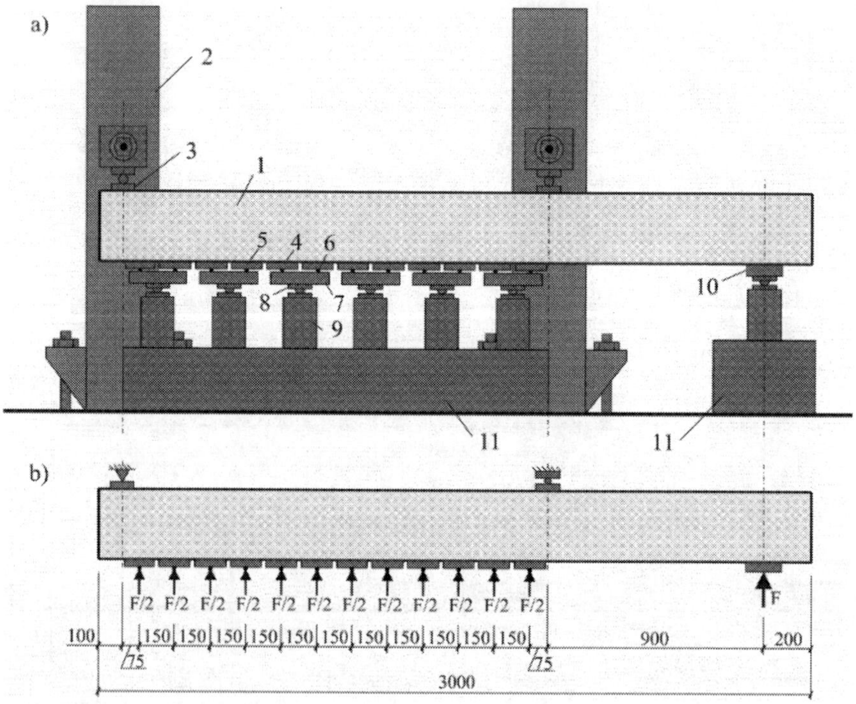

1 – experimental beam;
2 – 2 U-shaped steel frames connected to the strong floor;
3 – 4 steel supports (100x200x30 mm);
4 – 13 rubber bearings;
5 – 12 steel plates (130x130x30 mm);
6 – 12 steel bars (Ø14 mm);
7 – steel plates (245x155x50 mm);
8 – 7 steel hinges;
9 – 7 hydraulic jacks (250 kN);
10 – steel plates (150x150x40 mm);
11 – steel girder.

Fig.2. Test setup

Fig.3. General view of test setup

The following values of self-stresses in concrete on the depth of the cross-section gravity center to the self-stressing concrete expansion stabilization was obtained for beams:

(1) БІ-1СД18 – 0,62 MPa (span) and 0,79 MPa (cantilever part);

(2) БІ-2СД18 – 0,78 MPa (span) and 0,89 MPa (cantilever part);

(3) БІІ-3СД12 and БІІ-4УД12 – shrinkage compensation only.

As follows from the test results the first flexural cracks in the span for all series of beams developed under load equal 0,4..0,5q_u. It should be pointed out that for the beams of the series II (БІІ-3СД12 and БІІ-4УД12) in which self-stresses absent, the flexural cracks developed not only in the span but in the cantilever (see Figure 4). Flexural cracks were distributed in near the same distance along the length of the beam span. Depth of the cracks observed from 50% to 80% of the height of the cross-section. The basic difference between failure mode for beams series I and series II was following: for beams series I observed shear failure mode in the span but for beams series II on the cantilever. Diagonal cracks were developed at the distance $\geq d$ from support and the inclination angle was registered equal to $\theta \approx 60..70°$.

The results obtained within loading of self-stressed beams are presented in Tables 4–5 and shown in Figures 5-6. Observed cracking patterns and failure mode of the tested beams are illustrated in Figure 4.

As it can be seen from the static loading results listed in Table 1, initial volumetric stress-strain state obtained on the concrete self-stressing stage had a positive influence on the crack resistance moment increasing and on the reducing of the deflections (Fig. 4) as well crack opening.

The Table 4 presents the parameters of the observed diagonal cracks.

Most of the shear design incorporated in codes for the FRP reinforced concrete members have focused on modifying shear equations of steel reinforced concrete beams (see Table 5).

The shear resistance was calculated for various models based on the expressions given in [11–13].

2 Results and Discussion

All results obtained from experimental studies and predicted by the models from codes [11-13] are listed in Table 5. It is observed that the shear resistance predicted by prEN 1992-1-1 [11] is closer to the experimental values. However, it is found that the EN 1992-1-1 [12] and *fib* MC 2010 [13] under predicts the results compared to experimentally observed values. These will be discussed in detail in the further investigations.

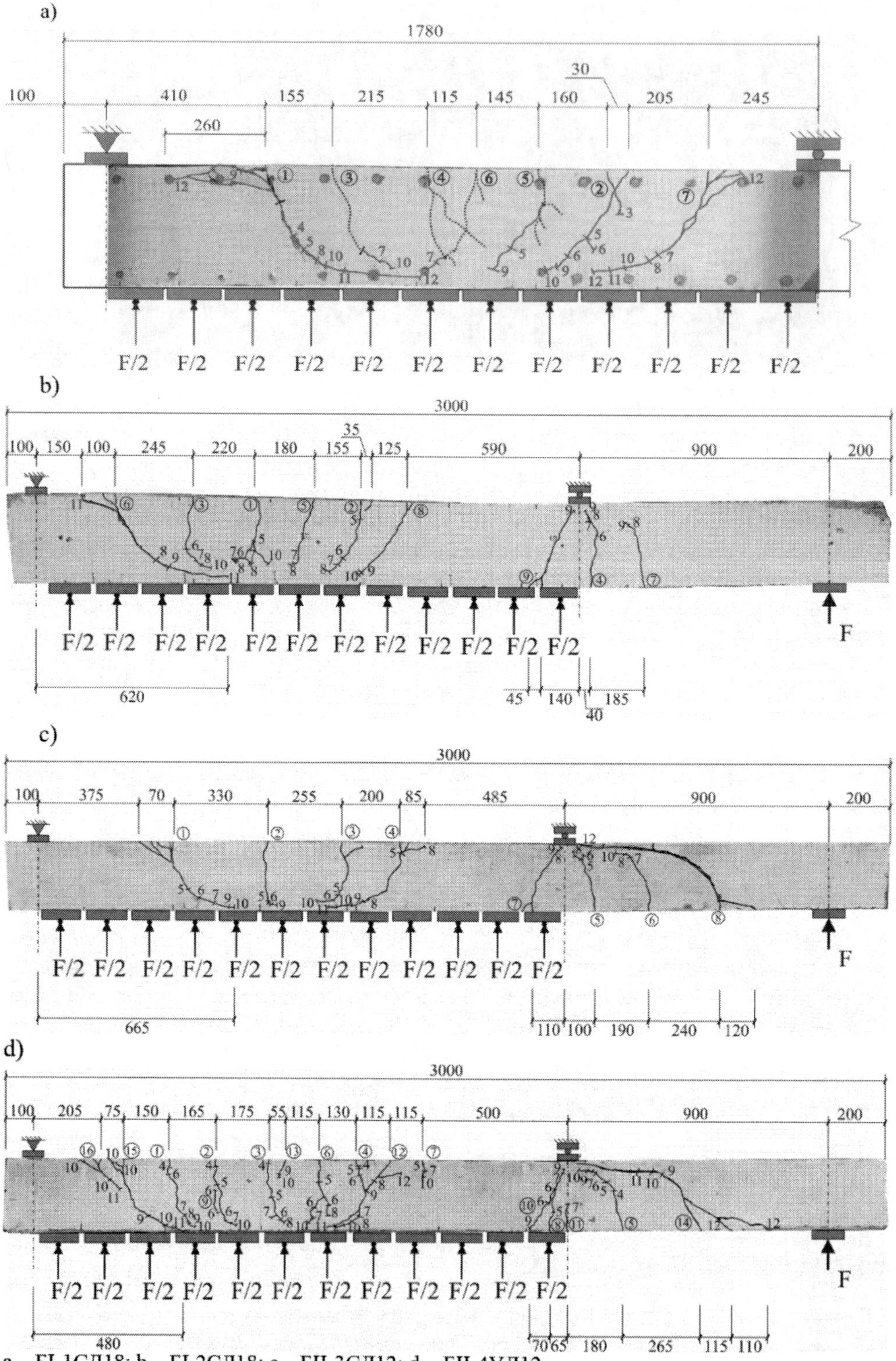

a – БІ-1СД18; b – БІ-2СД18; c – БП-3СД12; d – БП-4УД12

Fig.4. Cracks patterns for tested beams

 https://doi.org/10.1051/e3sconf/202021202017

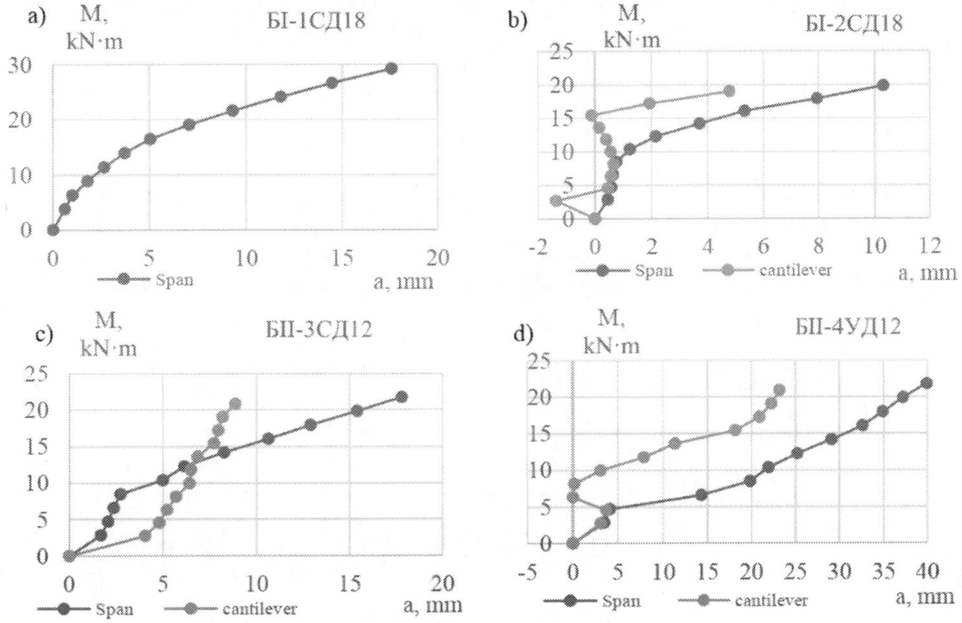

a – БІ-1СД18; b – БІ-2СД18; c – БІІ-3СД12; d – БІІ-4УД12

Fig.5. Relation “moment – deflection” for tested beams

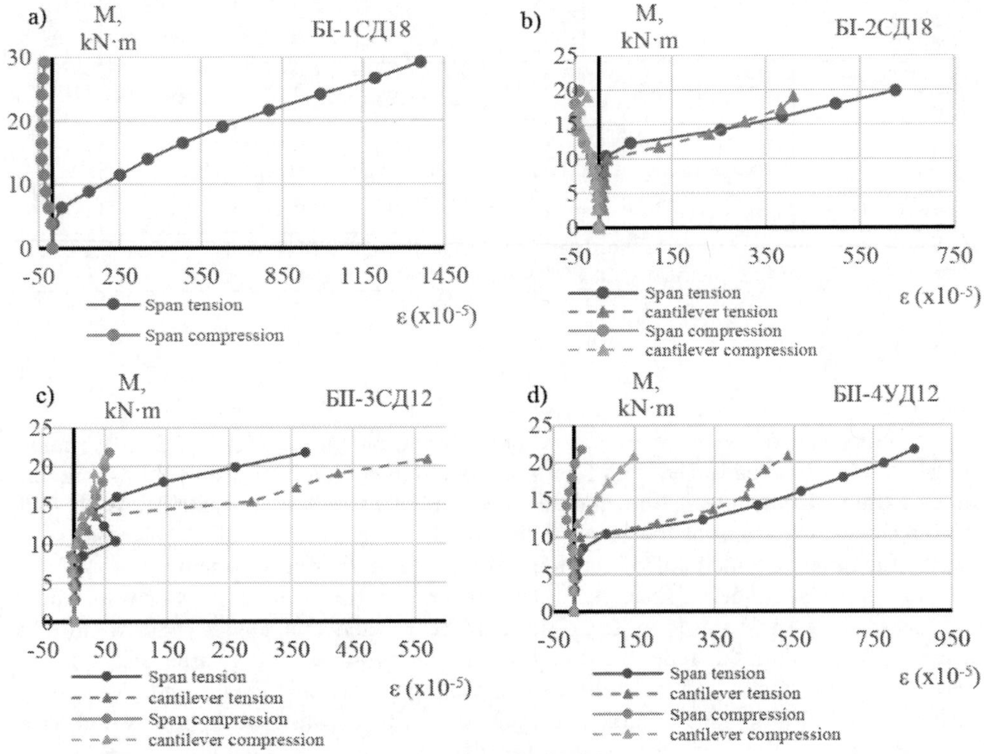

a – БІ-1СД18; b – БІ-2СД18; c – БІІ-3СД12; d – БІІ-4УД12

Fig.6. Relation “moment – strains” for tested beams

Table 4. Parameters of the diagonal cracks

Unit code	№ critical crack (see Fig. 4)	l_0, mm	l_{sp}, mm	$\frac{M}{V \cdot d}$	max w_k, mm	Inclination angle critical crack, θ°
БI-1СД18 (без нагружении консоли)	1	410	370	2,30	2,5	68
БI-2СД18	6	250	370	1,37	1,2	65
БII-3СД12	8	530	530	1,37	-	66
БII-4УД12	14	445	445	1,69	0,7	61

Notes:
l_0 – distance from support to critical diagonal crack. For beam series I – distance from support A, series II – from support B.
l_{sp} – length of the critical crack along beam axis;
$\frac{M}{V \cdot d}$ – ration in point of critical crack;
w_k – max. critical crack width.

Table 5. Parameters of V_{theo}/V_{exp} for the different design equations

Until code	V_{exp}, kN	V_{theo}, (EN 2), kN	V_{theo}/V_{exp}	V_{theo}, (MC 2010), kN	V_{theo}/V_{exp}	V_{theo}, (prEN 2), kN	V_{theo}/V_{exp}
БI-1СД18 (without loaded cantilever)	$\frac{50,2}{62,8}$	31,1	0.49	16,4	0,64	34,6	0,69
БI-2СД18	$\frac{34,6}{46,2}$	33,1	0,71	17,6	0,48	34,8	1,01
БII-3СД12	$\frac{25,2}{25,2}$	35,7	1,42	17,7	0,71	33,3	1,32
БII-4УД12	$\frac{25,2}{25,2}$	27,9	1,11	22,9	0,91	25,7	1,02

Note. Under the line – value of the shear force at the distance *d* from support, below the line – value of the shear force at the distance 0,5*d* from support (according to EN 1992-1-1).

Conclusion

The following conclusions can be drawn from the results of the present study:

– the mechanical model of shear resistance based on MCFT (*fib* MC 2010) cannot be applied to elements reinforced with fiberglass bars (GFRP), since this model has limitations on the strain of concrete in the middle of the element section ($\varepsilon_x \leq 0.003$). With such a reinforcement, the strain at the level of the middle of the section of the element will be greater due to the low modulus of elasticity of the fiberglass reinforcement;

– the empirical model of shear resistance presented in EN 1992-1-1 also has a limitation that does not allow its use in the design of beam elements loaded with a uniformly distributed load. With this type of loading, where the transverse force varies along the span, a critical inclined crack forms at a distance approximately equal to ≈*d* from the support, the design section in this model is taken at a distance of 0.5*d* from the support, which in turn will underestimate the shear resistance of the element;

– mechanical model CSCT (prEN 1992-1-1) has a fairly high level of accuracy. However, it should be noted that this model is not without its drawbacks.

– on the basis of studies [14-17], in rigid beams (at *L/d* <10), failure occurs when fracture of concrete above the tip of a critical crack, which was not observed when testing beams reinforced with polymer composites (see Figure 4). A detailed theoretical analysis of the results obtained is required.

References

1. ACI Committee 440 *Guide for the design and construction of structural concrete reinforced with FRP bars*. (2006)
2. ACI Committee 445, ACI J. Pr., **70** (1973)
3. R. Machial, M. S. Alam, A. Rteil, M. and Str., **45** (2001)
4. A. Ghowdhury, I. Mashfiqul, J. of C. and Env. Eng., doi: 10.4172/2165-784x.1000186. (2015)
5. T. Zhang, J. Oehlers, P. Visintin, J. of Comp. C., **18** (2014)
6. S. Lee, C. Lee, Eng. Str., **61**, 99 (2014)
7. L. C. Hoang, U. G. Jensen, M. of Concr. Res., **62**, 625 (2010)
8. L. C. Hoang, M. P. Nielsen, Cem. and Concr. Comp., **20**, 437 (1998)
9. U. G. Jensen, L. C. Hoang, M. of Concr. Res., **61**, 691 (2009)
10. V. Semianiuk, V. Tur, S. St. Phen., **292**, 230, doi:10.4028/ www.scientific.net/ ssp.292.230. (2018)
11. European Committee for Standardization. *Eurocode 2: Design of Concrete Structures, Part 1-1: General rules and rules for buildings*, 293 (2018)
12. *fib Model Code for Concrete Structures 2010*. – Ernst and Sohn, 389 (2013)
13. European Committee for Standardization. *Eurocode 2: Design of Concrete Structures, Part 1: General rules, rules for buildings, bridges and civil engineering structures*. Draft for committee works, 293 (2018)
14. N. A. Dassow, *Msc thesis. Austin (USA): The University of Texas*. 117 (2014)
15. A. Caldentey Perez, ACI Str. J., **109**, 595 (2012)
16. F. Cavagnis, *PhD thesis, Structural concrete laboratory (IBETON), École polytechnique fédérale de Lausanne (EPFL), Lausanne, Switzerland*. 201 (2017)
17. B. A. Podgorniak-Stanik, *Msc. Thesis. Toronto (Canada): The University of Toronto*. 711 (1998)

Shear resistance of self-stressing concrete elements expanded under 2D restraint conditions

Viktar Tur[1], and *Volha* Sannikava[1,*]

[1] Brest State Technical University, 224017, Moskovskaya 267, Brest, Belarus

Abstract. An analytical design model for estimation the shear resistance of plane self-stressing concrete elements with orthogonal reinforcement is presented. The design model is based on the strut-and-tie model concept and takes into consideration the restrained expansion strains and self-stresses as a result of self-stressing concrete expansion. The results of theoretical investigations were finally compared to the experimental ones showing the suitability of proposed design method.
Keywords: shear resistance, plane element, expansive concrete, design model.

1 Introduction

The usage of expansive concrete for the reinforcement prestressing during construction of monolithic structures reveals to be an efficient manner to cover both the effectiveness in compensating early-age shrinkage and the advance on durability and serviceability characteristics. One of the key problems during the design of monolithic self-stressing concrete structures concerns the assessment of shear resistance of ones. It's noticeable that in some cases (deep beams, corbels, beam ends, diaphragms) the shear design is essential for setting the geometry and reinforcement arrangement of self-stressing concrete elements.

Strut-and-tie model is a rational approach based on the traditional truss model [1, 2] and used for design of reinforced concrete structures in cases when shear stresses distribute nonuniformly over the depth of concrete section. According to the strut-and-tie model concept [3-9], reinforced concrete element represents an idealized system of compressive struts, tensile ties and nodal regions. Assessment of the resistance of reinforced concrete structures follows the requirements provided in actual codes [10-14]. When performing a design of self-stressing concrete elements, strut-and-tie model should take into account the presence of initial strains and stresses caused by prestressing of reinforcement during the concrete expansion.

*Corresponding author: volhasannikava@list.ru

In this paper a concept of calculation method for assessment the shear resistance of self-stressing concrete elements expanded under biaxial restraint conditions is presented. The proposed design model is based on the strut-and-tie model strategy counting for the initial strains and self-stresses as a result of self-stressing concrete expansion.

2 Design model for assessment the shear resistance of plane self-stressing concrete elements

Details of code requirements for design of reinforced concrete structures according to the strut-and-tie model [10–14] were compared and analysed in the authors' research papers [15–18]. It's noticeable that with respect to the resistance of tensile ties the code provisions [10–14] take into consideration stresses in both nonprestressed and prestressed reinforcement. At the same time concerning the compressive struts and node surfaces only requirements for nonprestressed elements are considered [10–14]. The experimental investigations [15, 17] also have shown that the most reliable provisions concerning the strut-and-tie model were presented in ***fib*** Model Code [12].

As in case of concrete elements prestressed in traditional ways (mechanical, etc.) the behavior of self-stressing concrete elements under the static load can be divided into two specific stages (Figure 1): before and after the moment of reduction the compressive stresses in concrete to zero (moment of decompression).

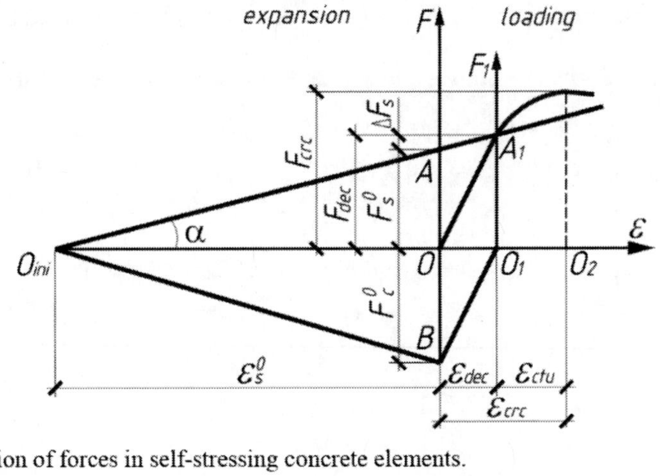

Fig. 1. Distribution of forces in self-stressing concrete elements.

In the Figure 1 F_S^0, F_C^0 are the forces in reinforcement and concrete (section OA and OB respectively) in the principle stresses direction by the beginning of load action; ΔF_S is the incremental force in reinforcement in the principle stresses direction at the stage preceding the moment of decompression; F_{dec} is the decompression force in reinforcement at the moment of reduction the compressive strains in concrete to zero in the principle stresses direction; F_{crc} is the cracking force in the principle stresses direction; ε_S^0 is the restrained expansion strain in the principle direction by the beginning of load action; ε_{dec} is the strain in self-stressing concrete element in the principle direction at the moment of decompression; ε_{crc} is the strain in self-stressing concrete element in the principle direction under loading at the moment of cracking; ε_{ctu} is the ultimate tensile strain in concrete.

The proposed design model for assessment the shear resistance of self-stressing concrete elements expanded under two-way restraint conditions differs from actual strut-

and-tie models by taking into consideration the shear resistance at the moment of reduction the compressive strains in concrete to zero. Thus, the total shear resistance of self-stressing concrete element can be presented as follows:

$$F_u = F_{u.dec} + F_{u.F} \quad (1)$$

where $F_{u.dec}$ is the shear resistance at the moment of decompression;

$F_{u.F}$ is the shear resistance that is calculated as in case of nonprestressed reinforced concrete elements following the code requirements for strut-and-tie models [12].

Such a division is convenient to consider the restrained expansion strains and self-stresses as a result of self-stressing concrete expansion in the strut-and-tie model solution.

On that basis the following assumptions related to the proposed strut-and-tie model with respect to self-stressing concrete elements can be postulated:

1) a model geometry is determined from the actual stress field during the elastic behavior of concrete;

2) forces in compressive struts and tensile ties are calculated according to the static scheme of element;

3) design of compressive struts, tensile ties and nodal regions are carried out based on the theory of plasticity;

4) from the applying of load until the moment of decompression the behavior of plane self-stressing concrete element is modelled as elastic, so the traditional stress-strain relationships are valid:

– strains and stresses in concrete can be described by corresponding Mohr's circles (Figure 2a, 2b);

– directions of principle strains and principle stresses coincide (Figure 2c).

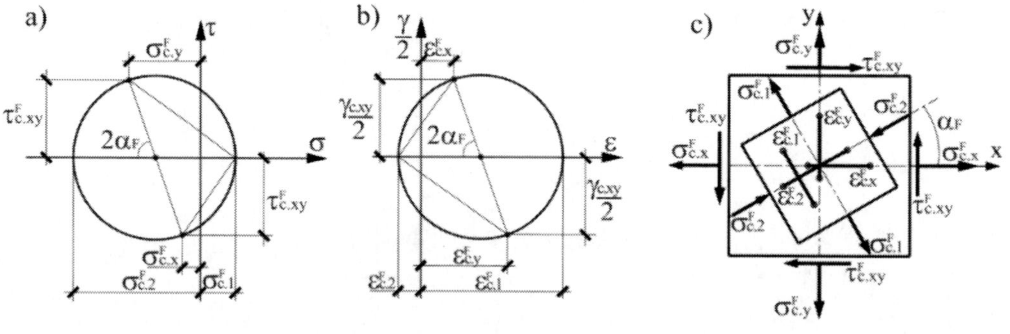

a) Mohr's circle for stresses; b) Mohr's circle for strains; c) directions of strains and stresses under static loading.

Fig. 2. Stress-strain relationship in plane self-stressing concrete element.

Calculation of resistance at the moment of decompression $F_{u.dec}$ is carried out based on the statement that initial compressive strains in concrete in the direction of normal to the strut longitudinal axis as a result of prestressing $\varepsilon^0_{c.y\alpha}$ equal to the principal tensile strains in concrete in the direction of principle tensile stresses under the action of concenctrated load $\varepsilon^F_{c.1}$.

The initial compressive strains in concrete in the direction of normal to the strut longitudinal axis $\varepsilon^0_{c.y\alpha}$ are defined by :

$$\varepsilon^0_{c.y\alpha} = \sigma^0_{c.y\alpha} / E_{cm.28} \tag{2}$$

where $E_{cm.28}$ is the Young's modulus of expansive concrete at the age of 28 days;
$\sigma^0_{c.y\alpha}$ are the initial compressive stresses in concrete in the direction of normal to the longitudinal axis of strut (or nodal surface):

$$\sigma^0_{c.y\alpha} = \sigma^0_{c.x} \cdot \sin^2\alpha + \sigma^0_{c.y} \cdot \cos^2\alpha \tag{3}$$

where $\sigma^0_{c.x}$, $\sigma^0_{c.y}$ are the initial compressive stresses in concrete in the orthogonal directions OX and OY respectively (it should be noted that reinforcing steel in self-stressing concrete element is located also in the orthogonal directions OX and OY);
α is the angle between axis of strut (or axis of nodal surface) and longitudinal axis of element.

Definition of initial compressive stresses in concrete $\sigma^0_{c.x}$ and $\sigma^0_{c.y}$ is carried out according to the modified early age strains development model for the case of two-way restraint conditions (2D MSDM) [19, 20].

The principal tensile strains in concrete in the direction of principle tensile stresses under the action of concentrated loads $\varepsilon^F_{c.1}$ are calculated according to expressions presented in [21]:

$$\varepsilon^F_{c.1} = \varepsilon^F_{c.x} + (\varepsilon^F_{c.x} - \varepsilon^F_{c.2}) \cdot \cot^2\alpha\,, \tag{4}$$

$$\varepsilon^F_{c.x} = \frac{M_{Ed}/d_z + 0{,}5 \cdot N_{Ed} + 0{,}5 \cdot V_{Ed} \cdot \cot\alpha}{E_s \cdot A_s}\,, \tag{5}$$

$$\varepsilon^F_{c.2} = \frac{\sigma^F_{c.2}}{E_{cm.28}} = \frac{V_{Ed}}{b \cdot d \cdot E_{cm.28}} (\tan\alpha + \cot\alpha)\,. \tag{6}$$

In formulas (4-6) $\varepsilon^F_{c.x}$ refers to the strains of extreme tensile reinforcement or in the section at a distance of $0{,}6d$ from the extreme top concrete fiber; $\varepsilon^F_{c.2}$ refers to the principle compressive strains in strut.

Calculation of the shear resistance $F_{u.F}$ in case of nonprestressed reinforced concrete elements is carried out according to the requirements for strut-and-tie models presented in ***fib*** Model Code 2010 [12]:

$$F_{u.F} = f_{\max} \cdot A_c \tag{7}$$

where A_c is the cross-sectional area of concrete strut (or nodal surface) that is defined based on the parameters of idealized bottle-shaped strut;
$f_{\max}$ is the maximum stress at a compressive concrete strut that is calculated taking into account the decrease of concrete compressive strength due to the tensile strains in concrete utilizing the reduction factor k_c :

$$f_{\max} = k_c \cdot f_{cd} \tag{8}$$

where f_{cd} is the design value of concrete compressive strength.

3 Validation of proposed design model

Validation of proposed design model for assessment the shear resistance of self-stressing concrete elements was carried out based on the experimental and theoretical investigations of three series of plane self-stressing concrete elements under the static loading. The plane self-stressing concrete element was performed as a part of compressive bottle-shaped strut and placed at a line between the support and point of application of concentrated load (Figure 3a). The plane specimens were under the axial concentrated static load according to the scheme in the Figure 3b.

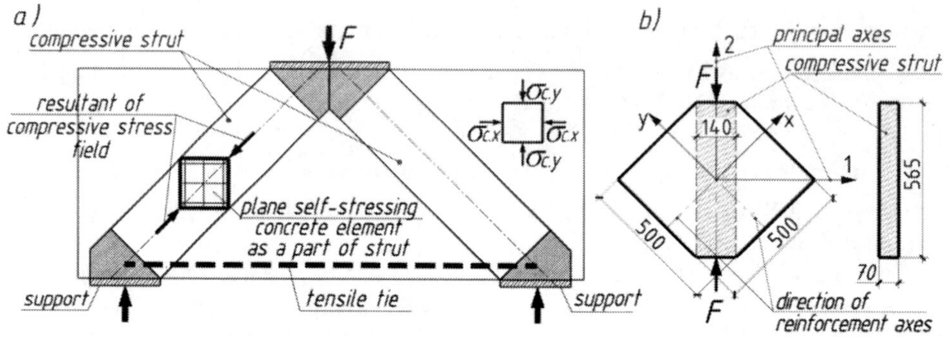

Fig. 3. Test specimen as a part of compressive strut (a) and under the concentrated load (b).

During the theoretical investigations of plane self-stressing concrete elements under the concentrated load the following parameters were calculated: the load at the moment of decompression $F_{dec.calc.\sigma}$ based on the values of initial compressive strains calculated according to the modified strain development model for the case of biaxial restraint conditions (2D MSDM) [19, 20]; the decompression load $F_{dec.\exp.\sigma}$ based on the experimental values of initial compressive strains, which were determined on the plane self-stressing concrete samples by the expansive concrete stabilization; the cracking load F_{crc} based on the ultimate tensile strains in concrete which characrerizes the behavior of element from the moment of decompression to the crack formation; the total cracking load $F_{crc.calc.\sigma}^{total}$ based on the calculated values of initial compressive stresses; the total cracking load $F_{crc.\exp.\sigma}^{total}$ based on the experimental values of initial compressive stresses.

It should be noticed that the total cracking load is calculated as a sum of the load at the moment of decompression and the cracking load which characterizes the behavior of element from the moment of decompression to the crack formation.

The results of investigations are presented in the Table 1. As it could be seen from the Table 1, a good agreement arises between the experimental and calculated values of cracking load.

Table 1. Results of theoretical and experimental investigations.

Series	$F_{crc.calc.\sigma}^{total}$, kN	$F_{crc.\exp.\sigma}^{total}$, kN	$F_{crc.\exp}^{total}$, kN	$\frac{F_{crc.calc.\sigma}^{total}}{F_{crc.\exp}^{total}}$	$\frac{F_{crc.\exp.\sigma}^{total}}{F_{crc.\exp}^{total}}$
I	383,44	382,90	340	1,127	1,126
II	335,66	316,43	340	0,987	0,931
III	278,02	271,41	300	0,927	0,905

4 Conclusions

A strut-and-tie model for assessment the shear resistance of self-stressing concrete elements expanded under biaxial restraint conditions has been developed. The distinctive idea of the proposed model consists in considering the initial strains and stresses as a result of concrete expansion. Suitability of the model was confirmed by experimental and theoretical investigations of plane self-stressing concrete elememts under axial concentrated load.

References

1. W. Ritter, Schweizerische Bauzeitung, **33**, 2 (1899)
2. E. Mörsch, *Concrete-Steel Construction* (McGraw-Hill, New York, 1909)
3. J. Schlaich, K. Schäfer, M. Jennewein, J. of Pr. C. Str, **32**, 76 (1987)
4. J. Schlaich, K. Schäfer, IABSE congress report, 8 (1984)
5. M.D. Brown [et al.], Research Report 0-4371-2 (University of Texas, Austin, 2006)
6. C. Williams, D. Deschenes, O. Bayrak, Final Report FHWA/TX-12/5-5253-01-1 (University of Texas, Austin, 2012)
7. A. Muttoni, M. Fernandez Ruiz, F. Niketic, ACI Str. J, **112**, 11 (2015)
8. M. Tyler Ley, Kyle A. Riding, Widianto, Sungjin Bae, ACI Str. J, **104**, 8 (207)
9. M. D. Brown, O. Bayrak, ACI Str. J, **105**, 9 (2008)
10. AASHTO LRFD Bridge Design Specifications, 8th Edition (Washington, 2017)
11. ACI Committee 318, Building Code Requirements for Structural Concrete (ACI 318 – 14) and Commentary (ACI 318R-14) (Farmington Hills, 2014)
12. ***fib*** Model Code 2010, Vol.1 (Federal Institute of Technology Lausanne, EPFL, Lausanne 2010)
13. PrEN 1992-1-1:2018 (Eurocode 2), Design of Concrete Structures – Part 1-1: General rules, rules for buildings, bridges and civil engineering structures (European Committee for Standardization, 2018)
14. EN 1992-1: 2001 (Eurocode 2), Design of Concrete Structures. General Rules and Rules for Building (European Committee for Standardization, 2001)
15. O. G. Sannikava, V. V. Tur, Materials of XXI PDDC, 10 (Brest, BSTU, 2018)
16. O. G. Sannikava, V. V. Tur, Vestnik PSU, Series F, **16**, 10 (Novopolotsk, RIO PSU, 2018)
17. O. G. Sannikava, V. V. Tur, Vestnik PSU, Series F, **16**, 6 (Novopolotsk, RIO PSU, 2018)
18. O. G. Sannikava, V. V. Tur, Materials of international academic readings "Safety of the construction fund in Russia. Problems and Solutions", 14 (Kursk, KSU, 2018)
19. V. Sannikava, V. Tur, Journal of sustainable architecture and civil engineering, 29 75–86 (2017), doi: http://dx.doi.org/10.5755/j01.sace.20.3.19082
20. O. G. Sannikava, V. V. Tur, Vestnik BSTU, **1,** 15 (2019)
21. V. V. Tur. A. A. Kondratchik, *Design of reinforced concrete structures for shear,* (Brest, BSTU, 2000)

Accidental actions values and combinations for key-elements checking

Viktar Tur[1*], *Andrei* Tur[1], *Aliaksandr* Lizahub[1], *Stanislav* Derechennik[1]

[1] Brest State Technical University, 224017, Moskovskaya 267, Brest, Belarus

Abstract. In the case of accidental design situations, if accidental actions can be identified, one of the possible design strategies is checking the key element. This strategy minimizes the possibility of local failure and subsequent progressive collapse. The paper considers accidental action combinations and values of identified accidental loads according the various codes. The combination of actions for accidental design situation for checking of the "key-element" resistance was proposed. In addition, the values of the combination factors for variable loads and partial factors for permanent loads in accordance with required reliability class RC for structural element and values of accidental loads was proposed.
Keywords: accidental design situation, accidental action, key-element, load combination, robustness.

1 Introduction

Most of the structural codes [1–4] provides a description of principles and application rules for the design of structural systems subjected to accidental action, including impact forces, actions due to internal explosions and due to local failure [5].

According to EN1991-1-7 [3] two groups of strategies are proposed in order to assess accidental design situations: those based on identified accidental actions and those based on limiting the extent of localized failure.

In the first group, it is proposed to design the structure to have sufficient minimum robustness, to prevent or reduce the effect of accidental action or to directly design the structural system to sustain the action.

The second group of strategies is based on limiting the extent of localized failure, either by increasing redundancy of structure or designing "key elements" to sustain notional accidental actions and applying some prescriptive rules like integrity or ductility.

Ellingwood [6] proposed a following formula to assess the probability of progressive collapse:

$$P(C) = P(C|DH)\cdot P(D|H)\cdot P(H) \qquad (1)$$

* Corresponding author: profturvic@gmail.com

where *P(C)* is the probability of progressive collapse; *P(H)* is the probability of the occurrence of a hazard *H*; *P(D|H)* is the probability of local damage *D* as a result of a hazard *H*; *P(C|DH)* – the probability of progressive collapse *C* of structural system as a result of local damage *D* caused by hazard *H*.

In [7] there is a good illustration of this expression (1) together with assigned appropriate terms (see Figure 1).

$$\left.\overbrace{\text{robustness}\;\;\begin{matrix}\text{element}\\ \text{behaviour}\end{matrix}\;\;\begin{matrix}\text{event}\\ \text{control}\end{matrix}}^{\text{collapse resistance}}\right\}\text{maximise}$$

$$P(C) = \underbrace{\overbrace{P(C|DH)}\cdot\overbrace{P(D|H)}}_{\text{vulnerability}}\cdot\underbrace{\overbrace{P(H)}}_{\text{hazard}}\;\}\;\text{minumise}$$

Fig. 1. Terms in the context of progressive collapse (from [7]).

Considering the above Eq. (1) and Figure 1, the probability of progressive collapse can be minimised in three ways, namely by: controlling abnormal events (term *P(H)*, controlling local element behaviour (term *P(D|H)*) and/or controlling global system behaviour (conventional probability *P(C|DH)*).

It is worth nothing in [7], that controlling abnormal events by structural engineers is normally very difficult, practically impossible. However, engineer can influence the local and global system behaviour, i.e. probabilities *P(D|H)* and *P(C|DH)*.

Conditional probabilities presented in Eq. (1) can be obtained by a probabilistic risk analysis (PRA), in which it is possible to model uncertainties, study their propagation and the effect on the required performance of the structural systems (with damaged elements). This approach is called structural reliability analysis and failure (collapse in the case in question) is considered achieved when demand *E* (i.e. the effects generated by the actions) exceeds collapse resistance *R*. In general case, the probability of failure is equal to:

$$p_f = \int F_R(x) f_E(x) dx \tag{2}$$

where $F_R(x)$ is the CDF (cumulative distribution function) of resistance *R* and $f_E(x)$ is the PDF (probability density function) of *E* (effect of actions).

The probability of disproportionate collapse can be defined according to EN1990 [4] as follows:

$$p_f = Prob\left[E \le R\right] \text{ or } p_f = \Phi(-\beta) \tag{3}$$

where β is the reliability index for structural system and $\Phi(\bullet)$ is a normal standard distribution function.

For a correct assessment of disproportionate collapse risk, it may be necessary to consider the presence of multiple hazard events and the initial stage of damage. In this case, Eq. (1) can be generalized as illustrated in the following equation (valid for independent event only):

$$P(C) = P(C|DH)\cdot P(D|H)\cdot \lambda_H \tag{4}$$

where: λ_H can substitute P(H) if occurrence probability is less that 10^{-2}/year. Values of λ_H are reported in [6].

As was shown in [8], if a performance based design approach is adopted, an acceptable value of risk tolerance has to be defined. In the case of a disproportioned collapse, which main consequence is the loss of human life, decision-makers can assume that the performance objective of safeguarding human life is achieved if the following relationship is verified:

$$P(C) \le p_{tag,h} \tag{5}$$

where $p_{tag,h}$ is the risk threshold defined as "de minimis" which in general case assumed values ranging from 10^{-5}/year and 10^{-7}/year. More detailed discussion presents in our publication [9].

Moreover, in particular case in which so called alternative load path method (ALP-method [6, 10]) is used in design phase, the collapse probability becomes P(C|DH), which in turn has to respect the following equation according to [6]:

$$P(C \mid DH) < \frac{p_{tag,h}}{\lambda_H} \tag{6}$$

Therefore, assuming λ_H equal to 10^{-6}/year..10^{-5}/year, the performance based target probability established by condition (6) requires that the conditional probability of collapse for the modified structural system be in the order of 10^{-2}/year..10^{-1}/year.

Consequently, as shown in [8], the reference reliability index β_0 for the limit collapse state of conditioned by the occurrence of the damage will be in order of 1.5. That is significantly lower than that assumed for ultimate limit state of new buildings for residential and office use in case of ordinary actions (i.e. $\beta_{tag} = 3.8$, which corresponds to reference probability for structural system collapse of the order of $\sim 10^{-4}$).

Currently, the combinations of accidental actions proposed in the design codes for checking structures in accidental design situations do not take into account that in some cases the accidental effect on the key element may be decisive (i.e., significantly higher than other effects). Therefore, a more objective model of the accidental combination of actions is needed, which will take into account the magnitude of the accidental effect and making it possible to obtain the resistance of the "key element" in accordance with the required reliability class.

2 Load combinations for accidental design situation according to various codes

According to [10, 11], a reliability based approach can be applied to determine reasonable loading combinations for accidental design situation. The actions to be combined reflect the small probability of a joint occurrence of the accidental action and design values of imposed (or live), snow, wind loads.

Hazard events, and mainly, malicious attack are a rare events and many of them suppressed early.

Focussing on the mechanical actions, these are traditionally subdivided into permanent actions and imposed (or variable) action according to [12]. Their variability with time is an aspect of particular relevance for checking of the structural system in accidental design situation. As was shown in [11], in partial factor design method (PFM) for normal conditions, the load variability is considered by a characteristic or design load with a low probability of being exceeded during the service life of the structure. This ensures that the building structure are designed both safety and economically, as in setting the design requirements a balance has been sought between the cost of premature failures and the cost of additional safety investment (see ISO2394 [1]).

Figure 2 shows that reliability indices (failure probabilities) are influenced by the efficiency of safety investments and consequences of failure [1]. The optimal reliability index β^* can be obtained by minimising the sum of investments in safety measures and the accompanying capitalised risk. The target reliability indices derived on the basis of economic optimisation might not be acceptable with regard to requirements concerning human safety, as it is stated in ISO2394 [1]. These reliability indices are denoted as $\beta_{tag,h}$.

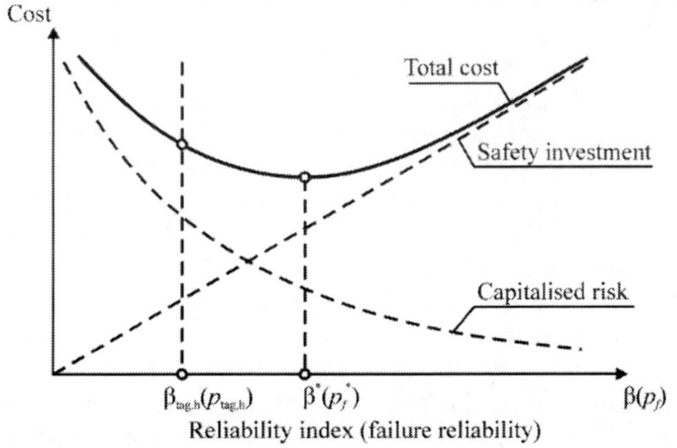

Fig. 2. Principles of cost minimisation, reliability optimum β^* and reliability minimum $\beta_{tag,h}$ according to ISO2394 [1].

It is clear, the day-to-day probability of occurrence of such high (design) load value is low, just as for the day-to-day probability of occurrence of a hazard event (accidental event). Simultaneously taking into account both events would result in very onerous design requirement for accidental design situation (in case of modified structural system robustness checking).

Hence, the reduced partial safety and combination factors in Eurocode [3], ASCE [13] and other codes [14–16] (see Table 1) lesser the required load under consideration for structural design in accidental design situation compared to normal design situation.

Table 1. Accidental action combinations according the various codes.

Standards	Load combinations
BS [14]	$D + L/3 + W/3$
Eurocode [3]	$\sum G + P + A_d + \psi_{1,1}Q_{k,1} + \sum \psi_{2,i} Q_{k,i}$ $\sum G + P + A_d + \psi_{2,1}Q_{k,1} + \sum \psi_{2,i} Q_{k,i}$
ASCE 7-05 [13]	($0.9D$ or $1.2D$) + ($0.5L$ or $0.2S$) + $0.2W_n$ – alternate load path method $1.2D + A_k$ + ($0.5L$ or $0.2S$) – specific local resistance method ($0.9D$ or $1.2D$) + $A_k + 0.2W_n$
GSA [15]	$2(D + 0.25L)$ – static analysis $D + 0.25L$ – dynamic analysis
UFC 4-023-03 [16]	($0.9D$ or $1.2D$) + ($0.5L$ or $0.2S$) + $0.2W_n$ – nonlinear dynamic analysis 2[($0.9D$ or $1.2D$)+ ($0.5L$ or $0.2S$) + $0.2W_n$] – static analysis
D – dead load, L – live load, W – wind load, S – snow load, A – accidental load.	

3 Accidental actions values for "key-element" checking (conditional probability P(D|H))

Recently, many standards [3, 13–16] do not define value for accidental loads, but leave them to the designer. As was noted in [7], after all, the inquiry on the Ronan Point disaster stated that there was only a risk of $3.5 \cdot 10^{-6}$/year of structural damage due to an urban gas explosion. The pressure due to the gas explosion at Ronan Point was estimated to be between 14 kPa and 83 kPa. Therefore, an accidental pressure of 34 kPa has been adopted by British Standards science 1970. In practice, it is used to determine a notional load applied to key-elements (KE), resulting in more robust construction against gas explosion and other loads as well. However, research [7] has shown, that this pressure may be too high. For example, Ellingwood [6] reports that for a typical residential apartment and venting, the mean maximum pressure generated by a gas explosion is about 17,3 kPa, with standard deviation of 3 kPa. Similary, Ellis and Currie stated that *"for appraisal of structures specially for the event of gas explosion, 17 kPa is considered to be over pressure that will act on all surfaces of the room simultaneously"*. It should be pointed, that this statement is based on housing units, where a maximum pressure of 13 kPa has been measured in gas explosion in single room. But, multi-room, cascading explosions, usually due to piped gas, can produce up to 90 kPa.

The EN1991-1-7 [3] allows a smaller accidental pressure (34 kPa), or value based on calculations that account for room volume, the presence of vents, the strength of windows, and less importantly, the properties of gas mixture. Moreover, the Eurocode [3] defined accidental actions due to the following events: impact from road vehicles; impact from forklift trucks; impact from train; impact from ships and hard landing of helicopters on roofs. According to [3], actions due to impact should be determined by dynamic analysis or represented by an equivalent static force. For buildings, actions due to impact should be taken into account for buildings used car parking; buildings in which vehicles of forklift trucks are permitted, and buildings that are locates adjacent to either road or railway traffic.

4 Load combinations for «key-element» checking in accidental design situation. Results and Discussion

As shown above [11], in general case, hazard events can be classified in two major types: unintentional but identified (Natural and Accidental) hazards and malicious attacks. According to [11], the distinct, nature of two types of hazard implies that the hazard associated uncertainties, severity and frequency of occurrences are significantly different. For unintentional hazards such as earthquake, wind, scour, vessel collision, random stochastic models are typically used to represent the hazard intensity and occurrence. However, for purposely plotted malicious destruction such as explosions and intentional collisions and purposely made accidents (criminal and terrorist attacks), the ordinary random stochastic model is not longer valid.

According to EN1990 [4] and EN1991-1-7 [3] the general format of effects of actions for the accidental design situations is analogous to the general format for STR/GEO ultimate limit states. Here, the loading action is the accidental action, and the most general expression of the design value of the effects of actions is the following:

$$E_d = E\{G_{k,j};\ P_i;\ A_d;\ (\psi_{1,1} \text{ or } \psi_{2,1})Q_{k,i};\ \psi_{2,i}Q_{k,i}\} \tag{7}$$

which can also be expressed as:

$$\sum_{j\geq 1} G_{k,j} \text{"+"} P \text{"+"} A_d \text{"+"} \left(\psi_{1,1} \text{ or } \psi_{2,1}\right) Q_{k,1} \text{"+"} \sum_{i>1} \psi_{2,i} Q_{k,i} \tag{8}$$

According to background to EN1990 [4], this combination considers that:

– accidents are unintended events such as explosions, fire or vehicular impact, which are of very short duration and have a low probability of accurence;

– a certain amount of damage is generally acceptable in the event of an accident;

– accidents generally occur when structures are in use.

Hence, to provide a realistic accidental combination, accidental actions are applied directly, with the frequent and quasi-permanent combination values used for the main (if any) and other variable actions respectively (see cl. 6.4.3.3 (3) EN1990 [4]).

Regarding the representative value (frequent and quasi-permanent) of a possible main variable action, EN1990 [4] states that discretion is left to national authorities for the reason that all accidental situations or events cannot be similarly treated. When the main variable action is not obvious, each variable action should be considered in turn as the main action.

The combination for accidental design situation either involve an explicit design value of accidental action A_d (e.g. impact) or refer to a situation after an accidental event ($A_d = 0$).

The partial factors for actions for ultimate limit states in the accidental design situations are normally taken equal to 1.0, in general, not only are the reliability elements for actions modified for the partial factors for resistances.

Based on results of our own investigations [17], we proposed to use for checking of the "key-element" resistance the following combination of actions for accidental design situation (combination comprises accidental action $A_d \neq 0$):

$$E_{d,A} = \sum_{j=1}^{n} (\gamma_{GA,j}) G_{k,j} \text{"+"} A_d \text{"+"} \psi_{A,1} Q_{k,1} \tag{9}$$

where $G_{k,j}$ is the characteristic value of a permanent action "j";

$Q_{k,1}$ is the characteristic value of the leading variable action;

A_d is the design value of the accidental action;

$\gamma_{GA,j}$ is the combination factor applied to a permanent action "j";

$\psi_{A,1}$ is the combination factor applied to the leading variable action according to Table 2.

In Table 2 we relate values of the combination factors ψ_A with required reliability class RC for structural element and factor k, which is determined as ratio:

$$k = \frac{A_d}{E_k} \tag{10}$$

with

$$E_k = \sum_{j=1}^{n} G_{k,j} + Q_{k,1} + \sum_{i>1} Q_{k,i} \tag{11}$$

where $Q_{k,i}$ is the characteristic value of the accompanying variable actions.

Proposed combination of actions for accidental design situation Eq. (9) gives resistance of the "key-element" in accordance with required reliability class. It was found that the application of the ψ_A coefficients to the accompanying (non-dominant) loads does not lead to a change in the reliability indices. Because of the significant accidental action A, the influence of all other loads decreases. Only the leading variable action have noticeable influence on the reliability indices. In this regard in Eq. (9), only dominant variable loads are taken into account in combination of actions for accidental design situation for the "key" system elements. For non-dominant variable action $\psi_{A,i,2} = 0$.

Table 2. Combination factors ψ_A and partial factors γ_{GA} for checking of the "key-element" resistance.

Reliability Class	k_A	**Values of the coefficient for variable actions**			
		Imposed (Q), $\psi_{A,Q}$	**Wind (W), $\psi_{A,W}$**	**Snow (S), $\psi_{A,S}$**	**Permanent (G), ψ_{GA}**
RC2	1.0	0.8	0.8	0.7	1.0
	1.5	0.6	0.6	0.55	
	2.0	0.5	0.5	0.4	
	2.5	0.35	0.4	0.3	
	3.0	0.2	0.3	0.2	
	3.5	0.1	0.2	0.1	
	4.0	0.05	0.15	0.05	
RC3	1.0	1.0	1.05	1.0	1.05
	1.5	0.9	0.95	0.85	
	2.0	0.8	0.8	0.7	
	2.5	0.7	0.7	0.6	
	3.0	0.55	0.6	0.5	
	3.5	0.45	0.5	0.4	
	4.0	0.4	0.45	0.35	

Conclusion

In the paper was considered accidental action combinations and values of identified accidental loads according the various codes. The combination of actions for accidental design situation for checking of the "key-element" resistance was proposed. The calibrated values of the combination factors $\psi_{A,i,1}$ for the key elements (see Table 2) depend on the required reliability class of structural elements and the factor k, which is defined as the ratio of the design value of the effect from the accidental action A_d on the element and the effect E_k from the total characteristic load on this element. This approach provides a more objective assessment of the resistance of the "key-element".

Depending on the magnitude of the accidental action, the values of the combination factors $\psi_{A,i,1}$ are for the imposed load from 0.05 to 1.0, for the wind load from 0.15 to 1.05, for the snow load from 0.05 to 1.0.

References

1. International Standards Organization, *ISO 2394: 2015: General principles on reliability for structures* (2015)

2. Joint Committee of Structural Safety, *JCSS Probabilistic Model Code. Part 2: Load Models* (2001)
3. European Committee for Standardization. *Eurocode 1 - EN 1991-1-7: Actions on structures - Part 1-7: General actions - Accidental actions* (2006)
4. European Committee for Standardization. *EN 1990:2002: Eurocode – Basic of structural Design* (2009).
5. R. Van Coile, D. Hopkin, N. Elhami Khorasani, D. Lange, T. Gernay, 3rd International Conference on Structural Safety under Fire and Blast, (2019)
6. B.R. Ellingwood, R. Smilowitz, D.O. Dusenberry, D. Duthinh, H.S. Lew, N.J. Carino, *Best Practices for Reducing the Potential for Progressive Collapse in Buildings*, (No. NIST Interagency/Internal Report (NISTIR)-7396) (2007)
7. S. Kokot, G. Solomos, *Progressive collapse risk analysis: literature survey, relevant construction standards and guidelines. Ispra: Joint Research Centre, European Commission* (2012)
8. B.R. Ellingwood, Journal of Fire Protection Engineering, **15(1)**, 43-61 (2005)
9. V.V. Tur, A.V. Tur, S.S. Derechennik, In Proceedings of the fib Symposium 2019: Concrete-Innovations in Materials, Design and Structures, 2126-2133 (2019)
10. Arup, *Review of International Research on Structural Robustness and Disproportionate Collapse* (London: Department for Communities and Local Government, 2011)
11. H. Gulvanessian, *Designers'guide to Eurocode: Basis of Structural Design* (2020)
12. CIB, *CIB Report, Publication 116: Actions on structures, live loads in buildings* (1989)
13. ASCE 7, *ASCE 7: Minimum design loads for buildings and other structures. American Society of Civil Engineers* (2005)
14. BS 6399, *BS 6399: Loading for buildings: Part 1: Code of practice for dead and imposed loads. British Standards Institute* (1996)
15. GSA Guidelines, *GSA Progressive Collapse Analysis and Design Guidelines for New Federal Office Buildings and Major Modernizations Projects. General Services Administration (GSA)* (2003)
16. DoD UFC Guidelines, *Design of Buildings to Resist Progressive Collapse, Unified Facilities Criteria (UFC) 4-023-03. Department of Defence (DoD)* (2005)
17. V.V. Tur, D.M. Markovskij, Stroitel'naya nauka i tekhnika, **2**, 23 (2009)

AUTHOR INDEX